Soft Nanoparticles for Biomedical Applications

RSC Nanoscience & Nanotechnology

Editor-in-Chief:
Paul O'Brien FRS, *University of Manchester, UK*

Series Editors:
Ralph Nuzzo, *University of Illinois at Urbana-Champaign, USA*
Joao Rocha, *University of Aveiro, Portugal*
Xiaogang Liu, *National University of Singapore, Singapore*

Honorary Series Editor:
Sir Harry Kroto FRS, *University of Sussex, UK*

Titles in the Series:

How to obtain future titles on publication:
A standing order plan is available for this series. A standing order will bring delivery of each new volume immediately on publication.

For further information please contact:
Book Sales Department, Royal Society of Chemistry, Thomas Graham House, Science Park, Milton Road, Cambridge, CB4 0WF, UK
Telephone: +44 (0)1223 420066, Fax: +44 (0)1223 420247
Email: booksales@rsc.org
Visit our website at www.rsc.org/books

Soft Nanoparticles for Biomedical Applications

Edited by

José Callejas-Fernández
University of Granada, Granada, Spain
Email: jcalleja@ugr.es

Joan Estelrich
Univeristy of Barcelona, Barcelona, Spain
Email: joanestelrich@ub.edu

Manuel Quesada-Pérez
University of Jaén, Jaén, Spain
Email: mquesada@ujaen.es

Jacqueline Forcada
University of the Basque Country, San Sebastián, Spain
Email: jacqueline.forcada@ehu.es

ROYAL SOCIETY
OF CHEMISTRY

THE QUEEN'S AWARDS
FOR ENTERPRISE:
INTERNATIONAL TRADE
2013

RSC Nanoscience & Nanotechnology No. 34

Print ISBN: 978-1-84973-811-8
PDF eISBN: 978-1-78262-521-6
ISSN: 1757-7136

A catalogue record for this book is available from the British Library

Published by The Royal Society of Chemistry,
Thomas Graham House, Science Park, Milton Road,
Cambridge CB4 0WF, UK

Registered Charity Number 207890

For further information see our web site at www.rsc.org

Preface

The book is focused on soft nanoparticles, a specific area in the field of nanotechnology. The activity in this field has grown exponentially worldwide during the past couple of decades, becoming a major interdisciplinary area of research. This growth is driven to a great extent by the integration of nanotechnology into medical science, where it is opening up many new possibilities. For instance, nanoparticles can be potential or actually used in diagnostic tests (detecting extremely low concentrations of pathogens and analytes), non-invasive imaging techniques and gene and drug delivery.

Among the existing soft nanoparticles, we have chosen those useful for actual or potential biomedical applications. More specifically, our choice is a compromise between well-established nanoparticles and promising candidates under testing. On the other hand, we have included some soft nanoparticles that have been used for decades (*e.g.* liposomes), which have been extensively described elsewhere.

Chapter 1 introduces us to the realm of soft nanoparticles and tells us how some of these nanoparticles began to be employed in pharmaceuticals. This chapter also outlines some colloidal aspects of these particles and summarizes the possibilities that computer simulations offer in this field.

Concerning the biomedical applications of soft nanoparticles, great emphasis has been placed to date on their chemistry. However, physical properties may also have a considerable influence. In fact, some authors suggest the importance of improving physics-derived solutions (including simulations) for controlling processes in which nanoparticles are involved. Trying to fill this gap, Chapter 2 focuses on key physical properties of nanoparticles (such as size, shape, surface charge and internal structure) and the methods and techniques used for their measurement.

RSC Nanoscience & Nanotechnology No. 34
Soft Nanoparticles for Biomedical Applications
Edited by José Callejas-Fernández, Joan Estelrich, Manuel Quesada-Pérez and Jacqueline Forcada
© The Royal Society of Chemistry 2014
Published by the Royal Society of Chemistry, www.rsc.org

Magnetoliposomes are hybrid systems formed by encapsulating iron oxide nanoparticles within liposomes. They are a platform that presents the advantages of being non-toxic and biocompatible and displaying magnetophoretic mobility. In other words, they can carry a drug within them and under the influence of an external magnetic field can be directed to a targeted area (tissues, cells, *etc.*) of the human body. The rationale for magnetoliposome-based targeting lies in the potential to reduce or eliminate the side effects concomitant with the administration of some drugs. Chapter 3 is devoted to the synthesis, characterization and applications of these soft nanoparticles.

Micro/nanogels are cross-linked colloidal particles that can swell by absorption (uptake) of large amounts of solvent, but they do not dissolve owing to the structure of the polymeric network, physically or chemically cross-linked. Nanogels are considered good candidates for gene and drug delivery. In their design, however, we should keep in mind that the cornerstone that controls the uptake and release of the load is the interaction between the molecules of the substance and the nanogel. If the cargo is an ionic species, the nanogel charge and the electrostatic interaction play an essential role. In any case, the encapsulation of a given solute inside the nanogel is a complex process that depends on many physical and chemical parameters. These are analysed in Chapter 4, which also presents a survey on the interactions that govern the colloidal stability of these nanoparticles.

Polymeric micelles can be considered a type of nanoparticles with a core–shell structure. The inner core is the hydrophobic part of a block copolymer, which usually contains the poorly water-soluble drug. The outer shell is the hydrophilic part of the block copolymer and its mission is to protect the drug from the aqueous environment. Polymeric micelles are interesting owing to their proved effectiveness for the specific delivery of anticancer drugs to tumours. Chapter 5 describes the synthesis of suitable block copolymers to design the desired micelles, their conversion into micelles, the characterization of the micelles formed and actual and future applications of these soft nanoparticles.

Gene therapy represents a new paradigm of therapy for diseases, in which the disease is treated at the molecular level by restoring defective biological functions or reconstituting homeostatic mechanisms within cells. Specific and efficient delivery of genetic material to diseased sites and to particular cell populations is a challenge that is being addressed using a variety of non-viral delivery systems, all of which have distinct advantages and disadvantages. Substantial progress has been made in binding DNA to nanoparticles or encapsulating DNA in and controlling the behaviour of these complexes. In Chapter 6, recent advances in the major colloidal delivery carriers are reviewed. The structure, synthesis, biological properties and the cellular transfection capabilities of the different colloidal systems are discussed.

Dendrimers are another class of highly branched polymers having low polydispersity. Usually, they have a core, branched units and surface groups that provide building blocks for the design of delivery systems for various applications. The molecular architecture of dendrimers can be controlled precisely through synthetic methods, resulting in well-defined nano-structures. Chapter 7 is devoted to these soft nanoparticles.

Bicelles are emerging as promising membrane models and, because of their attractive combination of lipid composition, small size and morphological versatility, they have become new targets in skin research. These nanoparticles are formed by two kinds of phospholipids, one with a large hydrocarbon chain that forms bilayers over a wide range of temperature and water content, and the other with a short hydrocarbon chain that generally forms micelles. Binary mixtures of these phospholipids at a suitable molar ratio tend to form disc-shaped particles known as bicelles (or bilayered micelles). The use of bicelles for skin applications is clearly advantageous: bicelles are constituted exclusively of lipids and have the ability to penetrate through the narrow intercellular spaces of the stratum corneum of the skin to reinforce its lipid lamellae. In addition, the bicelle structure allows for the incorporation of different molecules that can be carried through the skin layers. Bicelles are discussed in Chapter 8.

Hybrid nanoparticles consist of an inner part (core) and an outer part (shell) and both can be organic- or inorganic-based materials. Among the available core–shell nanoparticles, we have focused on those with a soft shell (hard nanoparticles such as quantum dots are not considered). Nowadays, a plethora of soft hybrid nanoparticles have been developed to target specific biomedical applications and to perform desirable diagnostic and therapeutic functions. As the reader can guess, there are also many methods that have been developed to synthesize such soft hybrid nanoparticles. The most important ones are briefly outlined in Chapter 9. Their relevance in biomedical applications such as cancer and gene therapy, diagnosis and bioimaging is concisely discussed in this chapter.

To finish, Chapter 10 offers an illustrative example of how coarse-grained simulations can provide valuable information about the behaviour of nanoparticles: the complexation of DNA-like polyelectrolytes and oppositely charged spherical nanoparticles. In the first part of the chapter, different computer simulation techniques are briefly discussed. In the second part, the effects of surface charge density, ionic strength, pH, ion valence and chain flexibility are considered.

The Editors express their gratitude to all the authors who have kindly contributed to this work. The editors also acknowledge the financial support from 'Ministerio de Economía y Competitividad, Plan Nacional de Investigación, Desarrollo e Innovación Tecnológica (I + D + i)', Projects MAT2012-36270-C04-01, -02, -03 and -04.

Contents

RSC Nanoscience & Nanotechnology No. 34
Soft Nanoparticles for Biomedical Applications
Edited by José Callejas-Fernández, Joan Estelrich, Manuel Quesada-Pérez and
Jacqueline Forcada
© The Royal Society of Chemistry 2014
Published by the Royal Society of Chemistry, www.rsc.org

Introductory Aspects of Soft Nanoparticles

JOAN ESTELRICH,*[a] MANUEL QUESADA-PÉREZ,[b] JACQUELINE FORCADA[c] AND JOSÉ CALLEJAS-FERNÁNDEZ[d]

[a] Departament de Fisicoquímica, Facultat de Farmàcia, Universitat de Barcelona, 08028 Barcelona, Spain; [b] Departamento de Física, Escuela Politécnica Superior de Linares, Universidad de Jaén, 23700 Linares, Jaén, Spain; [c] POLYMAT, Bionanoparticles Group, Departamento de Química Aplicada, UFI 11/56, Facultad de Ciencias Químicas, Universidad del País Vasco UP/EHU, Apdo. 1072, 20080 Donostia-San Sebastián, Spain; [d] Grupo de Física de Fluidos y Biocoloides, Departamento de Física Aplicada, Facultad de Ciencias, Universidad de Granada, 18071 Granada, Spain
*Email: joanestelrich@ub.edu

1.1 Nanoparticles

Nanotechnology is the science that deals with matter at the scale of 1 billionth of a metre (*i.e.* 10^{-9} m $= 1$ nm) and is also the study of manipulating matter at the atomic and molecular scale. A nanoparticle is the most fundamental component in the fabrication of a nanostructure and is far smaller than the world of everyday objects that are described by Newton's laws of motion, but larger than an atom or a simple molecule that are governed by quantum mechanics.

According to the definition of the International Organization for Standardization (ISO), a nanoparticle is a particle whose size spans the range between 1 and 100 nm.[1] Metallic nanoparticles have different physical and

RSC Nanoscience & Nanotechnology No. 34
Soft Nanoparticles for Biomedical Applications
Edited by José Callejas-Fernández, Joan Estelrich, Manuel Quesada-Pérez and Jacqueline Forcada
© The Royal Society of Chemistry 2014
Published by the Royal Society of Chemistry, www.rsc.org

chemical properties from bulk metals (*e.g.* lower melting points, higher specific surface areas, specific optical properties, mechanical strengths and magnetizations), properties that might prove attractive in various industrial applications. However, how a nanoparticle is viewed and is defined depends very much on the specific application. In this regard, for biomedical applications, structures and objects up to 1000 nm in size are included as nanostructured materials used in medicine.[2]

Of particular importance, optical properties are among the fundamental attractions and characteristics of a nanoparticle. For example, a 20 nm gold nanoparticle has a characteristic wine-red colour, a silver nanoparticle is yellowish grey and platinum and palladium nanoparticles are black. Not surprisingly, the optical characteristics of nanoparticles have been used for centuries in sculptures and paintings even before the fourth century AD. The most famous example is the Lycurgus cup (fourth century AD). This cup, at present in the British Museum in London, is the only complete historical example of a special type of glass, known as dichroic glass, that changes colour when held up to the light. When it is looked at in reflected light or daylight, it appears green. However, when light is shone into the cup and transmitted through the glass, it changes colour to red. This property puzzled scientists for decades and the mystery was not solved until 1990, when researchers in England scrutinized broken fragments under a microscope and discovered that Roman artisans were nanotechnology pioneers: they had impregnated the glass with a very small quantity of minute (~ 70 nm) colloidal silver and gold in an approximate molar ratio of 14:1, which gives it these unusual optical properties.

Gold suspensions were familiar to alchemists in the Middle Ages and the reputation of soluble gold was based mostly on its fabulous curative powers against various diseases, for example, heart and venereal diseases, dysentery, epilepsy and tumours. Metallic nanoparticles were used in mediaeval stained glasses. The mediaeval artisans trapped gold nanoparticles in the glass matrix in order to generate ruby-red colour in windows. They also trapped silver nanoparticles, which gave the glass a deep-yellow colour. Beautiful examples of these applications can be found in glass windows of many Gothic European cathedrals.

In the seventeenth century, the so-called Purple of Cassius was highly popular. It was a colloid made by reducing a soluble gold salt with stannous chloride. It was used as a colorant and to determine the presence of gold as a chemical test. The first scientific study of gold particles was carried out by Faraday in 1857.[3] He observed that gold suspensions with a ruby-coloured appearance, made by reducing an aqueous solution of chloroaurate ($AuCl_4{}^-$) with phosphorus in CS_2 (a two-phase system), changed their colour from red to blue upon heating or addition of salt. Faraday correctly attributed the colour change to an increase in the effective particle size caused by aggregation. Since that pioneering work, thousands of scientific papers have been published on the synthesis, modification, properties and assembly of metal nanoparticles, using a wide variety of solvents and other substrates. Nowadays, the most widely used nanotechnology product in the field of *in vitro* diagnostics is colloidal gold in lateral flow assays, which is used in rapid tests for pregnancy, ovulation, human immunodeficiency virus (HIV) and other indications. Gold nanoparticles were introduced into these tests in the late 1980s because gold conjugates have particularly high stability, which is critical for avoiding false positives.

In 1959, Richard Feynman gave a talk entitled 'There's plenty of room at the bottom', where he predicted the new things and new opportunities that one could expect in the very small world.[4] Norio Taniguchi of Tokyo University of Science was the first to propose in 1974 the term 'nanotechnology'. The age of nanotechnology had begun.

The activity in the field of the nanotechnology has grown exponentially worldwide during the past three decades, becoming a major interdisciplinary area of research. This growth has been driven to a great extent by the integration of nanotechnology into the field of medical science, since nano-structured materials have unique medical effects. The control of materials in the nanometric range not only results in new medical effects but also requires novel, scientifically demanding chemistry and manufacturing techniques. This definition does not include traditional small-molecule drugs as they are not specifically engineered on the nanoscale to achieve therapeutic effects that relate to their nanosize dimensions. Nanoparticles have numerous functional moieties on their surfaces capable of multivalent conjugation for diagnostic, targeting, imaging and delivery of therapeutic agents (Figure 1.1).

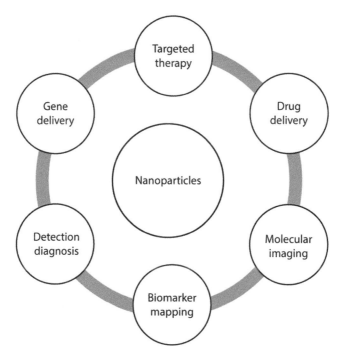

Figure 1.1   Biomedical applications of nanoparticles.

Owing to their unique characteristics, including large surface area, structural properties and long circulation time in blood compared with small molecules, nanoparticles have emerged as attractive candidates for optimized therapy through personalized medicine. Potential advantages of engineered therapeutic nanoparticles are the ability to convert unfavourable physicochemical properties of bioactive molecules to desirable biopharmacological profiles, to improve the delivery of therapeutic agents across biological barriers and compartments, to control the release of bioactive agents, to enhance therapeutic efficacy by selective delivery of drugs to biological targets and to perform theranostic functions by combining multimodal imaging and simultaneous diagnosis and therapy into multifunctional nanoplatforms.[5]

A handful of nanomaterials and nanoparticles are being studied in clinical trials or have already been approved by the US Food and Drug Administration (FDA) for use in humans and many proof-of-concept studies of nanoparticles in cell-culture and small-animal models for medical applications are under way.[6] Examples of such nanoparticles and nanomaterials are provided in Table 1.1.

Table 1.1 Nanoparticles and nanomaterials clinically approved, in clinical trials or in proof-of-concept research stages.

Nanomaterial/nanoparticle	Applications
Metallic	
Iron oxide	Magnetic resonance imaging/cancer therapy
Gold	*In vitro* diagnostics/cancer therapy
Gold (nanorods, nanoshells, nanocages)	Cancer therapy, diagnosis
Carbon structures	
Fullerenes	Cancer therapy (photodynamic therapy)
Carbon nanotubes	Fluorescence and photoacoustic imaging, antioxidant
Ceramic nanoparticles	
Silica	Cancer therapy, diagnosis
Alumina	Cancer therapy, diagnosis, computed tomography
Semiconductor	
Quantum dots	Fluorescent contrast, *in vitro* diagnostics
Organic	
Protein-based nanoparticles	Cancer therapy
DNA-based nanoparticles	Cancer therapy
Liposomes	Cancer therapy
Polymer nanoparticles	Cancer therapy
Polymer–drug conjugates	Cancer therapy
Polymeric micelles	Cancer therapy
Dendrimers	Microbiocides, cancer therapy
Nanogels	Gene and drug delivery
Bicelles	Topical delivery
Hybrid	
Magnetoliposomes	Magnetic resonance imaging/cancer therapy

As can be deduced from Table 1.1, there is a plethora of nanoparticles suitable for biomedical applications. This large number of nanoparticles is due to recent developments in synthetic methods, which mostly involve polymeric formulations, inorganic formulations or a combination of both. The resulting organic–inorganic hybrid materials inherit properties both of the polymers and of the metallic compounds.

Magnetoliposomes (liposomes encapsulating iron nanoparticles) are the most relevant example of hybrid nanoparticles. In this book, however, we will restrict ourselves to soft nanoparticles. More specifically, our choice is a compromise between well-known particles (*e.g.* polymeric micelles) and new particles whose potential is exciting but not yet widely proved.

1.2   Soft Nanoparticles

Classical micelles were the type of soft nanosystems used in pharmaceutical applications long before the emergence of nanotechnology. As a result of micelle formation, an organic compound that would normally be insoluble in water can be 'dissolved' in a surfactant solution because it can move into the oily interior of the micelle. This phenomenon, known as *solubilization*, has been used for solubilize drugs. As a few representative examples, phenolic compounds are frequently solubilized with soap to form clear solutions, which are widely used for disinfection. Non-ionic surfactants are efficient solubilizers of iodine. Such iodine–surfactant systems (referred to as iodophors) are more stable than iodine–iodide systems. On the other hand, the low solubility of steroids in water presents a problem in their formulation for ophthalmic use. The requirement for optical clarity precludes the use of oily solutions or suspensions. The use of non-ionic surfactants permits the production of clear solutions, which are stable to sterilization. This type of surfactant has also been used to solubilize essential oils and water-insoluble vitamins.[7]

Among the first nanotechnology drug-delivery systems were lipid vesicles, which were described in the mid-1960s and later became known as liposomes.[8] At first, they were used to study biological membranes; several practical applications, most notably in drug delivery, emerged in the 1970s (in this period, Donald Tomalia invented, named and patented the dendrimers, although they were not used as drugs until 1980s). Soon, however, it was observed that liposomes suffered an important drawback when used as carriers for therapeutically active compounds: they undergo rapid degradation due to the macrophage phagocyte system (MPS). As with all foreign colloidal particles, liposomes are quickly recognized as 'non-self' and taken up by the cells of the MPS, chiefly macrophages in the liver and spleen. This leads to the inability to achieve sustained drug delivery over a prolonged period of time.[9] The incorporation in the bilayer of cholesterol and, mainly, biocompatible, hydrophilic polymers with a flexible main chain, such as poly(ethylene glycol) (PEG), led to long-circulating liposomes, also known as sterically stabilized liposomes or Stealth liposomes.[10] Doxil, the first nanomedicine to secure regulatory approval by the FDA (for the treatment of AIDS-associated Kaposi's sarcoma in 1995 and in Europe in 1997 with the brand name Caelyx), was obtained by encapsulating doxorubicin within liposomes. At present, liposomes are the most commonly used soft nanoparticles for clinical applications, especially in the treatment of cancer and systemic fungal infections. The number of liposomal products on the market or in advanced clinical studies exceeds two dozen. Apart from liposomes, polymer-based nanoformulations constitute the majority of the nanoparticle therapeutic agents available for clinical use. Polymer–drug conjugates are another extensively studied nanoparticle drug-delivery platform currently in clinical practice. Many polymers have been proposed as drug-delivery carriers, but only a few of them with linear architecture have been accepted into clinical practice. PEG was first introduced into clinical use in the early 1990s. Today, there are around a dozen examples of PEGylated drugs in clinical practice. Other macromolecule–drug conjugates have also been developed as drug carriers, such as Abraxane, a 130 nm albumin-bound paclitaxel that was approved by the FDA in 2005 as a second-line treatment for patients with breast cancer.[11]

Biodegradable polymeric micelles with a size of 10–200 nm have attracted considerable attention as drug-delivery nanocarriers and have shown remarkable therapeutic potential. Polymeric micelles are formed by self-assembly of block copolymers consisting of two or more polymeric chains with different hydrophobicity. These copolymers spontaneously assemble into a core–shell micellar structure in an aqueous environment to minimize the Gibbs energy.[12]

When short-chain phospholipids were combined with long-chain phospholipids, structures closely related to liposomes and also with micelles were found: the bicelles. They were used in the 1990s as a model membranes well suited to magnetic resonance studies of membrane protein structure because of their ability to orient in a magnetic field, and have been used in most NMR structural studies of transmembrane proteins. As a biomedical tool, the use of bicelles has been proposed to favour the penetration of encapsulated drugs through the corneum stratus of the skin.[13]

Dendrimers have emerged as another novel class of drug-delivery soft nanoparticle platform because of their well-defined architecture and unique characteristics. The specific molecular structure of dendrimers enables them to carry various drugs using their multivalent surfaces through covalent conjugation or electrostatic adsorption. Alternatively, dendrimers can be loaded with drugs using the cavities in their cores through hydrophobic interaction, hydrogen bonds or chemical linkages.

As mentioned previously, there are promising candidates for nanoparticles whose clinical applications are still limited. Among them are polymer-based soft nanoparticles, which are very interesting for use as nanosized drug carriers. An ideal drug carrier needs to combine both the targeting property and the stimulus responsiveness to enhance the bioavailability of the drug together with the reduction of side effects. Therefore, the design of stimuli-responsive nanoparticles for drug delivery to release the drug in a controlled way when arriving at the targeted site is highly desirable. Among polymer-based soft nanoparticles there are stimulus-responsive nanoparticles or environment-sensitive nanoparticles having the ability to change their size or volume when exposed to external changes or signals. These nanoparticles are also known as micro/nanogels. Stimuli-responsive soft nanoparticles are classified depending on the stimulus (physical or chemical) into temperature, electric and magnetic field, light intensity, pressure, pH, ionic strength, specific (bio)molecules or enzymes and ultrasound-responsive nanoparticles. The development of soft stimulus-responsive nanoparticles for biomedical applications relies on the stimulus-sensitive polymer that constitutes their structure. Different polymer synthesis techniques are used and advanced polymerization processes have been developed to produce new stimuli-responsive nanoparticles, which are sensitive to other signals such as microwave, redox and different chemical substances, in addition to those already mentioned. Although the synthesis of new stimuli-responsive soft nanoparticles has attracted considerable interest in recent decades, in particular in the case of dual/multi-stimulus-responsive types, practical clinical applications remain limited except for thermo- and pH-sensitive nanoparticles with high sensitivities.[14]

In recent years, several attempts to prepare thermo-responsive hybrid micro/nanogels with inorganic silica cores have been reported.[15–18] These core–shell hybrid nanogels have interesting additional properties compared with polymer nanogels, being promising candidates as controlled drug-delivery vehicles in cancer therapies.[18,19] From the synthetic point of view, in almost all studies the preparation of these nanohybrids consisted of three steps. First, the silica nanoparticles are prepared. The second step is the functionalization of silica with a coupling agent, mainly 3-(trimethoxysilyl)propylmethacrylate (TPM). This coupling agent generates a hydrophobic surface on the silica particles and in addition can subsequently react by radical polymerization, allowing chemical coupling between the polymer network and the inorganic material. The last step consists of the well-known emulsion polymerization used to synthesize nanogels. Ramos *et al.*[20] reported a facile synthesis of thermo-responsive nanohybrids with a silica core and a poly(*N*-vinylcaprolactam) (PVCL)-based thermo-responsive shell. This was achieved by a batch emulsion polymerization of VCL with TPM and *N,N'*-methylenebisacrylamide (MBA) as cross-linkers. The hybrid character of these thermo-responsive nanogels together with the excellent biocompatibility conferred by silica and PVCL makes them suitable as carriers for drug delivery and biosensing.

As Lendlein and co-workers claimed,[21] we should conclude that, in order to bridge the gap between medical-grade material availability and material demand, there needs to be clear communication between the scientists who design and produce polymeric materials and those who implement them in biomedical applications and finally in the clinic.

Gene therapy is another field where soft nanoparticles might play a fundamental role in the near future. The concept of gene therapy appeared formally in the 1970s but the first proof of *in vitro* gene transfer was not reported until the 1980s, when white blood cells were extracted from the body of a girl.[22] Certain types of genes were implanted in such cells and then they were transferred back to the girl's body.[23] An improvement in the immune system of this person was reported. Since then and up to the present, trials on various diseases have continued.

As genes cannot be directly inserted into a patient's cells, they need a shuttle (carrier), named vectors. Usually, these are classified into viral and non-viral vectors. The first studies started with the use of a harmless virus altered in the laboratory. The genes were inserted into the virus and then they were mixed with the cells of the patient. The death of a patient in a clinical trial[24] and a blood disorder in children after another clinical trial caused a delay in the advancement of these techniques.[25] In any case, the interest in the development of non-viral vectors has grown considerably in recent times owing to this kind of risk. DNA particles are considered good non-viral vectors.

1.3   Colloidal Aspects of Soft Nanoparticles

In spite of constituting a motley assortment, nanoparticles share a common feature: they belong to the colloidal domain and we do not refer to size exclusively. The percentage of atoms at the surface of a material becomes significant when this material is split into nanosized particles. In fact, many interesting properties of nanoparticles can be mostly attributed to the large surface area of the material. The relevance of surface properties is a typical feature of colloids. Owing to this large surface-to-volume ratio, the Gibbs energy of a colloidal dispersion is generally higher than the Gibbs energy of the bulk material. Hence a colloidal dispersion will tend to lower its Gibbs energy unless there is some substantial energy barrier to prevent it, thus keeping the system in a metastable state. In any case, the colloidal nature of nanoparticles should not be neglected. In many cases, they are synthesized, stored and used in more or less concentrated suspensions.

A long-standing issue in colloid science is the effective force between colloidal particles,[26,27] which is the result of different supramolecular (or non-covalent) interactions. It should further be stressed that such interactions also affect the supramolecular constituents of soft nanoparticles (such as polymers, polyelectrolytes or lipids). In fact, the structure and properties of isolated nanoparticles and their suspensions depend on these constituents and the interactions between them. Some of these interactions depend strongly on the specific type of supramolecular entity involved. However, there are other forces that are rather non-specific, because they affect all (or almost all) kinds of supramolecular entities. For instance, van der Waals forces (due to the interaction between either permanent or induced dipoles) are always present (unless the refractive indexes of the dispersed and continuous phases are fairly similar). The excluded volume repulsion between hard entities (due to the impossibility of any overlap) can also be considered a non-specific interaction.

Polyelectrolytes and, in general, many macromolecules often carry an electrical charge and therefore attract or repel each other. However, the treatment of electrostatic forces between charged macromolecules (or nanoparticles) is not a trivial matter, for several reasons. On the one hand, the effective electrostatic interaction is screened by ions in aqueous solutions, but the role of ions goes beyond screening. For example, in the case of multivalent ions, strong ionic correlations can give rise to somewhat counterintuitive phenomena, such as electrostatic attraction between like-charged particles or (non-specific) charge inversion. On the other hand, the shape and charge distribution also affect electrostatic ion-mediated forces. For instance, the attraction between oppositely charged non-spherical proteins with complementary shapes is considerably increased compared with spherical particles.[26] The situation becomes even more complex when highly specific interactions are considered (such as hydrogen bonding). In some cases, the role of the solvent is not limited to a mere dielectric screening, as shown by a couple of examples: (1) water-mediated hydrophobic forces can be responsible for the collapse of polymer chains in aqueous solutions; and (2) water molecules play a key role in ion-specific effects (also known as Hofmeister's effects). Concerning nanoparticles, we should finally mention that steric forces between polymer-covered nanoparticles produce an additional repulsive force (which is predominantly entropic in origin), whereas solutions containing a non-adsorbing polymer can modulate interparticle forces, producing an attractive depletion force between particles.

In the case of charged soft nanoparticles, the role of electrostatic interactions goes beyond colloidal stabilization. It is well established that such interactions control processes in different areas of materials science. For instance, the electrostatic attraction between oppositely charged supramolecular entities is the foundation of assembly techniques (*e.g.* layer-by-layer assembly) that allow the synthesis of many nanoparticles.[28] However, electrostatic interactions are also essential in the function of biological entities. As a result of the charges of nucleic acids or proteins, the association of biomolecules into functional units is strongly dependent on local ionic concentrations. We should therefore expect that charge distributions will be important for developing functional soft nanoparticles. For example, asymmetric ionic profiles and interface charge excesses are crucial in the organization and function of ionic hydrogels. These interfaces generate a broad range of responses to external stimuli, including external electric fields.[29]

1.4 Measuring the Properties of Soft Nanoparticles

In a general sense, many of the fundamental chemical and physical properties of nanoparticles (NPs), such as size, shape, surface charge and internal structure, might have a strong influence on a number of biomedical applications such as drug delivery, targeting or gene therapy. Concerning the novel field of applications of NPs in medicine and biology, great emphasis has been put on the chemistry of new NPs in the past, as some authors have pointed out.[30,31] However, they also suggest the importance of improving physics-derived solutions (including simulations) for controlling processes in which NPs are involved. At present, it is clear that the physical properties mentioned above can have a great impact on the behaviour of NPs inserted in a biological environment. For that reason, Chapter 2 focuses on key physical properties of NPs and the methods and techniques used for their measurement.

Particle size is the first property of a nanoparticle that has to be known. Several important *in vivo* and *in vitro* functions of particles depend on their size, as can be widely corroborated through this book and further literature. A typical example can be found in Chapter 6, devoted to DNA particles and their use in gene therapy. In all the cases studied, the transfection capabilities of different DNA particles are strongly dependent on particles size. The literature also provides many examples of this kind. Transport and adhesion of nanoparticles in blood vessels, airways or gastrointestinal tract is governed by the size of NPs.[32,33] In passive and active targeting, as can be seen in Chapter 5, the size of polymeric micelles makes them suitable as drug carriers circulating in the bloodstream without losing their properties and extravasing to specific tissues. Chapter 7 offers another example related to dendrimers. The high level of control over the dendritic size makes them ideal systems for enhanced solubility of poorly water-soluble drugs.

The shape of NPs is another valuable property that has to be known. Particle shapes commonly observed include rods, spheres, filaments, toroids and globules. Recent studies confirm the important role that this property plays in biotechnology.[34,35] It is worth mentioning some examples. In drug delivery, phagocytosis by macrophages is strongly dependent on shape.[36] The aspect ratio of cylindrical particles influences their degree of internalization, as reported by Gratton *et al.*[37] Geng *et al.*[38] studied the effects of particle shape (comparing filaments and spheres) on flow and drug delivery. In comparing spheres ranging in size from 0.1 to 10 μm with an elliptical disc up to 3 μm, Muro *et al.*[39] concluded that the targeting efficiency of spheres is less than that of any ellipsoidal disc. Indeed, DeCuzzi *et al.* also emphasized the importance of particle geometry in processes such as endocytosis or intravascular delivery.[6,40] Chithrani *et al.*[41] determined the size and shape dependence of gold nanoparticle uptake into mammalian cells. In the present book, Chapter 8, devoted to bicelles and their application to the skin, provides another illustrative example: by varying the bicelle lipid composition and/or lipid ratios, the resulting morphologies and sizes may be varied to obtain more or less superficial effects. In the case of dendrimers, by varying the composition of the core, the interior branches and the surface functional groups, it is possible to obtain NPs with different shapes and sizes that, in consequence, could be good candidates for different applications. In DNA particles, when NPs are formed by lipids and DNA, the possible packing arrangements of the lipid bilayers lead to the formation of spherical, ellipsoidal or globular vesicles[27,42,43] with the corresponding influence on the overall transfection efficiency of these systems. Accordingly, the fundamental techniques used for the measurement of the size and shape of NPs are presented in Chapter 2. They can be divided into *direct* imaging techniques, such as scanning and transmission electron microscopy (SEM and TEM), and *indirect* techniques, particularly those based on the scattering of radiation and/or particles, such as light, neutrons or X-rays.

Surface charge is another basic physical property of outstanding interest for NPs. Several examples extracted from this book and from the literature support this statement. Regarding the magnetoliposome (ML) uptake in cells, the knowledge of the ML surface charge and the content of anionic-cationic lipids used are of paramount interest. What is more, the cytotoxicity of the ML is strongly influenced by the interaction of the cell medium with the ML surface charge (see Section 3.3.5). In addition, cationic dendrimers capable of condensing genetic material interact with negatively charged membranes of cells. Indeed, as can be seen in Section 7.3.3, there is strong evidence that the charged surface groups in cationic dendrimers play a key role in the solubilization of protein aggregates, responsible for Alzheimer's disease, Huntington's disease and Creutzfeldt–Jakob disease, among others.

These examples illustrate the important effects that electrostatic interactions derived from the surface charge of NPs have on their applications. In our opinion, as Editors, we feel that this aspect has not been soundly addressed in reviews and books devoted to the application of NPs in biomedicine. This is why we decided to include a large section (Sections 2.3 and 2.4) in which the ion cloud surrounding a colloidal particle and the methods for measuring the surface charge density of NPs are discussed.

In general, the internal and surface structure of NPs has been the focus of attention of considerable research in recent years, in order to study the relationship between the morphological characteristics of NPs and their functional activity in the different processes in which they are involved. Soft particles, as considered in this book, have non-homogeneous structures. MLs are pearl-like objects with a shell formed by phospholipids that includes grains of iron. Micro-nanogels are core–shell structures with the core formed by a cross-linker and polymer networks and the shell mainly formed by polymer chains. Polymeric micelles and dendrimers are more or less branched structures, mainly characterized by *holes* inside them. Hence in drug delivery they can bear a therapeutic agent that can be either dispersed in the polymer matrix or encapsulated by the polymer. Consequently, the knowledge of the internal and surface structures of the particles is a crucial issue. DNA particles are structures that can be considered as aggregates formed by a *platform* made of a colloidal particle plus nucleic acids (DNA, RNA), but what about their internal structure? Bicelles, in a rough image, are considered as disk-shaped aggregates of two kinds of phospholipids. Finally, core–shell hybrid particles can exhibit a broad variety of structures where a drug can be encapsulated either in the shell or in the core. Therefore, for experienced researchers and tyro students, it is of great interest to establish methods for determining the structure of NPs. Again, the scattering of light, neutrons or X-rays and also fluorescence and NMR techniques are valuable tools.

1.5 Computer Simulations and Soft Nanoparticles

Computer simulations of nanoparticles are one of the greatest challenges that soft matter science faces today owing to the time and length scales involved in these systems. For the same reason, the level of detail employed in computer simulations varies greatly. Two widely used representations of reality in simulations are atomistic models and coarse-grained (CG) models. The number of atomistic simulations, in which macromolecules and the surrounding solvent are explicitly modelled in full detail, has grown recently together with the size of computationally tractable systems. However, many questions regarding soft matter involve length and time scales that significantly exceed those that can be treated in atomistic simulations (not only nowadays but also for many years to come). Hence simulating the solvent explicitly is prohibitively expensive and avoided in many GC simulations. In addition, a single bead can even represent a small group of atoms in this kind of simulation (see Figure 1.2). In this way, the computational time is reduced by orders of magnitude (compared with atomistic simulations) and a meaningful exploration of the system is feasible. It should be stressed, however, that there is also a wide variety of different approaches in coarse-graining of soft materials, including integrated multiscale modelling, in which several models at different scales are interconnected.[44]

In any case, computer simulations are currently a powerful tool for achieving a better understanding of diverse aspects of soft nanoparticles, and some illustrative examples related to nanoparticles studied in this book are presented. Both atomistic and GC models have been employed in the simulation of lipid bilayers.[45] Their conclusions can be extremely useful for the comprehension of liposomes, MLs and bicelles. In fact, bicelles have been explicitly simulated.[46] For instance, the effect of the system composition on the edge structure, the edge composition and the line tension has been analysed.[47] A considerable excess of the short-chain lipids at the bilayer edge and a reduction in line tension were observed in bilayers predominantly formed of the long-chain component. This enrichment of short-chain lipids at the bicelle edge is corroborated by experiments. However, there is an excess of long-chain lipids near the edge and a slight decrease in line tension (compared with a bilayer of pure short-chain lipids) in bilayers predominantly composed of short-chain lipids. Bicelles have also been used as starting structures in GC simulations of the formation of vesicles.[48]

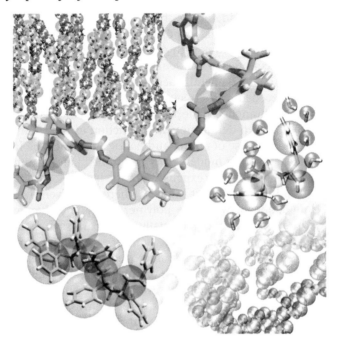

Figure 1.2   Coarse-grained superimposed with backmapped atomistic structures of a bisphenol A–polycarbonate chain and a polystyrene oligomer, of liquid crystalline compound 8AB8 (top left corner), of the aggregates formed by diphenylalanine (bottom right corner) and of a single diphenylalanine molecule with its water shell (bottom left).
Reproduced from Ref. 44 with permission from the Royal Society of Chemistry.

Computer simulations are very useful for the treatment of electrostatic interactions, which play a key role in many biological functions and self-assembly.[28] In Chapter 10, we analyze an illustrative example in depth: The complexation between nanoparticles and DNA-like polyelectrolytes is an essential mechanism in the binding of DNA to nanoparticles (see Chapter 6). In fact, electrostatic interactions control (to a great extent) the adsorption of charged polymers on charged surfaces and interfaces,[49] which is a long-standing issue in polymer physics but also a process used in the synthesis of many hybrid nanoparticles (Chapter 9). The electrostatic attraction between oppositely charged macromolecules is also the foundation of layer-by-layer assembly, a technique that provides a powerful tool for the synthesis of multicomponent coatings. Molecular dynamics simulations of the sequential adsorption of oppositely charged polyelectrolytes on a charged spherical particle have confirmed that the layer build-up proceeds through surface overcharging during each deposition step.[50]

Nanoparticles consisting of interconnected polymer chains are usually represented by a CG model known as the bead-spring model, in which monomer units are considered as hard spheres connected with their neighbours by harmonic bonds. Dendrimers have been simulated through these GC models.[28] Simulations with explicit counterions have shown that the phenomenon of counterion condensation has a significant effect on the conformations of charged dendrimers, namely the collapse of these macromolecules preceded by their swelling.[51] On the other hand, atomistic simulations of these branched nanoparticles have established the structural characteristics of charged dendrimers as a function of the solution pH.[52]

The swelling and collapse of micro- and nanogels can also be investigated with the help of bead-spring models.[46] As mentioned before, simulating the solvent explicitly is prohibitively expensive for the length scales of interest in charged nanogels. Implicit solvent models, where the solvent is replaced by an effective interaction between the polymer beads, are therefore of considerable importance, given the potentially significant savings in computational time. Even without an explicit solvent, the number of particles required for the simulation of a whole microgel could be extremely high. In fact, simulating microgel particles with diameters greater than 100 nm can become an unaffordable task. Therefore, simulating a small and inner piece of the microgel assuming that the rest of the gel behaves in the same manner becomes a feasible alternative. For this reason, infinite polyelectrolyte networks have mostly been simulated (replicating a small piece of them) during the last decade.

The assumption of an infinite network has been used by many workers when they try to apply the classical formalism for gels to microgels.[53] This could be reasonable for gels and even for microgels if their particle size is large enough, but it seems questionable for nanogels. In any case, the hypothesis of an infinite network involves two additional shortcomings: (1) the counterions that neutralize the gel charge must reside inside the network; and (2) it is not possible to consider the ionic atmosphere (the electric double layer) around the nanogel. If the network is small enough (of a few nanometres), the nanogel can be explicitly simulated, including the ionic atmosphere around it.[46,54]

However, temperature-sensitive micro- or nanogels have scarcely been simulated to date. It is worth considering the case of thermo-shrinking microgels in more detail as an illustrative example of how poor solvents can be accounted for in GC models through an effective interaction between the polymer beads. In water, a solvent-mediated hydrophobic attractive interaction between non-polar monomer beads is required. Admitting that the precise knowledge of hydrophobic forces and their functional form is far from complete, different functional forms have been proposed. For instance, it is possible to consider pairwise additive Lennard–Jones or square-well potentials.

In the case of thermo-shrinking micro- and nanogels [such as poly(*N*-isopropylacrylamide)-based microgels], the depth of this potential must increase with increase in temperature. This feature can be justified on molecular grounds as follows.[53] When non-polar molecules (or macromolecules) are inserted into an aqueous medium, the water molecules must rearrange their hydrogen bonds to form a structure (known as clathrate or iceberg) that surrounds the non-polar molecule like a cage. As a result of this phenomenon, non-polar molecules become soluble in water. When the temperature increases, the total number of water molecules structured around the hydrophobic solute decreases because this ice-like structure partially melts. As a result of this reduction in the clathrate, hydrophobic molecules aggregate (to maximize the volume-to-surface area ratio). Consequently, an increase in temperature promotes the hydrophobic attraction. This is just a qualitative conclusion, which is also obtained from thermodynamics. Kolomeisky and Widom derived a pair potential from a model encompassing the essential features of the hydrophobic mechanism (restricted orientations, ordering and energy release for solvent molecules near non-polar solutes).[55] This potential also predicts that hydrophobic forces must increase with increase in temperature, in agreement with our rationale. Unfortunately, our preliminary simulations revealed that the Kolomeisky–Widom (KW) potential overestimates the value of this interaction at high temperatures, yielding volume fractions larger than 1. For that reason their potential was replaced in subsequent work by an empirical expression, which is a sigmoid approximation of the square-well potential whose depth increases with increase in temperature (the main feature of the KW potential) but reaches a maximum at a given temperature.[56] This phenomenological potential can qualitatively and semiquantitatively reproduce swelling data for [poly(N-isopropylacrylamide-co-acrylic acid)]-based microgels with different proportions of charged groups per chain. In recent times, thermo-shrinking nanogels have been explicitly simulated with an effective interaction potential.[56] Concerning the thermal behaviour, the results obtained for nanogels agree qualitatively with those obtained previously for macroscopic gels. In addition, they confirm that the swelling decreases when monovalent counterions are replaced by divalent counterions. In relation to the charge profiles, these simulations reveal that, in general, charged beads and counterions are not evenly distributed in the inner region of the nanogel. Uncharged beads tend to form clusters when the temperature increases due to hydrophobic forces, whereas charged beads are segregated and mostly exposed to the solvent (see Figure 1.3). Simulations also show that the surface electrostatic potential increases when nanogels collapse.

Finally, the reader should keep in mind that the above-mentioned hydrophobic attraction between polymer beads is solvent mediated but these particles could also undergo direct (not mediated) attractive interactions. For instance, hydrogen bonds between polymer units can induce attractive forces between polymer chains and the collapse of the network. The rupture of such hydrogen bonds when the temperature increases would contribute in this case to the swelling of the gel (thermo-swelling behaviour), whereas the rupture of hydrogen bonds between water molecules of the clathrate contributes to the shrinking (as discussed previously). The competition between these two effects can give very rich behaviours.

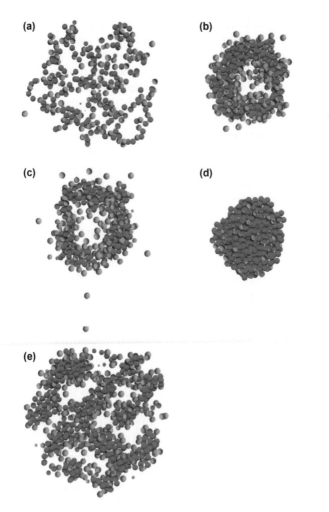

Figure 1.3 Snapshots of slices of the central part of the simulation box with dimensions $2 \times 20 \times 20$ nm to visualize cross-sections of the nanogels. (a) At 288 K with divalent counterions; (b) at 333 K with divalent counterions; (c) at 333 K with monovalent counterions; (d) at 333 K for the uncharged nanogel: (e) Nanogel with the same ratio of charged beads but larger chains in the presence of divalent counterions at 333 K. Blue, red, green and orange spheres symbolize uncharged beads, charged beads, cross-linkers and counterions, respectively.
Reproduced from Ref. 54 with permission from the Royal Society of Chemistry.

Acknowledgements

The authors acknowledge the support of the Spanish "Ministerio de Economía y Competitividad (MINECO)", Projects: MAT2012-36270-C04-01-02-03-04.

References

1. British Standards Institution, *Terminology for Nanomaterials, Publicly Available Specification BS PAS 136*, British Standards Institution, London, 2007; https://archive.org/details/bs.pas.136.2007.
2. V. Wagner, A. Dullaart, A.-K. Bock and A. Zweck, *Nat. Biotechnol.*, 2006, **24**, 1211.
3. M. Faraday, *Philos. Trans. R. Soc. London*, 1857, **147**, 145.
4. R. P. Feynman, *Eng. Sci.*, 1960, **23**, 22.
5. V. Sanna, N. Pala and M. Sechi, *Int. J. Nanomed.*, 2014, **9**, 467.
6. B. Y. Kim, J. T. Rutka and W. C. Chan, *N. Engl. J. Med.*, 2010, **363**, 2434.
7. D. Attwood and A. T. Florence, *Surfactant Systems, Their Chemistry, Pharmacy and Biology*, Chapman and Hall, London, 1983.
8. A. D. Bangham, M. M. Standish and J. C. Watkins, *J. Mol. Biol.*, 1965, **13**, 238.
9. D. Lasic, *Liposomes: from Physics to Applications*, Elsevier, Amsterdam, 1993.
10. V. P. Torchilin, *Nat. Rev. Drug Discov.*, 2005, **4**, 145.
11. L. Zhang, F. X. Gu, J. M. Chan, A. Z. Wang, R. S. Langer and O. C. Farokhzad, *Clin. Pharmacol. Ther.*, 2008, **83**, 761.
12. V. P. Torchilin, *Pharm. Res.*, 2007, **24**, 1.
13. U. N. H. Dürr, M. Gildenberg and A. Ramamoorthy, *Chem. Rev.*, 2012, **112**, 6054.
14. J. Ramos, J. Forcada and R. Hidalgo-Alvarez, *Chem. Rev.*, 2014, **114**, 367.
15. M. Karg, I. Pastoriza-Santos, L. M. Liz-Marzán and T. Hellweg, *ChemPhysChem*, 2006, 7, 2298.
16. V. Lapeyre, N. Renaudie, J.-F. Dechezelles, H. Saadaoui, S. Ravaine and V. Ravaine, *Langmuir*, 2009, **25**, 4659.
17. Z. Xing, C. Wang, J. Yan, L. Zhang, L. Li and L. Zha, *Soft Matter*, 2011, 7, 7992.
18. B. Chang, D. Chen, Y. Wang, Y. Chen, Y. Jiao, X. Sha and W. Yang, *Chem. Mater.*, 2013, **25**, 574.
19. B. Chang, X. Sha, J. Guo, Y. Jiao, C. Wang and W. Yang, *J. Mater. Chem.*, 2011, **21**, 9239.
20. J. Ramos, R. Hidalgo-Alvarez and J. Forcada, *Soft Matter*, 2013, **9**, 8415.
21. A. Lendlein, B. F. Pierce, L. Ambrosio and D. Grijpma, *Acta Biomater.*, 2012, **12**, 4199.
22. B. Lee and B. L. Davidson, *Hum. Mol. Genet.*, 2011, **20**(1), R1.
23. P. M. Patil, P. D. Chaudhari, S. Megha and N. J. Duragkar, *Int. J. Genet.*, 2012, **4**(1), 74.
24. S. Lehrman, *Nature*, 1999, **401**, 517.
25. E. Marshall, *Science*, 2002, **298**, 510.
26. V. Dahirel and M. Jardat, *Curr. Opin. Colloid Interface Sci.*, 2010, **15**, 2.
27. J. N. Israelachvili, *Intermolecular and Surface Forces*, Academic Press, London, 1992.
28. A. V. Dobrynin, *Curr. Opin. Colloid Interface Sci.*, 2008, **13**, 376.

29. M. Olvera de la Cruz, *Soft Matter*, 2008, **4**, 1735.
30. S. Mitragotri and J. Lahann, *Nat. Mater.*, 2009, **8**, 15.
31. M. Donkuru, I. Badea, S. Wettig, R. Veralll, M. Elsabahy and M. Foldvari, *Nanomedicine*, 2010, **5**, 1103.
32. V. R. S. Patil, C. J. Campbell, Y. H. Yun, S. M. Slack and D. J. Goetz, *Biophys. J.*, 2001, **18**, 1733.
33. A. Lamprecht, U. Schafer and C. M. Lehr, *Pharm. Res.*, 2001, **18**, 788.
34. J. A. Champion, Y. K. Katare and S. Mitragotri, *J. Control. Release*, 2007, **121**, 3.
35. P. Decuzzi and M. Ferrari, *Biophys. J.*, 2008, **94**, 3790.
36. J. A. Champion and S. Mitragotri, *Proc. Natl. Acad. Sci. U. S. A.*, 2006, **103**, 4930.
37. S. E. Gratton, P. A. Ropp, P. D. Pohlhaus, J. C. Luft, V. J. Madden, M. E. Napier and J. M. DeSimone, *Proc. Natl. Acad. Sci. U. S. A.*, 2008, **105**, 11613.
38. Y. Geng, P. Dalhaimer, S. Cai, R. Tsai, M. Tewari, T. Minko and D. E. Discher, *Nat. Nanotechnol.*, 2007, **2**, 249.
39. S. Muro, C. Garnacho, J. A. Champion, J. Leferovich, C. Gajewski, E. H. Schuchman, S. Mitragotri and V. R. Muzykantov, *Mol. Ther.*, 2008, **16**, 1450.
40. P. Decuzzi, R. Pasqualini, W. Arap and M. Ferrari, *Pharm. Res.*, 2009, **26**, 235.
41. B. D. Chithrani, A. A. Ghazani and W. C. Chan, *Nano Lett.*, 2006, **6**, 662.
42. G. Tresset, *PMC Biophys.*, 2009, **2**, 3.
43. A. González-Pérez, M. Schmutz, G. Waton, M. J. Romero and M. P. Kraft, *J. Am. Chem. Soc.*, 2007, **129**, 756.
44. C. Peter and K. Kremer, *Soft Matter*, 2009, **5**, 4357.
45. S. J. Marrink, A. H. de Vries and D. P. Tieleman, *Biochim. Biophys. Acta*, 2009, **1788**, 149.
46. J. Ramos, A. Imaz, J. Callejas-Fernandez, L. Barbosa-Barros, J. Estelrich, M. Quesada-Perez and J. Forcada, *Soft Matter*, 2011, 7, 5067.
47. J. de Joannis, F. Y. Jiang and J. T. Kindt, *Langmuir*, 2006, **22**, 998.
48. S. J. Marrink and A. E. Mark, *J. Am. Chem. Soc.*, 2003, **125**, 15233.
49. R. S. Pais and A. A. C. C. Pais, *Adv. Colloid Interface Sci.*, 2010, **158**, 48.
50. V. Panchagnula, J. Jeon and A. V. Dobrynin, *Phys. Rev. Lett.*, 2004, **93**, 037801.
51. A. A. Gurtovenko, S. V. Lyulin, M. Karttunen and I. Vattulainen, *J. Chem. Phys.*, 2006, **124**, 094904.
52. I. Lee, B. D. Athey, A. W. Wetzel, W. Meixner and J. R. Baker, *Macromolecules*, 2002, **35**, 4510.
53. M. Quesada-Perez, J. Alberto Maroto-Centeno, J. Forcada and R. Hidalgo-Alvarez, *Soft Matter*, 2011, 7, 10536.
54. M. Quesada-Perez and A. Martin-Molina, *Soft Matter*, 2013, **9**, 7086.
55. A. B. Kolomeisky and B. Widom, *Faraday Discuss.*, 1999, **112**, 81.
56. M. Quesada-Perez, J. Ramos, J. Forcada and A. Martin-Molina, *J. Chem. Phys.*, 2012, **136**, 244903.

CHAPTER 2

Experimental Techniques Used for the Characterization of Soft Nanoparticles

J. CALLEJAS-FERNÁNDEZ,[*a] J. RAMOS,[b] O. SANZ,[b]
J. FORCADA,[b] J. L. ORTEGA-VINUESA,[a] A. MARTÍN-MOLINA,[a]
M. A. RODRÍGUEZ-VALVERDE,[a] M. TIRADO-MIRANDA,[a]
A. SCHMITT,[a] B. SIERRA-MARTIN,[c] A. MALDONADO-
VALDIVIA,[c] A. FERNÁNDEZ-BARBERO,[c] R. PONS,[d]
L. F. CAPITÁN-VALLVEY,[e] A. SALINAS-CASTILLO,[e]
A. LAPRESTA-FERNÁNDEZ,[e] B. VÁZQUEZ,[f] M. R. AGUILAR[f] AND
J. SAN ROMÁN[f]

[a] Grupo de Física de Fluidos y Biocoloides, Departamento de Física
Aplicada, Facultad de Ciencias, Universidad de Granada, 18071 Granada,
Spain; [b] POLYMAT, Bionanoparticles Group, Departamento de Química
Aplicada, UFI 11/56, Facultad de Ciencias Químicas, Universidad del País
Vasco UPV/EHU, Apdo. 1072, 20080 Donostia-San Sebastián, Spain;
[c] Grupo de Física de Fluidos Complejos, Departamento de Física Aplicada,
Facultad de Ciencias, Universidad de Almería, 04120 Almería, Spain;
[d] Departament de Tecnologia Química i de Tensioactius, Institut de
Química Avançada de Catalunya, IQAC–CSIC, 08034 Barcelona, Spain;
[e] ECsens, Departamento de Química Analítica, Facultad de Ciencias,
Universidad de Granada, 18071 Granada, Spain; [f] Instituto de Ciencia y
Tecnología de Polímeros and CIBER-BBN, Juan de la Cierva 3, 28006
Madrid, Spain
*Email: jcalleja@ugr.es

RSC Nanoscience & Nanotechnology No. 34
Soft Nanoparticles for Biomedical Applications
Edited by José Callejas-Fernández, Joan Estelrich, Manuel Quesada-Pérez and
Jacqueline Forcada
© The Royal Society of Chemistry 2014
Published by the Royal Society of Chemistry, www.rsc.org

2.1   Introduction

This chapter is devoted to current techniques used to characterize soft nanoparticles (NPs). This task is directly related to the question of what kind of properties we want to measure. One may be interested in characterizing individual particles or in collective properties arising when particles form materials with a more or less compact internal structure in one, two or three dimensions.

Taking into account the final aims of this book, *i.e.* to describe applications of NPs in biomedicine and related fields, and potential readers (people mainly interested in these fields of knowledge), we decided to describe techniques that allow a basic set of fundamental properties of single NPs such as size, particle shape, surface charge and internal structure to be measured.

Obviously, the size of an NP is one of the first properties that has to be known. As has been recognized in recent years, there is no doubt that the particle shape is also of interest in biomedicine, owing to the important role that it plays in many aspects such as interaction between NPs and NPs or between NPs and living cells, the capability of NPs to encapsulate drugs or the possibility of NPs to cross cellular membranes. In addition to direct imaging techniques such as scanning electron microscopy and transmission electron microscopy (Section 2.2), we will also describe indirect techniques such as scattering of light, neutrons or X-rays (Section 2.5) to determine these parameters. The latter allow, for example, the molecular weight and the radius of gyration of the NPs to be assessed experimentally.

As the interested reader can check through the book, the surface charge of soft NPs and the ion clouds surrounding them are further outstanding properties that need to be known and characterized. These parameters play a crucial role in, for example, the interaction between NPs and living cells or macromolecules present in the human body. Hence we have devoted two sections to describe how these characteristics can be described and the corresponding parameters assessed experimentally. Section 2.3 tackles titration techniques and how they can be used to measure the particle surface charge. Electrokinetic techniques are described in Section 2.4. They assess mainly the electrophoretic zeta potential as one of the main parameters characterizing the electric double layer (EDL), *i.e.* the ion cloud surrounds charged particles in aqueous media.

Special attention is devoted to the internal structure of NPs, which refers to the spatial distribution of different molecules or macromolecules within the particles. It may be homogeneous or show inhomogeneities of different types, such as an inner core surrounded by an outer shell or a structure formed by the superposition of several layers of cross-linked or branched polymers. With respect to this issue, the properties and localization of particular chemical groups within the particles are of outstanding interest. Sections 2.5, 2.6 and 2.7 explain how scattering, fluorescence and NMR techniques can be employed to characterize the internal structure of NPs.

In a chapter devoted to experimental techniques and their applications, it is difficult to find an equilibrium between a thorough description of the corresponding theoretical background, the technical specification and details and data evaluation methods and procedures. After having reviewed the literature on NPs, the Editors came to the conclusion that imaging, fluorescence and NMR techniques are more frequently and deeply discussed than scattering techniques or techniques for characterizing the particle surface charge. Given the importance of the latter, the authors have put special emphasis on describing the physicochemical background and usefulness of these experimental techniques for characterizing NPs employed in biomedicine and related fields (Sections 2.3, 2.4 and 2.5).

2.2   Imaging Techniques

2.2.1   Electron Microscopy

Electron microscopy (EM) is an important asset to soft NP technology. It is well known that the particle shape, size and distribution of an emulsion or latex control the properties and end-use applications. Complex NP systems can be directly characterized with different EM techniques, such as transmission electron microscopy (TEM) and scanning electron microscopy (SEM).

EM offers numerous possibilities for the visualization and elemental analysis of colloidal systems. Solid and hydrated samples can be characterized after appropriate preparation. Analytical EM methods allow not only for the characterization of the spatial structure of different samples but also for their chemical composition.

An electron microscope uses electrons instead of light for the imaging of objects. The resolving power of a microscope is a linear function of the wavelength. Thus, the use of electrons, which have wavelengths about 100 000 times shorter than photons, allows a resolution of better than 50 pm.[1] The development of the electron microscope was based on theoretical work of Louis de Broglie, who found that the wavelength is inversely proportional to momentum.[2] In 1926, Hans Busch discovered that magnetic fields could act as lenses by causing electron beams to converge in a focus. Six years later, the first operational electron microscope was presented by Ernst Ruska and Max Knoll. The original form of the electron microscope was a TEM instrument that uses high-energy electrons for sample imaging. An SEM instrument, however, collects secondary electrons for obtaining an image.

The different EM techniques employed for analysis of soft NP systems are briefly presented in the following sections.

2.2.2   Transmission Electron Microscopy

Conventional transmission electron microscopes are electron optical instruments analogous to light microscopes.[3] The specimen, however, is not illuminated by light but by an electron beam. This requires operation in vacuum since gas molecules would deflect the electrons. At the top of the microscope column there is an electron gun and a system of electromagnetic lenses that focuses the electron beam on the sample. The image contrast in TEM is a consequence of electron scattering, *i.e.* the interaction of the electrons with the sample material. The resolution in TEM is directly proportional to the acceleration voltage of the electrons. High resolution is obtained due to the short wavelength of the electrons when the voltage increases. An increasing acceleration voltage, however, leads to poorer contrast since the scattering of the electrons is decreased at higher velocity.[4] Typical instruments are capable of voltages from 40 to 120 kV and microscopes in the range 200–400 kV are becoming more common. In TEM studies of colloidal systems, voltages between 80 and 200 kV are usually employed.[4]

The resolution in TEM is limited by the properties of the specimen rather than by those of the instrument. The present limit is at 0.3–1.0 nm.[5] Absorption of electrons by the specimen is unusual. Electrons scattered under larger angles, however, do not contribute to the image in the usual bright field mode and, therefore, appear to be absorbed. In the case of ordered or crystalline sample material, this gives rise to diffraction contrast which is strongly dependent on the crystal orientation. In amorphous materials, a mass thickness contrast is obtained given that the image brightness depends on the local mass thickness. Hence darker regions in the bright field image mode are regions of higher scattering.[6] For more detailed information on EM, the general literature is recommended.[7]

In the case of hydrated samples, such as soft NPs, cryoscopic methods are among the most suitable sample preparation techniques.[8] If cryo-TEM is not available, a conventional negative staining analysis with or without dilution can be performed on soft NPs, allowing certain basic information to be obtained. Staining techniques are frequently employed for imaging of colloidal systems with TEM since they are easy, fast and universally applicable.[4,9] The most common staining agents are salts of heavy metals such as molybdenum, tungsten and uranium with atomic numbers between 42 and 92. These agents must be benign to wet specimens, form a thin glassy layer upon drying and resist electron beam radiation damage to a satisfactory extent.[10] In NP analysis, phosphotungstic acid or its salt solutions[9,11] or uranyl acetate[12,13] are most frequently employed. During sample preparation, a droplet of the nanoemulsion is placed on a carbon-coated grid on which it is rapidly adsorbed. Subsequently, an aqueous solution of a heavy metal salt is applied for staining. The sample is then left to dry and finally observed by TEM at room temperature. This technique allows for the identification of the dehydrated shells of the nanoemulsion droplets, which are stabilized by a surfactant. The strongly scattering metal ions form an amorphous shield around the weakly scattering oil droplets in order to enhance the imaging contrast.[14] A high reverse contrast is thus seen in bright field TEM images, obtaining clearly visible structures against a darker background. Negative staining can be used to visualise the size, shape and internal structure of the sample. The negative stain provides not only contrast for weakly scattering specimens, but also physical support against collapse of the sample structure during drying and protection against electron beam damage.[10] The sample may also be coated with a carbon film under ambient conditions before analysis.[15]

However, it should be kept in mind that conventional EM is prone to artefacts in the case of surfactant solutions, *i.e.* hydrated colloidal dispersions.[16] For hydrated samples such as nanoemulsions, the factors responsible for the preservation of the structural integrity are identical in TEM and SEM.[17] Both drying and staining techniques can affect the structure and morphology of the sample, hence great care should be taken when interpreting the images obtained.[18] Severe shrinkage or even complete collapse, selective dimensional modification and aggregation of the constituents of colloidal systems due to complete dehydration and drying usually cause strong structural changes. In addition, the use of heavy metal salts leads to a selective sample appearance since only structures that can be reached by or react with the staining agents can be detected.[19] Consequently, the uppermost surface of the sample and also the compounds of the system that can react chemically with the staining agents are clearly observed while many compounds of the sample remain practically invisible. As a result, the final EM images may describe completely modified structures that have almost nothing in common with the original morphology. Moreover, conventional negative staining on continuous carbon support films bears the risk of sample distortion due to adsorption and flattening during the drying of the thin aqueous films of negative stain or evaporation in the TEM instrument.[20] Apart from adsorption artefacts, variable spreading and incomplete specimen coverage by the staining agent can lead to non-uniform staining results. Specimen distortion due to surface tension forces during evaporative drying or the formation of a saturated salt solution before the final drying may also occur.[10]

As noted above, TEM is only suitable for analyzing the particles in a dry state and therefore gives no information about the swollen state achieved at various temperatures. In order to overcome this problem, cryogenic techniques have been developed. Here, the aqueous sample is shock-frozen at the temperature of liquid nitrogen and the aqueous phase is converted into hyperquenched glassy water.[21,22] The particles are thus embedded in an amorphous water phase and no crystallization effects disturb their original structure.[23,24] The freezing-in of the sample proceeds very rapidly (10^4 K s^{-1}) and even states at elevated temperature may be frozen in.[25] As an applied example of cryo-TEM, Crassous *et al.* reported the first study of the transmission behaviour of thermoresponsive core–shell particles by cryo-TEM.[26] Figure 2.1 shows the cryo-TEM images for the core–shell microgel particles quenched from room temperature and from 45 °C. The volume transition in the shell from an expanded state at 23 °C to the shrunken state at 45 °C can be clearly seen. Moreover, the shell assumes a homogeneous state that is tightly wrapped around the core. This micrograph reflects also the state in which the shell is affixed to the cores.[27]

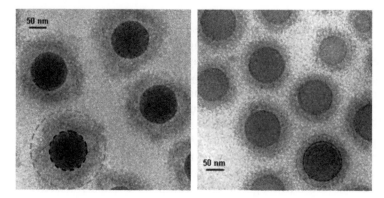

Figure 2.1   Cryo-TEM images of polystyrene/poly(*N*-isopropylacrylamide) core–
shell particles. The samples were kept at 23 °C (left) and 45 °C (right)
before vitrification. The circle around the core marks the core radius
determined by dynamic light scattering (DLS) in solution. The circle
around the entire particle gives the hydrodynamic radius of the core–
shell particles as determined by DLS.
Reprinted with permission from Ref. 26. Copyright 2006 American
Chemical Society.

Most soft matter systems, however, give comparatively poor contrast in
cryo-TEM. The inherently low contrast of unstained vitrified specimens can
be compensated to a large extent by an optimal use of phase contrast.[28] A
variation in focus can serve to reveal different parts of a structure while a
total defocus creates optical artefacts.[4] Additional staining techniques can
be employed to produce images with improved contrast.[29]

As a conclusion, EM techniques are an essential tool to obtain precise
information about basic structural properties of nanoparticle systems. Although both conventional SEM and TEM can be employed for this task,
more representative and artefact-free images can be obtained with cryogenic
methods of sample preparation.

2.2.3   Scanning Electron Microscopy

Scanning electron microscopy (SEM) has been used by researchers since
1935 to examine micro- and nanometre-sized structures.[30] The image is
formed point by point by scanning a focused electron beam across the
surface of a solid specimen. Electrons are dislodged from the atoms at the
surface of the samples and are attracted to a positively charged detector grid.
Various signals are generated as a result of the impact of the incident
electrons. The electrons that are collected by the detector system are registered for image formation or analyzed to obtain specific information about
the sample surface giving rise to different types of contrast.

The most common mode of operation is the detection of secondary electrons emitted by atoms excited by the impinging electron beam. When a set
pattern of primary electron beam scanning is used over the surface,

recording of the secondary electrons (SEs), *i.e.* all emitted electrons with exit energies below 50 eV, allows the surface topology to be displayed and a quasi-three-dimensional image to be recorded. Secondary electron detectors are standard equipment in all SEM instrumentation. SEM analysis is faster, less expensive and easier to perform than TEM. The main disadvantage of SEM, however, is the relatively complex sample preparation procedure.

For conventional SEM imaging, specimens must be electrically conductive, at least at the surface, and electrically grounded to prevent the accumulation of electrostatic charge at the surface. Metallic objects require little preparation for SEM. Non-conductive specimens such as polymeric NPs, however, are usually coated with an ultrathin layer (a few nanometres) of an electrically conducting material such as gold, platinum, osmium, tungsten or graphite that is deposited by means of low-vacuum sputtering or high-vacuum sublimation. Moreover, conventional SEM requires samples to be imaged under vacuum given that a gas atmosphere would rapidly spread and attenuate the electron beam. As a consequence, samples that produce a significant amount of vapour, *e.g.* NP suspensions, must be either dried or cryogenically frozen. Hence it is not always clear whether SEM images of polymeric NPs in the dispersed state are an exact representation of the original suspension. Xu *et al.* developed an improved specimen preparation technique for obtaining the morphology and size of colloidal NPs (see Figure 2.2).[31] In this study, dilution of the samples with an appropriate volatile solvent (ethanol) and different drying methods (freezing) were studied without destroying the natural particle morphology. The method described in their study, however, is not suitable for systems where chemical reactions may occur due to the presence of ethanol.

One form of EM that permits the imaging of samples in wet state is environmental scanning electron microscopy (ESEM), which employs a scanned electron beam and electromagnetic lenses to focus and direct the

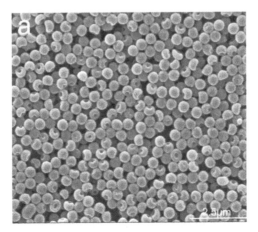

Figure 2.2   SEM images of poly(styrene–acrylic acid) microspheres.
Reprinted with permission from Ref. 31. Copyright 2011 John Wiley & Sons.

beam on the specimen surface in the same way as in conventional SEM. The ESEM instrument must have, just as in conventional SEM, a high-vacuum region for the generation and focusing of the electron beam. In SEM, the high-vacuum region also contains the sample whereas in ESEM it is placed in a separate region.[32] In this way, the chamber can be held at pressures of up to 20 Torr[33] while the gun and column remain at pressures of 7.5×10^{-7} Torr. Moreover, by using a correct pumpdown procedure[34] and by controlling the temperature of the sample usually by a Peltier element, dehydration can be inhibited and hence samples can be imaged in their natural state without the need for a conductive coating. Through adequate control of the temperature or the pressure, it is furthermore possible to produce evaporation conditions within the specimen chamber that allow examination of film formation processes.[35]

In the case of colloidal lattices, samples are not only wet but are also totally submerged in a liquid phase. In this case, only the top surface of the liquid is imaged under classical ESEM conditions with poor contrast and possible drifting of objects. Bogner *et al.* proposed the use of scanning transmission electron microscopy (STEM) modes that allow transmission observations of wet samples in an ESEM setup.[36] Since the particles are immerged in a layer of liquid, the specimens can be maintained in their original wet state. In this way, they are neither deformed nor collapsed and the size can be correctly estimated. Moreover, the fairly low voltages used in SEM (30 kV) in comparison with TEM give rise to a stronger electron–sample interaction, thus increasing the contrast of the images obtained.[37]

Another valuable SEM imaging method is the detection of backscattered electrons (BSEs). When the electrons of the incident beam (primary electrons) collide with the nucleus of an atom, especially a heavy atom, they may be scattered in any direction without losing much of their energy. Those electrons scattered backwards are referred to as 'backscattered' electrons and registered by a detector. BSE energy is directly related to the weight density of the atoms. Since heavy elements with a high atomic number backscatter electrons more strongly than light elements with a low atomic number, they appear brighter in the image. Consequently, the images obtained allow the variations of surface composition to be visualized.

When elemental analysis becomes necessary, energy-dispersive spectrometry (EDS) has to be used. In this technique, electrons are bombarded on the sample surface, knocking out electrons from inner atomic shells that are subsequently filled with electrons from higher energy levels.[38] This means that an electron beam is finely focused on to the specimen, resulting in characteristic X-ray photons produced from a microvolume of approximately 1 μm^3 of sample. These X-rays are detected by EDS and the results are plotted as a spectrum. Each element has its own 'fingerprint' of peaks, which allows for qualitative and quantitative determination of the elements present in the selected region of the sample. EDS is an essential tool for Pickering emulsion polymerization when solid emulsifiers are used.[39] It also has been used to localize species such as NaCl, which is a substance historically employed in latex production and added to control the ionic strength of water.[35]

2.3 Techniques for the Determination of the Surface Charge Density of Soft Nanoparticles

2.3.1 Foundations of Titration

One of the most important parameters that characterize any colloidal particle is its surface charge density, σ_0.[40-43] The role played by σ_0 in the colloidal stability of hydrophobic particles in aqueous solutions was explained by Derjaguin, Landau, Verwey and Overbeek six decades ago. As predicted by this theory, the lack of stability may derive from a low σ_0 value. In addition, charged and polar groups located at the water/particle interface do not only control the surface charge of colloidal nanostructures, but also affect the hydrophobic/hydrophilic character of the surface, which is another crucial parameter to be taken into account in order to understand the colloidal stability of a given system controlled by repulsive hydration forces.[42,44,45] In the following sections, the determination of the surface charge density of different types of colloidal particles by potentiometric and conductimetric titrations is explained.

2.3.1.1 *Potentiometric Titrations*

An acid–base titration is based on the neutralization of a base dissolved in an aqueous medium with an unknown concentration by adding a 'standard' acid solution, *i.e.* a solution in which the acid concentration is exactly known. The same concept can be applied using a standard basic solution to calculate the unknown concentration of an acid. The standard solution is usually composed of a strong acid (or base) diluted in water. A 'strong acid' is an electrolyte that dissociates completely in water releasing protons with a dissociation constant $K_a \approx \infty$ (*e.g.* HCl, HNO$_3$, H$_2$SO$_4$), and a 'strong base' generates OH$^-$ ions with a dissociation constant $K_b \approx \infty$ [*e.g.* NaOH, KOH, Ba(OH)$_2$]. In a typical titration experiment, the standard solution is gradually added to the unknown solution using a calibrated pipette or a dosimetric device. The pH of the solution is continuously monitored and recorded in real time using a pH electrode. We will explain first how to calculate the neutralization point of a strong acid solution (HCl) by using a standard solution of a strong base (NaOH) as reactant. Figure 2.3 shows the pH evolution when the NaOH solution, with an exactly known concentration, is added little by little to the unknown acidic solution. The base reacts rapidly with the acid yielding water molecules and sodium chloride, which is completely dissociated in the aqueous environment due to its intrinsic high solubility:

$$NaOH(aq) + HCl(aq) \rightarrow NaCl(aq) + H_2O(l) \qquad (2.1)$$

The concentration of protons in solution is reduced as soon as they react with the OH$^-$ ions to form water. At the same time, the pH increases step by step. The neutralization point is defined as the moment at which each proton released by the acid molecule (HCl) is coupled to an OH$^-$ donated by the base. At this instant, the number of moles of HCl that were dissolved in the original solution exactly coincides with the number of moles of NaOH added from the standard solution; this point is also known as the 'stoichiometric point'. The equivalence point can easily be determined from a pH plot similar to that shown in Figure 2.3, because neutralization coincides with the inflection point of the pH curve. Since we have a strong acid and a strong base, the pH at the equivalence point in the first example has to be 7, taking into account that (i) the ionic product of water is $K_w = 10^{-14}$ in a situation in where $[H^+] = [OH^-] = 10^{-7}$ and (ii) the NaCl formed is a completely dissociated electrolyte that does not react with water molecules any longer. The inflection point gives the volume of the standard solution needed to neutralize the acid exactly. Considering the exactly known concentration of the standard solution, one can then calculate the number of moles of NaOH added to the sample. Since the numbers of moles of acid and base have to be equal at the equivalence point and knowing the volume of the acid solution used at the beginning of the experiment, the acid concentration in the original solution can be easily obtained. Once the equivalence point is exceeded, the addition of more NaOH molecules produces an increase in pH.

In the next example a weak acid solution [*e.g.* CH$_3$COOH(aq)] is titrated with a standard solution composed by a strong base (*e.g.* NaOH). At the

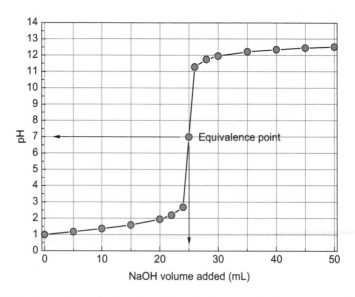

Figure 2.3   Potentiometric titration of a solution containing a strong acid (HCl) with a standard solution of a strong base.

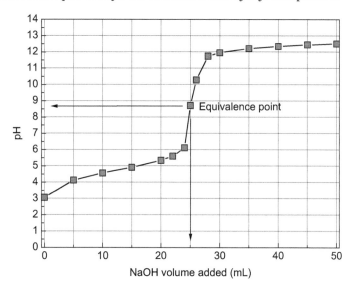

Figure 2.4   Potentiometric titration of a solution containing a weak acid (CH_3COOH) with a standard solution of a strong base.

equivalence point, the numbers of moles of both substances (acid and base) coincide, generating $NaCH_3COO(aq)$. However, the just-formed electrolyte reacts with water molecules yielding to a hydrolysis reaction – *i.e.* a 'breakdown' of water molecules – shifting the pH from neutral to basic values at the equivalence point, according to the following equilibrium:

$$CH_3COO^-(aq) + H_2O(l) \rightleftharpoons CH_3COOH(aq) + OH^-(aq) \qquad (2.2)$$

This explains why the pH curve changes when a weak acid is titrated with a strong base, giving rise to a solution that has an excess of OH^- ions at the equivalence point – *i.e.* a pH >7 – as can be seen in Figure 2.4.

Similarly, the titration of a weak base (*e.g.* NH_3) with a strong acid (*e.g.* HCl) leads to an acidic pH at the equivalence point. The NH_4Cl formed during titration completely dissociates, making ammonium ions that react with water according to the following hydrolysis reaction:

$$NH_4{}^+(aq) + H_2O(l) \rightleftharpoons NH_3(aq) + H_3O^+(aq) \qquad (2.3)$$

This equilibrium is governed by the K_b constant of the classical reaction for ammonia in water:

$$NH_3(aq) + H_2O(l) \rightleftharpoons NH_4{}^+(aq) + OH^-(aq) \qquad (2.3a)$$

The protons released by the $NH_4{}^+$ ions undergo the indicated hydrolysis reaction, leading to a pH <7 at the neutralization point. Figure 2.5 shows the curve that is expected when a weak base solution is titrated against a strong acid.

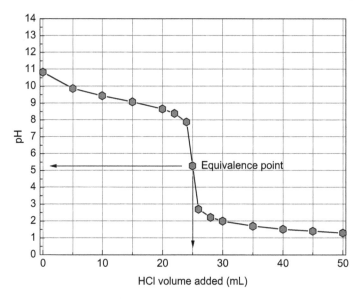

Figure 2.5 Potentiometric titration of a solution containing a weak base (NH_3) with a standard solution of a strong acid.

2.3.1.2 *Conductimetric Titrations*

The equivalence point in a titration experiment can also be determined by measuring the electrical conductivity of the solution. The conductivity (κ_e) of an electrolyte solution gives information about the charges transported by the ions dissolved in the medium under the effect of an applied electric field. The conductivity depends on three factors: the ion concentration (c_i), the ion valence (z_i) and the ionic mobility (u_i). The last term takes into account the nature of the ion (size and shape) and the drag forces that act when a given ion moves in a viscous environment such as water. The smaller the ion size, the higher is its ionic mobility. This explains why protons are the ions with the highest u_i value in water and why the ionic mobility of OH^- is higher than that of Cl^-. The theoretical expression for the conductivity of a solution containing a totally dissociated electrolyte is

$$\kappa_e = F(|z_+|c_+u_+ + |z_-|c_-u_-) \qquad (2.4)$$

where F is the Faraday constant. The subscripts '$+$' and '$-$' refer to the cation and anion, respectively.

The conductivity of a solution can easily be measured with inexpensive conductimetric devices that are available in most laboratories. An electric probe with two electrodes that face each other at a given distance is immersed in the solution and detects the amount of electric charge that is transported between the electrodes after applying an electric potential

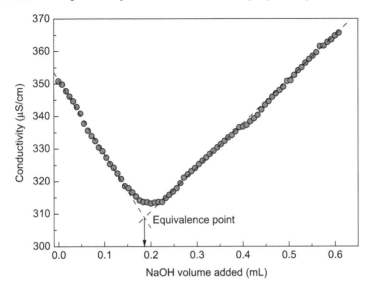

Figure 2.6   Conductimetric titration of a solution containing a strong acid (HCl) with standard solution of a strong base.

difference. During a titration experiment, the conductivity usually changes linearly, showing different slopes before and after the neutralization point. If a strong acid (*e.g.* HCl) is titrated with a strong base (*e.g.* NaOH), the conductivity of the solution will behave as shown in Figure 2.6.

In the first part, when there is an excess of acid, the conductivity decreases as the basic solution is added little by little. The original conductivity at the beginning of the experiment is caused by Cl^- and H_3O^+ ions. When some NaOH is added, neither the chloride concentration (c_-) nor the concentration of cations (c_+) changes given that the protons that react with the added OH^- forming neutral water molecules are replaced by positively charged sodium ions. Therefore, the ionic concentration does not change before achieving the equivalence point, *i.e.* c_- and c_+ remain constant in eqn (2.4). The ionic mobility of the cationic species, however, does change. Protons with a very high mobility u_+ are interchanged with larger Na^+ ions with a lower mobility. Consequently, the global conductivity of the solution decreases slightly according to eqn (2.4). Once the neutralization point has been reached, any excess of base increases the concentration of dissolved ions, since now both Na^+ and OH^- species increment their concentration continuously and no water molecules are formed any longer in the absence of protons to react with. This explains why the conductivity of the solution increases once the equivalence point has been exceeded. In conductimetric titrations, the neutralization point can easily be derived from the intersection of the two lines that fit best the two linear parts of the curve.

2.3.2 Titration of Non-Penetrable Particles with Only Strong Acid Groups on Their Surface

A classical example of this type of particle can be found in polystyrene nanospheres stabilized by sulfate or sulfonate groups at the particle/water interface. Polystyrene particles have been used extensively in biomedicine and biodiagnostic applications, *e.g.* in latex immunoagglutination assays (LIAs) and latex agglutination tests (LATs).[46,47]

In order to preserve electroneutrality, each charged anionic group deriving from a strong acid molecule attached to the polystyrene particle surface is counterbalanced by a dissolved cation located in a diffuse electrolyte layer near the interface. Owing to the extremely high K_a values of these strong acid groups, they are always completely deprotonated. For a correct σ_0 determination, it is essential that only protons are coupled to the sulfate or sulfonate groups on the particle surface. The reason is that these groups cannot be determined directly by a titration procedure but their complementary protons can. Hence, the first and may be the most important step before determining the surface charge density of this kind of colloidal particle is to carry out an exhaustive cleaning process in order to remove any other cations such as Na^+ from the bulk solution. This guarantees that each proton detected by an acid–base titration corresponds as counterion to one charged anionic group on the surface. When the proton concentration is known by titration, the surface charge density σ_0 can easily be calculated taking into account the total surface area of the particle/water interface used for the experiment. The latter can be determined from the average particle size and shape and their concentration in the solution.

Among the different cleaning processes, the use of ion-exchange resins is usually recommended.[48] The initial sample volume, the concentration of the standard titration solution and the amounts to be added in each titration step are variables that must be optimized and adapted for each colloidal system. Potentiometric and conductimetric measurements can be performed simultaneously. It is also recommended to bubble nitrogen through the sample solution in order to remove traces of dissolved CO_2 that reacts with water to form carbonic acid. The use of automatic dosimeters simplifies the process and ensures that the titrant is added properly. Sometimes the addition of an inert electrolyte (*e.g.* KBr) is recommended in order to facilitate pH measurements. The reason is that very clean colloidal solutions have an initial conductivity near to that of pure water (<1 μS cm^{-1}) and therefore the signal detected by a pH probe is very poor and strongly affected by experimental noise. Under typical experimental conditions, pH meter readings are stabilized by increasing the initial sample conductivity to 400–500 μS cm^{-1} by adding, for example, KBr. Figure 2.7 shows the result of a typical titration experiment carried out with non-penetrable particles having only strong acid groups on their surface.

Figure 2.7 shows that the neutralization points determined by conductimetry and potentiometry, are very similar although they do not match

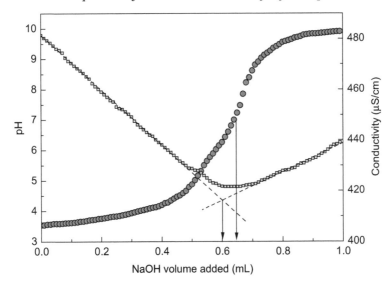

Figure 2.7   Potentiometric (●) and conductimetric (□) titration of a polystyrene
latex with only sulfonate as charged surface groups. Total area titrated,
12.5 m^2; NaOH concentration, 26.2 mM.

exactly. It should be noted that in both cases the exact localization of the
inflection point (on the potentiometric curve) and the crossing points (for
the conductimetric data) depend to a certain extent on the experimenter.
This means that both methods should be considered as independent and
complementary, and an average of the results obtained should be used as
final datum.

2.3.3   Titration of Non-Penetrable Particles with Weak Groups on Their Surface

2.3.3.1   Acidic Groups

There are many organic and inorganic groups with a $K_a \neq \infty$. Among them,
the most usual group in many colloidal suspensions is that derived from
carboxylic acids (R–COOH). Particles with lactic, glycolic, acrylic, methacrylic
or deoxycholic acids located on their surface are examples in which this type
of weak acid group is present in colloids with biotechnological and biomedical applications.[49–52] Titration of particles with only weak acid groups
differs completely from those having only strong acid groups attached to
their surface. In this case, the cleaning process must not acidify the suspension at all. If the starting point was an acidic pH, protonation of the
initially charged COO$^-$ groups due to the acid–base equilibrium governed by
the K_a value would yield uncharged particles that would spontaneously aggregate owing to a lack of electric repulsion. Since titration of an aggregated
system makes no sense, it is not possible to obtain the surface charge density

of particles with only weak acid groups from direct titrations, *i.e.* using a strong base such as NaOH as titrant. In order to overcome this problem, so-called 'back' titrations are recommended for determining σ_0.

The procedure for a back-titration starts with a comprehensive cleaning process of the colloidal suspensions with repeated cycles of centrifugation–decantation–redispersion if that is possible for the experimental system used. Alternatively, a serum replacement method with pure deionized water can be employed. In contrast to ion-exchange resins, these cleaning methods do not replace surface charge counterions by protons, thus avoiding any undesirable charge cancellation and the subsequent aggregation of the colloidal system before characterization. Once the cleaned colloidal suspension has been placed in the titration device and the potentiometric and conductimetric probes have been immersed in the solution under nitrogen, the pH of the medium must be increased to ~10 by adding NaOH. It is not necessary to collect the data for this 'direct' titration since they would be useless. When the pH value is set at around 10, the particles are fully charged owing to the basic environment and, consequently, the system remains completely stable. Now back-titration can start, using a standard solution of a strong acid (*e.g.* HCl). It may sound strange to titrate particles with weak acid groups on their surface with an acidic solution, but actually what is titrated is the weak base group COO^-, which is the conjugated base of the weak acid COOH. Figure 2.8 shows typical results obtained for such a process.

Back-titration of particles with only weak acid groups generates conductimetric curves with three different slopes. The first part of the curves has

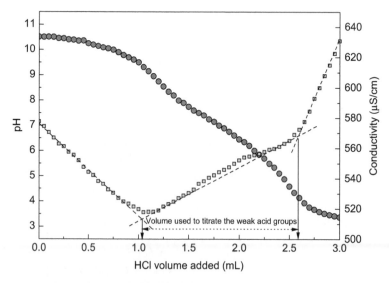

Figure 2.8   Potentiometric (●) and conductimetric (□) titration of a polystyrene latex with only carboxylic groups on the particle surface. Total area titrated, 14.9 m^2; HCl concentration, 21.6 mM.

a negative slope and corresponds to the titration of the excess of the previously added NaOH. Addition of HCl replaces OH^- ions by Cl^- ions when OH^- ions react with the added protons to form water. Given that anions with a higher ionic mobility (OH^-) are replaced by a less mobile species (Cl^-), the sample conductivity decreases, despite the fact that the concentration of anions in the solution remains constant. Once the excess of NaOH has been eliminated, titration of the weak acid groups takes place. The addition of the titrant (HCl) now increases the conductivity, since the values of the ion concentrations c_+ and c_- increase. Nevertheless, part of the added protons do not remain in the bulk because they associate with the carboxylate groups on the particle surface according to the corresponding chemical equilibrium governed by K_a. The conductivity increase in this part of the titration experiment comes to an end when all the COO^- groups have been completely protonated and the particle surface is completely uncharged. Any further addition of HCl gives rise to a sharp increase in conductivity since there is no removal of added protons from the bulk by the surface groups. This means that the third part of the conductimetric titration curve corresponds to the excess of the titrant. Evidently, the amount of HCl used for titrating the weak acid charges of the particles corresponds to the central part of the curve shown in Figure 2.8.

It can also be seen in Figure 2.8 that the slopes of the different parts of the conductimetric curves change at the inflection points of the potentiometric data. As explained before, the first inflection point corresponds to the titration of the initial excess of NaOH and, therefore, should be observed at pH 7. This neutralization point is, however, masked by the protonation of the carboxylate groups because the titration of a strong base (NaOH) overlaps with that of a weak base ($-COO^-$), which is the conjugated base of the weak acid $-COOH$. This is the reason why the neutralization point of NaOH is not observed at neutral pH, as seen before in Figure 2.3. The second inflection point corresponds to the neutralization of the weak base, *i.e.* the $-COO^-$ groups. It takes place at an equivalence point located at a pH <7 similar to that shown in Figure 2.5.

It should be noted that at the end of the back-titration of a colloid with only weak acid groups, the sample will be completely aggregated due to the lack of surface charge. This fact should be taken into account by the experimenter, since the sample usually has to be discarded after this kind of titration.

2.3.3.2   Basic Groups

The chemical agents that supply positive charges to most of the cationic colloids derive from amine or amidine groups. Throughout this book, poly(amidoamine) (PAMAM) dendrimers, positive-lipid nanocapsules or colloids coated with chitosan molecules are mentioned as examples of cationic particles that may have biomedical applications, *i.e.* they belong to the weak base species.[45,53–55] Although it is possible to quantify the surface density of amine groups by other chemical methods, classical acid–base titrations may also be employed. In this case, the procedure is similar to that explained in the previous section, setting the initial pH to an acidic value and using a standard base solution for back-titration. After performing the cleaning process with a serum replacement device and/or with centrifugation–decantation–redispersion cycles, the colloidal suspension with cationic particles must be acidified, *e.g.* with HCl, until a pH of about 3–4 is reached. The back-titration using a standard NaOH solution will produce a graph similar to that shown in Figure 2.8, *i.e.* a conductimetric curve with three different slopes and with a potentiometric sigmoidal curve with two inflection points. Evidently, the starting pH will be acidic and the pH at the end of the titration process will be basic. If only weakly basic groups exist on the particle surface, the sample will have aggregated at the end owing to charge cancellation and, consequently, should be discarded for further experimental studies.

2.3.4 Titration of Non-Penetrable Particles with Strong and Weak Acid Groups on Their Surface

Particles with a heterogeneous distribution of different anionic groups on their surface can also be found and employed for biomedical applications. Examples are colloids with sulfate or sulfonate groups that coexist with carboxylic terminal groups. When only two kinds of negatively charged chemical groups are present on the particle surface and both of them derive from acids differing in their respective K_a values, it is possible to determine the surface concentration of each of these chemical species. As an example, we will assume that one of them corresponds to strong acid with $K_a \approx \infty$ and the other to a weak acid. In order to quantify the concentration of strong acid groups, it is essential to clean the sample thoroughly with ion-exchange resin as explained previously. The resin can be used after an initial cleaning process consisting of centrifugation cycles or serum replacement procedures. In this case, a direct titration will be performed starting from an acidic pH and using an NaOH solution as titrant. Although all weak acid groups on the particle surface will be protonated, *i.e.* uncharged at the beginning of the experiment, the sample will remain stable without suffering any aggregation phenomenon thanks to the charge provided by the strong acid groups. Potentiometric and conductimetric titrations will be carried as usual under nitrogen, adding an indifferent electrolyte in order to stabilize the pH reading. Typical conductimetric and potentiometric data are plotted in Figure 2.9. The former plot shows three linear regions and the latter two inflection points.

The first region of the conductimetric curve corresponds to the titration of the strong acid groups. The reason for the negative slope was already justified in Section 2.3.3. The amount of the standard solution used for the initial part of the experiment serves to calculate the surface charge density

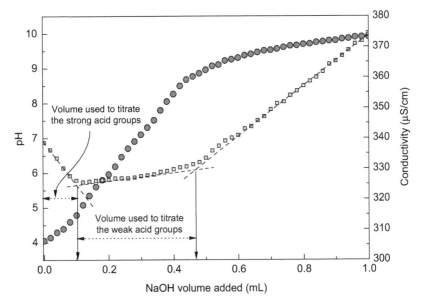

Figure 2.9   Potentiometric ($\bullet$) and conductimetric ($\square$) titration of a polystyrene latex with both sulfate and carboxylic groups in the particle surface. Total area titrated, 10.7 m^2; NaOH concentration, 22.5 mM.

exclusively supplied by this kind of groups, *i.e.* the sulfate and/or sulfonate groups. The second region corresponds to the amount of NaOH solution used to titrate the weak acid groups. The conductivity increase observed in the third part of the curve is caused by the excess NaOH once all acid groups on the particle surfaces have been neutralized. As mentioned previously, the crossing points of the lines fitted to the conductimetric curves fall close to the inflection points given by the potentiometric measurements and so again we recommend the use of the average of the two sets of data as the best value for further evaluation.

2.3.5   Dependence of Surface Charge Density on pH

In those cases where the particles have only weak acid (or basic) groups on their surface, it is possible to determine the corresponding equilibrium constant K_a (or K_b) by titration. In addition, such an experiment allows the surface charge density of the particles to be determined as a function of the pH. The method explained in the previous paragraphs gave only the maximum value for σ_0, *i.e.* when all surface groups are completely charged. Particles with weak acid groups, however, have a charge density that depends on the pH of the medium. They may be almost or even totally uncharged at acid pH values and fully charged at basic pH values. Shirama and Suzawa[56,57] explained how to quantify the surface charge density of polystyrene latex as a function of pH by means of potentiometric titrations.

In order to perform this type of experiment, it is necessary to titrate two identical solutions, *i.e.* two aliquots differing just in the presence or absence of colloidal particles. We will treat a case of a colloid consisting only of particles with weak acid groups. The dependence of σ_0 on pH can be obtained from the titration of a blank sample of NaOH without colloidal particles and the titration of the same basic solution containing the nanospheres. The potentiometric curves of the two titration experiments will look like those shown in Figure 2.10.

Since the two solutions differ only in the presence or absence of particles, the difference between the two curves at a given pH provides information about the amount of the titrant (HCl) employed to neutralize the weak acid groups of the particles. At acidic pH values, the difference between the two curves decreases and, consequently, the σ_0 obtained diminishes. For the data shown in Figure 2.10, a volume of 0.7 mL of HCl was needed to titrate the particles at pH 8, whereas a volume of only 0.3 mL was required at pH 5. It should be pointed out that both curves must match at the pH where the particles are completely uncharged, *i.e.* at their isoelectric point. For this reason, it is almost unavoidable to determine previously the isoelectric point of the particles by other techniques such as electrophoretic mobility measurements. This will allow both titration curves to be shifted appropriately, making them coincide exactly at the correct pH value. It should be noted that the isoelectric point of the samples corresponding to Figure 2.10 (pH ≈ 3.5) was determined using zeta potential data measured with an electrophoretic

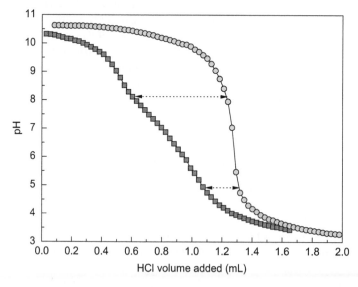

Figure 2.10 Potentiometric titration of polystyrene particles with only carboxylic groups on their surface to obtain a σ_0 *versus* pH dependence. ($\bigcirc$) Titration of an NaOH blank; ($\blacksquare$) titration of the same blank plus latex particles added. The dotted horizontal arrows are referenced in the main text.

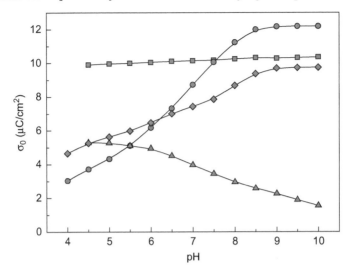

Figure 2.11 Dependence of the surface charge density on pH for different polystyrene latex samples varying in the nature of their surface charged
groups. (■) Particles with only strong acid groups (sulfonate); (●)
particles with only weak acid groups (carboxylic); (▲) particles with
only weak base groups (amidine); (♦) particles with both strong and
weak acid groups.

instrument as a function of pH and not from the titration experiment itself.
Figure 2.11 shows the surface charge as a function of pH obtained by this
titration method for four polystyrene latex samples with different surface
charge groups.

2.3.6   Titration of Soft Colloidal Particles

As an example of the titration of soft colloidal particles, we described the
case of micro- and nanogels in Chapter 4. This kind of particle can be
thought of as swollen colloidal systems in which ions can penetrate into
their inner part simply by diffusion. Hoare and Pelton[58] proposed a time-
dependent titration method to estimate the number and approximate distribution of functional charged groups in polymeric gels of mesoscopic size.
This gives an idea of how complex the titration of soft particles can become
when ions can penetrate inside. Their study was focused in microgels
composed of poly(N-isopropylacrylamide) (PNIPAM) microgels, containing
carboxylic groups. The degree of hydration of gels made of this polymer
depends strongly on the temperature, given that PNIPAM undergoes a reversible deswelling volume phase transition in the temperature range 32–35 °C. The
interested reader is referred to Chapter 4 for further information about
phase transitions in gels. Using different synthesis protocols for the polymerization of PNIPAM with methacrylic acid and acrylamide molecules, the
authors managed to prepare three microgels with varying distributions of

carboxyl groups. In fact, they did not only synthesize two samples, referred to as MAA-NIPAM and H-AM-NIPAM-60, where most of the weak acid groups resided on or near the microgel surface, but also produced systems, called H-AM-NIPAM-30, in which about 35% of the total number of carboxyl groups were located within the bulk of the microgel. The determination of this distribution and the quantitative estimation of its concentration were carried out by non-equilibrium conductimetric titrations. When performing titration experiments for microgel systems, one has to bear in mind that ions are not restricted to the aqueous bulk phase as happens in conventional colloids. Given that penetration of titrant ions into the microgel bulk is diffusion limited, much longer stabilization times for full equilibration of the titrant in the aqueous and the gel phases are required.[59] If the titration speed is too high, the equilibration distribution is not reached and an incorrect result will be obtained. Hence the determination of the charged group concentration obtained from titration methods for penetrable colloidal particles becomes apparently time dependent if performed at an inadequate titration speed. It should be noted that there are workers[51,54] who did not use any time-dependent titration analysis to characterize their ion-penetrable particles. Nevertheless, we consider that such a simplification is not recommendable at all and should only be used if there is clear evidence for equilibrium.

For the sake of clarity, in this section we will only compare the results obtained by Hoare and Pelton[58] with H-AM-NIPAM-30 and MAA-NIPAM microgels. Table 2.1 shows the influence of the titration speed on the measured number of titratable carboxylic groups in both colloidal systems. A significant time dependence is observed in the titrations when carboxylic groups are located not only in the external part of the particle but also along the whole mesh of the microgel. This occurs in the H-AM-NIPAM-30 sample where fast titrations gave degrees of acrylamide hydrolysis (that generated carboxylic groups) below 50%. When the titration speed was reduced, the measured degree of hydrolysis exceeded 70%, since longer titration times

Table 2.1 Carboxylic acid functional group content of H-AM-NIPAM-30 (centre column) and MAA-NIPAM (last column) microgels as determined by base-into-acid conductimetric titrations (see Figure 2.12) at various speeds.

Titration speed (min per pH unit)	Measured H-AM[a] incorporation (% of feed)	Measured MAA[a] incorporation (% of feed)
2.2	46.6 ± 2.3	98.9 ± 4.3
4.4	43.4 ± 2.6	101.4 ± 4.4
6.7	54.0 ± 2.5	99.8 ± 4.3
33	70.8 ± 3.1	105.2 ± 4.5
67	71.2 ± 2.8	103.8 ± 4.5

[a]H-AM refers to acrylamide molecules that were hydrolyzed into carboxylic groups; MAA refers to methacrylic acid groups. Adapted and reprinted with permission from T. Hoare and R. Pelton, *Langmuir*, 2004, **20**, 2123. Copyright 2004 American Chemical Society.

facilitate a complete equilibration of the titrant ions in the aqueous and the gel phases. Only at very long stabilization times of 33 and 67 min per pH unit did the titration results give similar values, suggesting that the inter-phase pH equilibrium was eventually achieved while non-equilibrium results were observed at faster titration speeds.

Figure 2.12 shows the conductimetric titrations for (left) the H-AM-NIPAM-30 sample and (right) the MAA-NIPAM sample. All data included in Table 2.1 were calculated from the conductimetric titrations as shown in Figure 2.12.

Four linear sections can be distinguished in the curves for the H-AM-NIPAM-30 microgels at faster titration speeds (6.7 min per pH unit), while the three classical linear zones reappeared when titration was slowed to 67 min per pH unit. A similar three-slope curve was observed for the MAA-NIPAM system at all titration speeds. The additional transition region in Figure 2.12 (left) is denoted V_3 and has an intermediate slope between that of zones V_2 and V_4. It is ascribed to a non-equilibrium process limited by ion diffusion. The volume obtained from zone V_2 was used for calculating the data shown in Table 2.1, since this region actually corresponds to the total number of titrated carboxylic groups within the particle according to the slope data.[58] The volume of zone V_2 increased as the titration time was slowed and the intermediate zone V_3 tended to disappear, suggesting that it was of kinetic origin.

The kinetic process based on ion diffusion was corroborated by potentiometric titrations. Forward (base into acid) and back (acid into base) titrations were carried out using a stabilization time of 2.2 min per pH unit. The interested reader can find these potentiometric data in Figure 3 in reference 58, where the back-titrations gave further insight into the differences between the functional group distributions of the microgels.

For the sample in which carboxyl groups are mainly located on the surface (MAA-NIPAM), good overlap was found between the forward and backward titration even at the shortest stabilization time studied (2.2 min per pH unit).

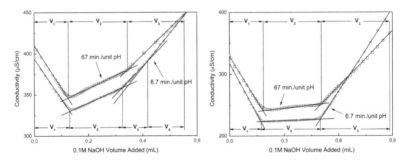

Figure 2.12   Left: a base-into-acid (charging) conductimetric titration of H-AM-NIPAM-30 copolymer microgel using varying titration speeds. Right: a base-into-acid (forward) conductimetric titration of MAA-NIPAM copolymer microgel using varying titration speeds.
Adapted and reprinted with permission from T. Hoare and R. Pelton, *Langmuir*, 2004, **20**, 2123. Copyright 2004 American Chemical Society.

This is a clear indication that the titrant ions are fully equilibrated in the aqueous and gel phases before the addition of the next volume of titrant. For the H-AM-NIPAM-30 system, however, a hysteresis-like pattern is observed. All these results allow us to conclude that the functional groups in the H-AM-NIPAM-30 system are much less accessible to titrant ion diffusion than those in the MAA-NIPAM sample. This means that the methacrylic acid–NIPAM copolymer microgels are characterized by a core–shell morphology in which most carboxylic groups are located at or near the particle surface, whereas in the case of H-AM-NIPAM-30 a significant number of carboxylic acid groups are located within the bulk of the microgel. Further information about the usefulness of titrimetric methods for characterizing other carboxylic-functionalized microgels can be found elsewhere.[60]

Summarizing, the basic principles of conductimetric and potentiometric titrations have been explained in this section as clearly as possible. We hope that it will be helpful for any researcher who needs to characterize colloidal systems.

2.4   Electrokinetic Techniques

2.4.1   Introduction

Soft particles have a surface layer or tangle of *electrically charged* polymers or polyelectrolytes that is usually fixed to a *charged* solid core.[61] With respect to its electrical state, a soft spherical particle will behave like a solid spherical particle (*hard particle*) when there is no polyelectrolyte layer and as a *spherical polyelectrolyte* in the absence of a core (see Figure 2.13).

As is emphasized throughout this book, the electrical charge of soft particles is an important factor for their colloidal stability and consequently also for their applications.[62] As will be explained in more detail in Chapter 4, the charge can derive from the initiator, the cross-linker, the surfactants employed and the ionic groups spread out and inside the particle. In the previous section, the experimental procedure to determine the charge of soft particles by means of conductimetric and potentiometric titrations was

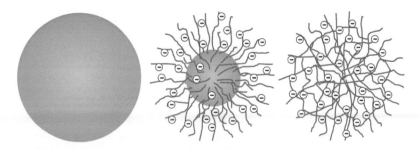

Figure 2.13   Schematic illustration of a solid spherical particle (left), a soft spherical particle (centre) and a spherical polyelectrolyte without particle core (right).

described. There are, however, other possibilities for determining the charge of soft particles. For instance, the functional group distribution of carboxylic acid groups within PNIPAM-based microgels has been calculated theoretically using a kinetic model and determined experimentally by means of *dynamic light scattering* (DLS), electrophoresis and *small-angle neutron scattering* (SANS) measurements.[62] The charge of polyvinylcaprolactam (PVCL) nanogels has been measured with *nuclear magnetic resonance* (NMR) spectroscopy. Most of these techniques are rather sophisticated and their use is limited. Therefore, this section is focused on the study of the electrical properties of soft particles based on fairly simple electrophoretic techniques. The section is organized in three parts concerning (1) the basic principles of electrophoretic light scattering, and its application to (2) charged hard spheres and (3) polyelectrolyte nanogels.

2.4.2   Electrophoretic Light Scattering

The electrostatic interactions among colloidal particles dispersed in a liquid have their origins mainly in the presence of surface charge, which plays an essential role in colloidal stability. One way to quantify the electrostatic interaction is by means of electrophoresis, *i.e.* the motion of charged particles dispersed in a liquid upon application of an external electric field (see Figure 2.14). Using electrokinetic techniques, the main parameter to be obtained is the electrophoretic mobility μ, which is defined as the observed migration velocity of the particle v divided by the electric field strength E.[63]

Electrophoretic light scattering techniques are based on the analysis of the light scattered by a suspension of moving colloidal particles mixed with the reference beam coming directly from the light source. The scattered light is frequency shifted by the Doppler effect and its superposition with the unshifted reference beam leads to beating at a frequency that depends on the speed of the moving particles. Traditional commercial devices perform a spectral analysis of the shifted frequency spectrum. This procedure,

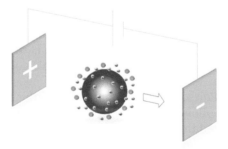

Figure 2.14   Schematic illustration of the electrophoresis technique. Movement of a charged particle (immersed in an electrolyte solution) induced by an external electric field.

however, is not very precise for the low electrophoretic mobility region. This is the case, for instance, for electrolyte solutions with high ionic concentrations or non-polar media. Trying to overcome this limitation, some electrophoretic devices use *phase analysis light scattering* (PALS) to measure the electrophoretic mobility. Phase analysis takes place over many cycles of the respective waveforms given that the optical phase of the scattered light is characterized by means of the *amplitude-weighted phase difference* (AWPD) function instead of a simple correlation treatment. This function improves the statistical behaviour because the detected signal usually fluctuates in amplitude due to relative particle movements and concentration fluctuations. Hence PALS is able to measure electrophoretic mobility values that are at least two orders of magnitudes lower than those determined with traditional light scattering methods. A complete discussion of the PALS method and the AWPD treatment can be found elsewhere.[64]

For the case of charged particles immersed in an electrolyte solution, μ generally increases with increase in the charge of the particles and decreases as the electrolyte concentration increases because the smaller electrolyte ions tend to screen the particle surface charge and reduce the range of their electrostatic interaction. Electrophoretic techniques are suitable for studying the electrical properties of a charged particle since the electrophoretic mobility is related to the electrostatic potential, which depends directly on the structure of the ion layer surrounding the charged colloidal particle. For evident reasons, the latter is commonly termed the *electric double layer* (EDL). Historically, the classical *Gouy–Chapman* model[65–67] supposes that the electric charge on the particle surface is counterbalanced by a diffuse layer of point-like ions. Using the Poisson–Boltzmann equation, the electrostatic interaction potential for molecular solutes in salty, aqueous media can be described by

$$\nabla^2 \psi(r) = -\frac{e}{\varepsilon_r \varepsilon_0} \sum_i z_i \rho_i^0 \exp[-z_i e \beta \psi(r)] \tag{2.5}$$

where $\psi(r)$ is the electrostatic potential, expressed as function of the radial distance from the particle surface (r), e is the elementary electric charge, $\varepsilon_r \varepsilon_0$ is the permittivity of the dielectric continuum, ε_0 is the permittivity of vacuum, z_i is the valence of the ionic species i, ρ_{2i}^0 is the density of species i in the bulk and $\beta = 1/k_B T$ where k_B is Boltzmann's constant and T is the absolute temperature.

Following a pure Poisson–Boltzmann approach, the local ionic concentration close to the colloidal surface reaches unrealistic values. This theoretical failure motivated the development of the *Stern model*, where the total ionic charge in the medium is separated into two parts: an adsorbed layer at the locus of the hydrated ions known as the *Stern layer* or *Helmholtz plane* and a *diffuse layer* beyond it reaching out to the bulk. Additionally, the Stern model supposes that the ions in the diffuse layer interact with the charged surface only *via* electrostatic interactions, whereas the ions at the Stern layer

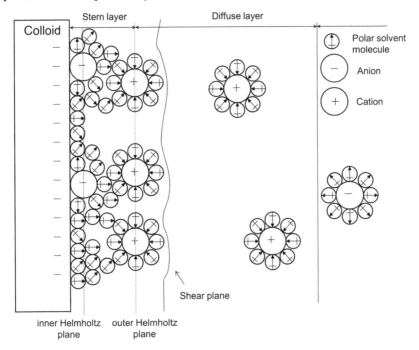

Figure 2.15   Stern model of a colloidal EDL.

can have a chemical affinity for the surface that allows them to be specifically adsorbed *via* covalent bonds or van der Waals forces. This model was further refined by Grahame with the triple-layer model (see Figure 2.15), wherein a distinction is made between the location of the Stern layer localized at the *inner Helmholtz plane* and the location where the diffuse layer begins (*outer Helmholtz plane*). The ions that are specifically adsorbed at the inner Helmholtz plane are known as specific ions, whereas the point ions that are in the diffuse layer beyond the outer Helmholtz plane are called indifferent ions owing to their electrostatic-only interaction with a charged colloid.[68]

The question that arises now is how electrophoretic data, *i.e.* the experimentally obtained values for the electrophoretic mobility μ, are linked to the electric properties of the particles and their EDL. The answer to this question is the *zeta* or ζ-*potential*, which is defined as the electric potential at the plane where the liquid velocity relative to the particle is zero. This plane, which may be up to two or three water molecule diameters away from the particle surface, is called the *slipping* or *shear plane*. Several numerical expressions have been developed to calculate the ζ-potential from μ measurements. Since the ζ-potential is related to the electrostatic potential $\psi(r)$, it provides information about the *effectiveness* of the electrostatic interaction among the particles due to their surface charge. This property therefore has important applications for studying the potential distribution and stability at interfaces involving surfactants, multivalent ions, polymers and even more complex biomolecules such as DNA or proteins, as the reader will see throughout this book.[69]

In addition, several estimations of the *effective charge* carried by charged particles, *i.e.* their *electrokinetic charge*, can be derived from ζ-potential measurements using different models for the EDL.[70] In general, the charge obtained from titration data (see the previous section) is much greater than the values determined from electrokinetic measurements. This is mostly due to the different approximations used to simplify the theoretical description of the EDL such as ion size, discreteness of the surface charge distribution and the molecular nature of the solvent[70,71] and also because not all the charges of the particle *really* contribute to its electrokinetic movement. As will be discussed later; these features are especially significant for soft particles.

The ζ-potential depends on the electrical charge of the particle, any adsorbed layer at the interface and the nature and composition of the surrounding medium in which the particle is suspended.[72,73] As a result, different expressions for the conversion of μ into values of ζ-potential can be derived according to the system considered. Before studying the case of soft particles, we will start with the case of charged hard particles.

2.4.3   Charged Hard Sphere Limit

The conversion of electrophoretic mobility into ζ-potential of a hard solid/electrolyte solution interface must take into account all possible mechanisms of double-layer polarization. The term polarization implies that the EDL around the particles is regarded as being distorted from its equilibrium shape by the motion of the particle. The classic problem of establishing a quantitative relationship between the steady-state electrophoretic mobility and the ζ-potential was pioneered by von Smoluchowsky, Hückel and Henry[74-76] and was developed further into the classical theory for the electrophoretic motion of a spherical macroion by Wiersema *et al.*[77] and O'Brien and White.[78] In all these theories, the *standard electrokinetic model* (SEM) is employed and it considers hard particles of radius a covered by a uniform charge density (σ) that are submerged in a continuum solution containing a binary point ion electrolyte.[79] Apart from the size of the particle, it is useful to know the relative extension of the EDL regarding the particle size. The dimensionless particle radius κa provides this information, κ being the well-known *Debye–Hückel* parameter. Its reciprocal represents the characteristic thickness of the EDL and is called the *screening Debye length*. The parameter κ can be calculated as a function of $\kappa^2 = 4\pi l_B \sum_i \rho_i z_i^2$, where $l_B = e^2\beta/(4\pi\varepsilon_0\varepsilon_r)$ is known as the *Bjerrum length* and for water at room temperature is ~ 0.7 nm. Within the classical SEM, there are two different asymptotic cases as functions of κa. In the limiting case where the particle size is much smaller than the thickness of the EDL ($\kappa a \ll 1$), the *Hückel equation* can be written as

$$\mu = \frac{2\varepsilon_0\varepsilon_r}{3\eta}\zeta \tag{2.6}$$

where η is the viscosity of the bulk solution. The applicability of this equation is restricted in aqueous systems. In non-aqueous media, however, the ion concentration may be very low and the Hückel equation does apply.[80] In the limiting case where the particle radius is much larger than the thickness of the EDL ($\kappa a \gg 1$), the *Helmholtz–Smoluchowski equation* relates the electrophoretic mobility as a function of the ζ-potential:

$$\mu = \frac{\varepsilon_0\varepsilon_r}{\eta}\zeta \tag{2.7}$$

This is the case, for instance, for large particles and/or solutions with a high ionic strength. For arbitrary values of κa, the full electrokinetic equations for the electrophoretic mobility of spherical particles have to be solved numerically. This is done with the *Henry equation*, which provides an integral expression relating μ to ζ for equilibrium EDLs and low electric potentials (<25 mV). O'Brien and White reported a well-known method based on the conjugation of linear hydrodynamics with a classical account of the equilibrium EDL.[78] This method takes into account the distortion of the EDL due to the external electric field (*relaxation effect*) during the movement of the charged particle. The resulting polarization of the ionic atmosphere induces an extra contribution to the dragging force over the particle. More recently, a more realistic model was proposed by Lozada-Cassou and González-Tovar considering also the ion size, which is relevant for the description of concentrated electrolytes and/or highly charged particles.[79] In all these cases, the equations obtained do not lead to simple analytical expressions relating the ζ-potential to the experimentally accessible electrophoretic mobility. Consequently, they have to be solved numerically using computer programs.[81] Approximate analytical expressions, however, have been proposed by several authors. Without any doubt, the mostly commonly used approximate equations are those proposed by Ohshima, who used the EDL classical models as a starting point to derive approximate analytical solutions for numerous different situations.[73] Further, Ohshima also provided analytical equations to calculate the electrokinetic charge of colloids from electrophoresis experiments, *i.e.* the measured electrophoretic mobility μ and the corresponding estimated ζ-potentials. For the case of spherical particles with radius a immersed in a symmetrical electrolyte solution of valence z, the electrokinetic charge density can be calculated from ζ using the following empirical expression:

$$\sigma = \frac{\varepsilon_0\varepsilon_r\kappa}{\beta ze}\left[2\sinh\left(\frac{\beta e\zeta}{2}\right) + \frac{4}{\kappa a}\tanh\left(\frac{\beta e\zeta}{4}\right)\right] \tag{2.8}$$

The maximum error of this equation is about 20%. It assumes that the shear plane coincides with the outer Helmholtz plane, *i.e.* $\zeta \approx \psi_d = \psi(a)$, where ψ_d is the electric potential at the closest approach of the hydrated ions to the surface that is known as *diffuse potential*. This latter approximation is valid for model colloids with smooth and homogeneous surfaces.[73]

The Ohshima formalism will be used in the next subsection for describing basic notions about the relationship between the electrophoretic mobility and the ζ-potential in a suspension of soft particles and, in particular, charged spherical polyelectrolytes.

2.4.4 Charged Spherical Polyelectrolytes

The electrophoresis of soft particles is completely different from that of hard particles because the concept of ζ-potential loses its physical meaning. Instead, the *Donnan potential* (ψ_{DON}) plays an essential role in the electric potential distribution of the soft particle and as a consequence its motion due to an applied external electric field.[82] The Donnan potential is defined as the electric potential difference between two solutions separated by an ion-exchange membrane in the absence of any current flowing through the membrane. A schematic representation of the surface of a soft particle and its corresponding electric potential distribution is illustrated in Figure 2.16.

According to this representation, a soft particle has a hard core of radius a covered by an ion-penetrable surface layer of polyelectrolytes of thickness d.

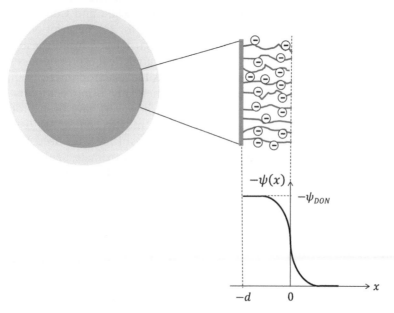

Figure 2.16 Schematic representation of a soft surface and its electric potential distribution.

Logically, this feature requires changes in the classical EDL model described so far. Nevertheless, analytical expressions for the potential distribution around a soft particle determined on the basis of the Poisson–Boltzmann equation have also been developed.[61] Ohshima published a general theory for the electrophoresis of soft particles and approximate analytical expressions for their mobility that include theories for hard particles and polyelectrolytes.[83] Ohshima assumed that the core particle is uncharged and that ionized groups of valence Z are distributed within the polyelectrolyte layer at a uniform density N so that the polyelectrolyte layer is uniformly charged at a constant density of $\rho_{\text{fix}} = ZeN$. In this scenario, approximate expressions for μ are derived for limiting values of κa, λa and λd, where λ characterizes the degree of friction exerted on the liquid flow in the surface layer, $1/\lambda$ being the *electrophoretic softness length*.[83] For the case of a symmetrical electrolyte with valence Z and bulk concentration n, the following expression is derived in the limit of $\kappa d \gg 1$, *i.e.* a polyelectrolyte layer thicker than the EDL:

$$\mu = \frac{2\varepsilon_0 \varepsilon_r}{3\eta} \left(1 + \frac{a^3}{2b^3}\right) \frac{\psi_0/\kappa_m + \psi_{\text{DON}}/\lambda}{1/\kappa_m + 1/\lambda} + \frac{ZeN}{\eta \lambda^2} \tag{2.9}$$

with

$$\kappa_m = \kappa \left[1 + \left(\frac{ZN}{2zn}\right)^2\right]^{\frac{1}{4}}$$

where $b = a + d$ and ψ_0 is the potential at the boundary between the surface layer and the surrounding electrolyte solution. The expressions for ψ_{DON} and ψ_0 can be derived for $\kappa d \gg 1$ as follows:[82]

$$\psi_{\text{DON}} = \frac{1}{\beta z e} \ln \left\{ \frac{ZN}{2zn} + \left[\left(\frac{ZN}{2zn}\right)^2 + 1\right]^{\frac{1}{2}} \right\} \tag{2.10}$$

$$\psi_0 = \psi_{\text{DON}} - \frac{1}{\beta z e} \tanh \left(\frac{\beta z e \psi_{\text{DON}}}{2}\right) \tag{2.11}$$

The former expression for μ converts into the Helmholtz–Smoluchowski equation in the limit $\lambda a \gg 1$, *i.e.* the case without any polyelectrolyte layer, assuming ψ_0 to be equal to ζ and $\kappa a \gg 1$. These results have been successfully applied to analyse electrophoretic data for cells and model particles.[84–86]

Finally, for the case of a general electrolyte solution of ions with valences z_i, another analytical expression can be derived for a spherical polyelectrolyte ($\lambda a \ll 1$):

$$\mu = \frac{\rho_{\text{fix}}}{\eta \lambda^2} \left(1 + \frac{\lambda^2}{3\kappa^2} A + \frac{\lambda^2 \kappa^2}{3} \frac{1 + \dfrac{1}{\kappa b}}{\dfrac{\lambda^2}{\kappa^2} - 1} B \right) \tag{2.12}$$

where A and B are parameters that depend on a, b and κ. This expression uses approximate solutions for ψ_{DON} and ψ_0 for the limit of spherical poly-electrolytes.[83,87] Furthermore, Ohshima derived other analytical expressions for different geometries and experimental conditions.[83]

Finally, we will discuss an example of the usefulness of the electro-phoretic mobility data for a particular case of soft particles. Electro-phoresis experiments for complexes formed by cationic liposomes and DNA (cationic lipoplexes) are shown in Figure 2.17.[88] Here, the electro-phoretic mobility of DOTAP:CHOL–DNA lipoplexes is measured as a function of the lipid/DNA mass ratio (L/D). The experimental data for μ_e as a function of L/D can be divided into three different ζ-potential regions: (i) a region where lipoplexes show a net negative and almost constant elec-trophoretic mobility; (ii) a region where the isoelectric point is reached and an inversion of the electrophoretic mobility sign takes place; and (iii) a region of net positive electrophoretic mobility that tends towards the value for the pure liposomes. The change of sign of the electrophoretic mobility is explained in terms of charge inversion or overcharging phenomena. Accordingly, electrophoresis experiments allow the isoelectric point of this system to be calculated also when liposomes covered by DNA exhibit a negative or positive net charge. This feature has a direct application in gene therapy in which DNA molecules have to be inserted into biological

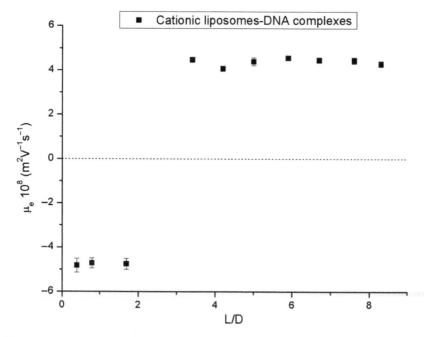

Figure 2.17   Electrophoretic mobility of DOTAP:CHOL–DNA lipoplexes at differ-ent lipid/DNA mass ratios(L/D) in HEPES buffer (10 mM, pH 7.4). DOTAP:CHOL molar ratio, 1:1; DNA concentration, 0.05 mg mL^{-1}.

cells (see Chapter 6). Since biological cell membranes are negatively charged under physiological conditions, positive DNA complexes are expected to be electrostatically attracted to cells. To this end, the results shown in Figure 2.17 provide the *L/D* ratios needed to form positively charged complexes.

2.5   Scattering Techniques

2.5.1   Introduction

Scattering is a general physical process where some form of radiation, such as light, sound or moving particles, is forced to deviate from a straight trajectory by one or more localized non-uniformities in the medium through which they pass. The aim of scattering methods applied to soft NPs is the determination of some key properties such as size, shape and internal structure. The characteristic lengths of the particles to be analysed range from a few ångstroms (~ 5–10×10^{-10} m) to the order of the microns ($\sim 10^{-6}$ m). To cover this wide range of lengths, X-rays, neutrons and light radiation have to be used. Light scattering is suitable for length scales greater than 40–50 nm, whereas X-rays or neutrons allow a wide range from ångstroms to microns to be measured. In static experiments, the key is to send a beam of radiation through a sample containing solid particles immersed in a fluid, and to measure and analyse the scattered intensity $I_s(\theta)$ as a function of the scattering angle θ, where θ is defined as the angle between the forward direction and the direction where the scattered ray is detected.

The incident radiation could be an electromagnetic wave in the case of X-rays or light or a flux of particles in the case of neutrons. According to the de Broglie hypothesis, neutrons can be also considered as a wave of wavelength $\lambda = h/p$, where h is Planck's constant and p the linear momentum of the neutrons. X-rays and light are scattered by the electrons in the particle whereas neutrons are scattered by the nuclei of the atoms. Therefore, irrespective of their nature, we can speak of an incoming radiation characterized by a wavevector with a modulus of $|\vec{k}_i| = 2\pi/\lambda$ (Scheme 2.1) where λ is wavelength in the medium. The outgoing radiation is also characterized by a wavevector $\vec{k}_s$ and it is measured by a detector that counts neutrons or registers the intensity of the impinging electromagnetic radiation.

The key parameter in a scattering experiment is the scattering vector $\vec{q}$, defined as $\vec{q} = \vec{k}_s - \vec{k}_i$ with a modulus

$$q = \frac{4\pi}{\lambda} \sin \frac{\theta}{2} \qquad (2.13)$$

Apart from the detailed mathematical development that can be found in more specialized books, the fundamental idea is that the scattered intensity $I_s(q)$ for a dilute system containing n monodisperse particles per unit

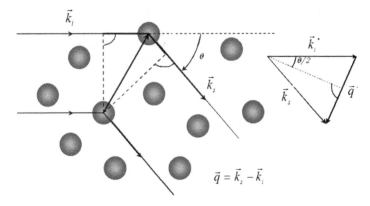

Scheme 2.1 Picture containing scattering geometry. Incident and scattered rays and vector q are represented.

volume is $I_s(q) = nP(q)$, where $P(q)$ is known as the particle form factor, defined as

$$P(q) = \left| \int_V A(r)e^{i\vec{q}\cdot\vec{r}} dV \right|^2 \qquad (2.14)$$

where V refers to the volume of the particles, $\vec{r}$ covers this volume and the function $A(r)$ takes different analytical expressions depending on the radiation employed, as will be seen below. Physically, $P(q)$ is a result of the interference between wavelets emanating from different points within the particles and this is the meaning of the integral in eqn (2.14). The key idea is that $P(q)$ measurements make it possible to determine the spatial point distribution that originates this particular $P(q)$, *i.e.* to know the internal particle structure. As indicated earlier, this method is called static because it supposes the particles to be at rest. In reality, however, particles immersed in a suspending medium undergo Brownian motion and so the scattered intensity $I_s(q)$ changes continuously and fluctuates in time. Hence one should rather talk about $I_s(q,t)$. The monitoring of $I_s(q,t)$ as a function of time and the corresponding treatment are the basis of dynamic or correlation methods such as dynamic light scattering (DLS) in the case of visible light.

2.5.2 Light Scattering

Absorption and scattering are the two fundamental ways in which electromagnetic radiation can interact with colloidal particles. Absorption means that the atomic shell electrons use the impinging electromagnetic radiation energy to jump to higher electronic levels. If this energy is re-emitted afterwards at the same or lower frequency, spontaneous emission or fluorescence is taking place. On the other hand, scattering means that the incident radiation which forces the electrons to perform an oscillatory motion is re-emitted with the same frequency. Hence absorption and scattering are different concepts. The absorption–emission process is slower $(\sim 10^{-12}$ s) than scattering $(\sim 10^{-14}$ s). Using the term photon, emission means that the electrons absorb a photon from the incident radiation and, after a brief period, emit another photon of the same or lower energy. In scattering processes, however, the same photon that approaches departs. The terms elastic and Rayleigh–Brillouin scattering are employed when the angular frequency of both the incident and scattered radiation are the same. If a third particle, such as a phonon, is involved, inelastic or Raman scattering is employed. In this case, an additional energy shift will be observed. For the sake of simplicity, only elastic scattering is discussed in this section and emission, fluorescence and other related phenomena, such as Raman scattering, are excluded.

From a physical point of view, light scattering occurs when light interacts with matter. In this scenario, the optical properties of the medium play a crucial role. Since light does not change its propagation direction in a medium with a uniform refractive index, this would only happen in our case when the refractive indices of the particles and liquid match. When the particles and liquid have different refractive indices and the particle concentration fluctuates, the electromagnetic radiation is scattered in all directions.

In typical elastic light scattering experiments (see Figure 2.18), a light source generates a beam of electromagnetic radiation that is directed towards the cell containing the particle dispersion. In conventional devices, this radiation is a monochromatic and vertically polarized laser. A small portion of the incident energy is scattered and the rest is transmitted. The scattered light is collected by a detector sensitive to intensity, *i.e.* energy flux

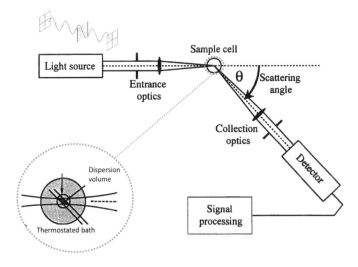

Figure 2.18   Scheme of a standard light scattering setup.

per unit area. The intersection between the scattered and the incident beams defines the scattering volume. The plane that contains both the incident and the scattered beams is known as the scattering plane, which usually is perpendicular to the polarization plane of the electric field of the incident electromagnetic wave. The detector is placed at a distance from the scattering volume that is much larger than the light wavelength. These experimental details are considered in this section.

When light hits matter, the incident electric field $\vec{E}_i(\vec{r},t)$ induces a polarization in the particles and the electric charges in the illuminated volume are accelerated.[89] According to classical electromagnetism, the accelerated charges radiate light and therefore they become secondary sources of light. If the dispersion volume is divided into sub-regions where the distance between charges is smaller than the incident wavelength, all charges inside a sub-region feel the same incident electric field. Hence the scattered electric field in one direction is the sum of the electric fields scattered in that direction from all sub-regions (interference phenomenon). If all sub-regions have the same refractive index, the scattered light would be cancelled except along the incident direction due to destructive interference. Nevertheless, when light hits a particle with a refractive index different to that of the surrounding liquid, the amplitudes of the electric field scattered from different sub-regions are not identical. This produces a constructive interference and the light is scattered in all directions.

A plane electromagnetic wave involves an oscillating electric field that is expressed in complex number notation as $\vec{E}_i(\vec{r},t) = \vec{E}_i \exp[i(\vec{k}_i \cdot \vec{r} - w_i t)]$, where $\vec{E}_i(\vec{r},t)$ is the electric field of the incident light, $\vec{r}$ is the position vector, t is the time, $\vec{E}_i$ is the wave amplitude vector, w_i is the angular frequency and $\vec{k}_i$ is the incident wavevector. Usually, it is assumed that the electric field inside the scattering volume is constant and equal to the incident electric field. This approximation, known as the first Born approximation,[90] is valid only when scattering is weak. The Maxwell equation resolution under these conditions gives a simplified expression for the scattered electric field, $\vec{E}_s(\vec{R},t)$, at a far field point, $\vec{R}$ (such as the detector position):[89]

$$\vec{E}_s(\vec{R},t) = |\vec{E}_i| \frac{\exp\left[i\left(\vec{k}_s \cdot \vec{R} - w_s t\right)\right]}{|\vec{R}|} \sum_{j=1}^{N} \vec{b}_j(\vec{q},t) \exp\left[-i\vec{q} \cdot \vec{R}_j(t)\right] \qquad (2.15)$$

where $|\vec{E}_i|$ is the amplitude vector of the incident wave, w_s and $\vec{k}_s$ are the angular frequency and wavevector of the scattered wave, respectively, $\vec{R}_j(t)$ is the position of the jth particle mass centre, $\vec{q}$ has been previously defined and the sum extends to the numbers of scatterers N. Therefore, $\vec{b}_j(\vec{q},t)$ is the wave amplitude scattered by the jth particle that depends on both the geometry and optical properties of the particle. Since the angular frequency w_s of the temporary fluctuations of the scattered wave is of the order of 10^{14} s^{-1} for visible light, it is very far from the threshold of temporary detection of any photodetector (10^{-7} s), so the first term on the right-hand side of eqn (2.15) can be considered as constant. Focusing on the second term of eqn (2.15) that modulates the amplitude of the scattered wave, $\vec{E}(\vec{q},t)$, we observe that it is independent of the initial amplitude, $|\vec{E}_i|$ and the position of the field point, $\vec{R}$:

$$\vec{E}(\vec{q},t) \equiv \sum_{j=1}^{N} \vec{b}_j(\vec{q},t) \exp\left[-i\vec{q} \cdot \vec{R}_j(t)\right] \qquad (2.16)$$

Therefore, as $\vec{R}_j(t)$ changes in time due to the Brownian motion of particles, the resulting scattered field $\vec{E}_s(\vec{R},t)$, with its dependence on scattering amplitudes $\vec{b}_j(\vec{q},t)$, fluctuates in time also due to this random motion of particles.

The scattered intensity can be obtained from the scattered electric field. Both quantities are determined by the size, shape and motion of particles inside the scattered volume. The instantaneous intensity of scattered light is proportional to the square of the electric field according the following expression:[89]

$$I(\vec{q},t) \propto |\vec{E}(\vec{q},t)\vec{E}^*(\vec{q},t)| = |\vec{E}(\vec{q},t)|^2 \qquad (2.17)$$

where the asterisk (*) denotes the complex conjugate. A change in the configuration of the particles (reorientation and/or translation) modifies the resulting scattered electric field and thereby the instantaneous scattered intensity. The scattered intensity therefore fluctuates due to the Brownian motion of the particles (rotational and/or translational) around a mean value. Precisely, the detector collects the instantaneous scattered intensity during a light scattering experiment.

There are two types of techniques, static light scattering (SLS) and dynamic (DLS) light scattering, depending on how the intensity is measured and analysed. In SLS, the time-averaged total intensity is measured as a function of scattering angle θ or scattering vector $\vec{q}$. These experiments give information on the internal structure, molecular weight and shape of the particles. In DLS, the temporal variation of the intensity is measured and the diffusion coefficient and size distribution of the particles can be obtained. Both techniques assume single scattering, this means that the radiation is only scattered by one localized particle. There is no multiple scattering.

2.5.2.1 Static Light Scattering

In SLS experiments, the intensity should be collected over a time interval that is much larger than the time required for the Brownian particles to probe all accessible configurations. When the magnitude of the characteristic dimension of the scatterer r_0 (for example, the radius a of a spherical particle) is compared with the wavelength of the incident radiation, λ, several light scattering regimes and the corresponding theories can be distinguished: Rayleigh $(r_0 \ll \lambda)$, Rayleigh–Gans–Debye $(r_0 \approx \lambda)$ and Mie $(\lambda \ll r_0)$.

2.5.2.1.1 Rayleigh Scattering. The Rayleigh model[91] assumes that the particles have a size much smaller than the wavelength of the incident radiation, are isotropic, are non-absorbing and have a refractive index that is not too large. Under these conditions, the mean intensity $\langle I(\vec{q},t)\rangle_{\text{Ray},N}$ of the scattered wave per unit volume by N particles in a dilute solution of volume V at a distance R is given by the following expression:[91]

$$\langle I(\vec{q},t)\rangle_{\text{Ray},N} = \frac{I_0 N}{V}\frac{16\pi^4 a^6}{R^2\lambda^4}\left(\frac{m^2-1}{m^2+2}\right) \tag{2.18}$$

where I_0 is the incident light intensity and m is the relative refractive index (*i.e.* the ratio $n_{\text{particle}}/n_{\text{medium}}$). This expression considers that the incident beam is perpendicularly polarized with respect to the scattering plane. As can be seen, the scattered intensity in the Rayleigh limit is independent of the scattering angle, so all directions in the scattering plane are equivalent. The strong dependence on both the size of the scatterer and the wavelength of incident light is notable. The intensity scattered by a larger particle is 64 times more intense than for a mid-sized particle. Therefore, high-power and low-wavelength lasers are more suitable for measuring small particles with low contrast in the refractive index $(m\approx 1)$.

A common Rayleigh scatterer for calibration is toluene. The average scattered intensity as a function of the scattering angle for a toluene sample is shown in Figure 2.19. The intensity fluctuates around a constant value over the wide scattering angle range of 10–150°. As a rule of thumb, a light scattering device is well calibrated when it provides a curve similar to that in Figure 2.19.

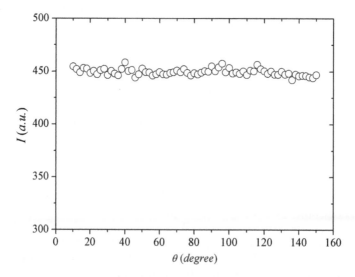

Figure 2.19   Scattered intensity *versus* scattering angle for a toluene solution.

Rayleigh scattering may be used for measuring the particle radius in solution by comparing the intensity scattered from two samples: a known sample containing ρ_1 particles of radius a_1 per unit volume and a sample containing ρ_2 particles of unknown radius a_2 per unit volume. From eqn (2.18), the ratio of the scattered intensities from the two samples provides the value of a_2 to be determined if both a_1 and the particle densities ρ_1 and ρ_2 are known. This method does not require the incident light intensity or the distance to the scattered light detector, which usually are not known.

2.5.2.1.2 Rayleigh–Gans–Debye Scattering.

Rayleigh also extended his theory to particles of arbitrary shape and small relative refractive index $|m-1| \ll 1$. Further contributions were made by Debye and Gans.[90] Particles with dimensions comparable to the wavelength of incident light, $r \ll \lambda/|m-1|$, and/or low optical contrast can be studied with the Rayleigh–Gans–Debye (RGD) theory. In the RGD theory, a particle of arbitrary shape is subdivided into different regions that behave as Rayleigh scatterers excited by the incident field. Because the distances between these various scattering centres are of the same magnitude as the wavelength of light, there will be interference between the light scattered by different parts of the same particle. The interference effect may be constructive or destructive, depending on the value of the scattering angle.

It can be demonstrated that the scattering intensity averaged over all possible orientations of two scattering centres, $\langle I(\vec{q},t) \rangle_{\mathrm{RGD},2}$, is given by the following expression:[92]

$$\langle I(\vec{q},t) \rangle_{\mathrm{RGD},2} = I_s \left[1 + \frac{\sin(qr)}{qr} \right] \tag{2.19}$$

where r is the distance between the two scattering centres and I_s is the scattered intensity without interference, *i.e.* the scattered intensity in the Rayleigh regime. The subsequent step is to extend eqn (2.19) to a single particle with N scattering centres. After doing so, we obtain[92]

$$\begin{cases} \langle I(\vec{q},t) \rangle_{\mathrm{RGD},N} = \dfrac{I_s}{N^2} \sum_i^N \sum_j^N \dfrac{\sin(qr_{ij})}{qr_{ij}} = I_s P(q) \\[12pt] P(q) = \dfrac{1}{N^2} \sum_i^N \sum_j^N \dfrac{\sin(qr_{ij})}{qr_{ij}} \end{cases} \tag{2.20}$$

where r_{ij} is the distance between the i and j scattering centres. This equation assumes the scattering centres to be identical. $P(q)$ is known as the particle form factor and physically describes the interference between the electric fields scattered from the different volume elements within each single particle. Therefore, $P(q)$ represents the variations of the intensity due to the finite size of the particle and contains information about the particle size, as will be seen below. Theoretically, the form factor has been evaluated for certain standard geometries such as homogeneous spheres, spherical shells and rigid rods. For optically homogeneous spheres of radius a, Rayleigh proposed the following expression for the form factor:[89]

$$P(q) = \left[3 \frac{\sin(qa) - qa\cos(qa)}{(qa)^3} \right]^2 \qquad (2.21)$$

By fitting the experimental scattered intensities using eqns (2.20) and (2.21), the particle radius a can be determined. Figure 2.20 shows the experimental form factor as a function of the scattering angle for spherical particles of 50 nm radius. The continuous line shows the best fit, obtaining a fitting value of 49.5 nm. The form factor shows a clear dependence on both scattering angle and scatterer size.

For more complex geometries such as a spherical shell of thickness t, the form factor is written as[93]

$$P(q) = \left[3 \frac{\sin(qr_{int}) - qr_{int}\cos(qr_{int}) - \sin(tqr_{int}) + tqr_{int}\cos(qr_{int})}{(1 - t^3)(qr_{int})^3} \right]^2 \qquad (2.22)$$

where r_{int} is the inner radius of the shell. Figure 2.21 shows the experimental normalized form factor for a sample of liposomes (see Chapters 3 and 6). The solid line represents the best fit to the experimental points according to eqn (2.22) and taking into account the polydispersity *via* a Schulz distribution.[93] The optimum fits gives $r_{int} \approx 92$ nm and a shell thickness of 5 nm.

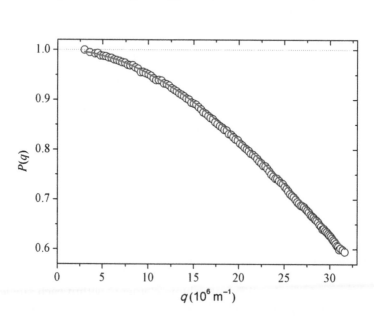

Figure 2.20   Experimental form factor $P(q)$ as a function of the scattering vector for an aqueous suspension of polystyrene spherical particles (100 nm in diameter). The solid line corresponds to the best fit according to RGD theory.

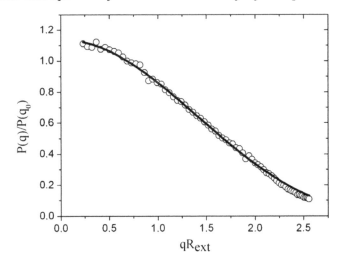

Figure 2.21 Normalized form factor $P(q)/P(q_0)$ as a function of the adimensional quantity qR_{ext}. Open circles represent experimental values and the solid line denotes the best fit. Note that $R_{ext} = r_{int} + t$.

Hence good agreement between experiment and theory is observed. One can definitely claim that RGD theory and SLS are suitable methods for measuring the size and shape of soft NPs.

Radius of Gyration. In biotechnology and materials science, particles with different and unknown shapes and geometries are usually found, and revealing information about the molecules or atoms inside the particle can be of interest. The radius of gyration provides average information. The most commonly used definition of the radius of gyration of a colloidal particle is the mean square value of the distances r_{ij} over all statistical configurations:[92]

$$\overline{R_g^2} = \frac{1}{2N^2} \sum_i^N \sum_j^N r_{ij}^2 \tag{2.23}$$

From eqn (2.20), the form factor of nanoparticles may be expressed as a power series for the case in which qr_{ij} is small. In this case and also when the modulus of the scattering vector q is small regardless of the distance between the scattering centres r_{ij}, the form factor reads[92]

$$P(q) = 1 - \frac{q^2}{6N^2} \sum_i^N \sum_j^N r_{ij}^2 = 1 - \frac{q^2 \overline{R_g^2}}{3} \tag{2.24}$$

where the mean square value of the radius of gyration of one particle has been introduced. This equation is valid only for $q^2 \overline{R_g^2} \ll 1$. As the second term on the right-hand side of eqn (2.24) is much smaller than unity, the inverse of form factor is given by

$$\frac{1}{P(q)} \approx 1 + \frac{1}{3}\overline{R_g}^2 q^2 \tag{2.25}$$

In the RGD regime, a plot of $1/P(q)$ as a function of q^2 allows the direct determination of $\overline{R_g}^2$ in scattering experiments at small q values. Nevertheless, the radius of gyration obtained from this method gives a mean value for the size distribution of the particles. For very small particles, the decrease in intensity over the entire experimentally accessible wavevector range is insufficient for an accurate determination of the particle radius.

The radius of gyration for structures of standard geometries can be expressed as follows:

Sphere:

$$R_g = \sqrt{\frac{3}{5}}a$$

where a is the radius of a spherical particle;

Cylinder:

$$R_g = \left(\frac{L^2}{12} + \frac{a^2}{2}\right)^{\frac{1}{2}}$$

where L and a are the length and radius of a cylindrical particle;

Random chain:

$$\overline{R_g}^2 = \frac{Nb^2}{6}$$

where N represents the number of segments of length b of a random chain.

Molecular Weight. Rayleigh and RGD theories also enable the molecular weight of soft particles or macromolecules in solution, such as proteins, synthetic polymers, liposomes, micelles and colloids, to be obtained.

In a light scattering experiment, the incident light intensity I_0 and the distance to the detector R are fixed but unknown. From a practical point of view, these quantities and the scattered intensity measure can be combined into one quantity known as the Rayleigh ratio, R_θ, defined as[90,92]

$$R_\theta = \frac{\langle I(\vec{q}, t)\rangle_{\text{Ray},N} R^2}{I_0} \tag{2.26}$$

Hence eqn (2.18) can be rearranged in terms of the Rayleigh ratio and the resulting expression becomes

$$R_\theta = \frac{N}{V}\frac{16\pi^4 a^6}{\lambda^4}\left(\frac{m^2 - 1}{m^2 + 2}\right) \tag{2.27}$$

It is known that the intensity of scattered light from colloidal dispersions or solutions of macromolecules depends on the average of the square of the polarizability[92] and this magnitude depends on the molecular weight M of the scatterer, the second virial coefficient B and other experimentally measurable quantities, such as the refractive index of the solution n and the refractive index gradient dn/dc. After some algebra, the Rayleigh ratio can be finally written as[91,92]

$$R_\theta = \frac{Kc}{\dfrac{1}{M} + 2Bc} \tag{2.28}$$

where c is the solute concentration in g L^{-1} and the numerical constants (N_A), optical properties $(n, dn/dc)$ of the solution and wavelength of the incident light have been included in the constant K as follows:

$$K = \frac{2\pi^2 n^2}{\lambda^4 N_A}\left(\frac{dn}{dc}\right)^2 \tag{2.29}$$

Given that the solvent molecules can also scatter light in colloidal dispersions, an excess Rayleigh ratio is defined as $\Delta R_\theta = R_\theta - R_{\theta,\text{solvent}}$ (see Chapter 5) and replaces R_θ in the calculations. In practice, to determine the average molecular weight of Rayleigh nanoparticles, we have to use a reference sample which a known Rayleigh ratio. Usually, this reference sample is toluene. From the right-hand side of eqn (2.27), $R_\theta^{\text{toluene}}(\theta)$ depends mainly on the wavelength, because its r_0, N, V and m values are known. In consequence, the $R_\theta^{\text{toluene}}(\theta)$ data are tabulated for different λ values. Using eqn (2.26), $R_\theta^{\text{sample}}(\theta)$ is determined experimentally using the relationship

$$R_\theta^{\text{sample}}(\theta) = \left[\langle I(\vec{q},t)\rangle^{\text{sample}} / \langle I(\vec{q},t)\rangle^{\text{toluene}}\right] R_\theta^{\text{toluene}}(\theta)$$

which is independent of the incident light intensity and detector position. From the plot of $Kc/R_\theta^{\text{sample}}$ as a function of c, the average molecular weight of the NPs M, evaluated from the intercept at infinite dilution, and the second virial coefficient B, calculated from the slope, are obtained. For a polydisperse suspension, the average molecular weight in the Rayleigh eqn (2.28) corresponds to the weighted-average molecular weight.

In the RGD regime, the theoretical scattered intensity is obtained multiplying the Rayleigh intensity by the particle form factor [see eqn (2.20)]. Hence eqn (2.28) reads[91,92]

$$\frac{Kc}{R_\theta^{\text{RDG}}} = \frac{1}{P(q)}\left(\frac{1}{M} + 2Bc\right) = \left(\frac{1}{M} + 2Bc\right)\left[1 + \frac{16\pi^2 n^2 \bar{R}_g^2}{3\lambda^2}\sin^2\left(\frac{\theta}{2}\right)\right] \tag{2.30}$$

where eqn (2.25) and the definition of vector $\vec{q}$ have been used. Now, the R_θ^{RDG} ratio is defined as

$$R_\theta^{\text{RDG}} = \frac{\langle I(\vec{q},t)\rangle_{\text{RDG},N} R^2}{I_0}$$

and determined experimentally with the same procedure as in the Rayleigh regimen.

From eqn (2.30) at the limit of both zero concentration and zero scattering angle, $Kc/R_\theta^{RGD} = 1/M$, so the molecular weight can be obtained. In the limit of zero scattering angle where $P(q) = 1$, B can be evaluated from the slope of the plot of Kc/R_θ^{RGD} as a function of c once M is known. Accordingly, in the limit of zero concentration, Kc/R_θ^{RGD} is proportional to $\sin^2(\theta/2)$, so the radius of gyration can be calculated.

For particles with a characteristic length of $r_0 > \lambda/20$, a double extrapolation to zero angle and zero concentration is necessary to obtain the molecular weight and the radius of gyration, as mentioned before. This is usually performed on a Zimm plot.[92] For small particles ($r_0 \leq \lambda/20$), no angular dependence is observed and eqn (2.30) reduces to eqn (2.28). In this case, no angular measurements are necessary to obtain M.

As a rule of thumb, in an experimental Zimm plot, Kc/R_θ^{RGD} is plotted as a function of $\sin^2(\theta/2) + k'c$, where k' is a constant to be chosen arbitrarily. Experiments are performed at several angles that satisfy the condition $qR_g < 1$ and at high enough particle concentrations. Figure 2.22 shows a representative Zimm plot.

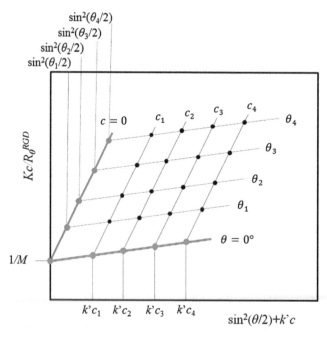

Figure 2.22   Schematic representation of a typical Zimm plot. Black points could be the experimental data and blue and red points could be the extrapolations at $c = 0$ and $\theta = 0$, respectively.

2.5.2.1.3 Mie Scattering. Gustav Mie described the scattering of electromagnetic radiation by dielectric absorbing spherical particles of arbitrary size and refractive index.[89,92] Currently, the term Mie scattering is used in many contexts, for example for scattering by stratified spheres or other geometries such as infinite cylinders or clusters of spheres. In order to calculate the scattered intensity theoretically, Maxwell's equations are solved considering the effects of particle geometry, angular dependence of the incident beam and optical properties of the particle with respect to the medium. The solutions are expressed in terms of infinite series of vector spherical harmonics from which cross-sections, the efficiency factor and the intensity distribution for a particle can be derived. These solutions are implemented in codes written in different computer languages available *via* different web pages. The average-time scattered intensity can be expressed as

$$\langle I(\vec{q}, t) \rangle = \frac{\lambda^2}{4\pi^2 r^2} |S_1(q)|^2 \tag{2.31}$$

where $S_1(q)$ is the amplitude function that depends on the Bessel functions. Rayleigh and RGD scattering theories are particular cases of the Mie solutions.

Figure 2.23 shows the experimental normalized form factor for a stable sample of polystyrene particles. The particles are highly monodisperse spheres with an average diameter of 600 ± 10 nm. The solid line was

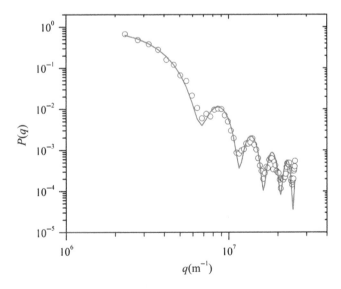

Figure 2.23 Experimental form factor $P(q)$ as a function of the scattering vector for an aqueous suspension of polystyrene spherical particles (600 nm in diameter). The solid line corresponds to the best fit according to Mie theory.

calculated according to Mie's theory for spheres with an average radius of 620 nm and a refractive index of $n = 1.49$.

2.5.2.2 Dynamic Light Scattering

Dynamic light scattering (DLS) is one of the most popular methods used to determine the size of particles. In DLS, the temporal variation of the scattered intensity is measured. In general, the intensity accessible in a scattering experiment for a fixed scattering angle depends on time, since the scattering particles are in constant random motion due to their kinetic energy. It is known that the intensity scattered by small particles fluctuates faster than in the case of large particles. This behaviour can be explained by the faster diffusive motion of smaller particles. Hence the scattered intensity fluctuates around an average value and the temporal fluctuations contain information on the Brownian motion of the particles.[94]

Brownian motion of colloidal particles is the result of collisions between the particles and the solvent molecules. The Brownian path is usually described by the mean quadratic displacement and related to the diffusion coefficient of the particles.[94,95] This parameter can be obtained by DLS, as will be seen later. For simple geometries, there is a direct relationship between the diffusion coefficient and the particle size.

In light scattering experiments, a detector collets the scattered intensity over a period of time[94,96,97] and amplifies the small single-photon signals. The detector output signal consists of a set of pulses than can be treated in different ways. Current electronic correlators provide simultaneously the averaged intensity and the autocorrelation or cross-correlation functions.

In DLS, the temporal variation of the scattered intensity over a period of time is contained in the intensity autocorrelation function. Correlation quantifies how two dynamic functions are related over a period of time. Likewise, autocorrelation relates the value of a function at two different times t and $t + \tau$. The autocorrelation function of the scattered intensity,[94,95] $G_I(q,\tau)$, describes the rate of change of the scattered intensity by comparing the intensity at time t with the intensity at a later time $(t + \tau)$. $G_I(q,\tau)$ is written mathematically as an integral over the product of intensities at an arbitrary time with a certain delay time τ:

$$G_I(q,\tau) = \langle I(q,t)I(q,t+\tau) \rangle = \lim_{T \to \infty} \frac{1}{T} \int_t^{t+T} I(q,t')I(q,t'+\tau)\mathrm{d}t' \qquad (2.32)$$

The brackets denote time averaging, t is the time at which a measurement is initiated and T is the time over which it is averaged. This definition is useful only if T is greater than the period of fluctuations of $I(q,t)$; ideally, T should be ∞. If τ is very small compared with the period of fluctuations, $I(q,t)$ and $I(q,t+\tau)$ will be very close. For large τ values, $I(q,t+\tau)$ will probably be very

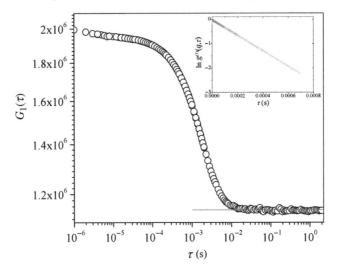

Figure 2.24 Intensity autocorrelation function *versus* delay time τ for a dilute solution of polystyrene spheres of diameter 46 nm in water. Inset: logarithm of the normalized autocorrelation function $g^{(1)}(q,\tau)$ as a function of τ. The solid line corresponds to a linear fit of the experimental data.

different from $I(q,t)$. It can also be verified that $G_I(q,\tau) \to \langle I(q,t)^2 \rangle$ when $\tau \to 0$ and $G_I(q,\tau) \to \langle I(q,t) \rangle^2$ when $\tau \to \infty$. This function is plotted in Figure 2.24. It can be seen how at shorter times the scattered intensity signals are correlated (same values) because the particles have almost no time to move. At longer times, the particles lose their correlation (different intensity values) because of the Brownian motion of the particles.

The ratio of $G_I(q,\tau)$ to its asymptotic value is known as the normalized autocorrelation function of scattered intensity[96]:

$$g^{(2)}(q,\tau) = \frac{\langle I(q,t)I(q,t+\tau) \rangle}{\langle I(q,t) \rangle^2} \qquad (2.33)$$

Like $g^{(2)}(q,\tau)$, the normalized autocorrelation function of the scattered electric field $\vec{E}_s(q,t)$ is defined as[94]

$$g^{(1)}(q,\tau) = \frac{\langle \vec{E}_s(q,t)\vec{E}_s(q,t+\tau) \rangle}{\langle \vec{E}_s(q,t) \rangle^2} \qquad (2.34)$$

This function reveals the motion of the particles relative to each other. The relationship between the particle motion contained in the $\vec{E}_s(q,t)$ and $g^{(1)}(q,\tau)$ functions and the measured intensity fluctuations, $g^{(2)}(q,\tau)$, is given by the Siegert equation[98]:

$$g^{(2)}(q,\tau) = 1 + \beta |g^{(1)}(q,\tau)|^2 \qquad (2.35)$$

where β is an experimental constant $(0 \leq \beta \leq 1)$ that depends on the coherence of the scattered field at the detector. This equation is valid when the electric field scattered by colloids is Gaussian with zero mean.

For dilute and monodisperse samples, $g^{(1)}(q,\tau)$ is a simple exponential function with a constant decay rate, Γ. The decay constant is related to the particle translational diffusion coefficient, D_0, by $\Gamma = D_0 q^2$. In this case, the normalized electric field autocorrelation function is written as[94]

$$g^{(1)}(q,\tau) = \exp(-\Gamma\tau) = \exp(-D_0 q^2 \tau) \qquad (2.36)$$

Large particles diffuse more slowly than small particles, hence the correlation function decays more slowly.

Using the Stokes–Einstein equation,[94,99] the translational diffusion coefficient of a spherical particle can be related to the hydrodynamic particles radius, r_H:

$$D_0 = \frac{k_B T}{6\pi\eta r_H} \qquad (2.37)$$

where k_B is the Boltzmann constant, T is the absolute temperature and η is the viscosity of the sample. The radius r_H measured using this equation reveals the size of the moving particles.

For polydisperse suspensions, *i.e.* mixtures of non-identical particles, data analysis becomes very complex and the correlation function $g^{(1)}(q,\tau)$ can be expressed as a sum or integral over different $e^{-\Gamma\tau}$ terms.[100,101] In this way, $g^{(1)}(q,\tau)$ describes jointly the motion of individual particles:

$$g^{(1)}(q,\tau) = \int_0^\infty G(\Gamma) e^{-\Gamma\tau} d\Gamma \qquad (2.38)$$

where $G(\Gamma)d\Gamma$ is the fraction of the scattered intensity due to particles with decay rates from Γ to $\Gamma + d\Gamma$. As Chu reported,[101] obtaining $G(\Gamma)$ from $g^{(1)}(q,\tau)$ becomes a non-trivial task. For this reason, several methods are used: cumulant expansion, regularization (Contin), maximum entropy, *etc.* For simplicity, we describe only the cumulant expansion method, which was first described by Koppel.[102] Essentially, the logarithm of the correlation function $g^{(1)}(q,\tau)$ is fitted by a polynomial:

$$\ln |g^{(1)}(q,\tau)| = -\mu_1 \tau + \frac{1}{2!}\mu_2 \tau^2 - \frac{1}{3!}\mu_3 \tau^3 + \dots \qquad (2.39)$$

where the coefficients $\mu_1, \mu_2, \mu_3 \dots$ are called cumulants. In practice, it is difficult to determine cumulants higher than μ_2. The initial slope of the $\ln|g^{(1)}(q,\tau)|$ function gives μ_1, which may be interpreted as an average decay rate $\langle\Gamma\rangle$, related to the average particle radius. The quantity μ_1/μ_2^2 characterizes the variance of the particle size distribution. This method is easy to implement and its results are often used as a starting point for a more detailed analysis.

In summary, we describe the procedure to obtain the particle diffusion coefficient from a DLS experiment as follows. (1) The correlator collects and integrates the scattered intensities at the different delay times, in real time.

Therefore, the intensity autocorrelation function $G_I(q,\tau)$ is obtained directly from a correlator device. (2) Next, $G_I(q,\tau)$ is normalized with respect to the average squared scattered intensity at large time, $\langle I(q,t)\rangle^2$ for $\tau \to \infty$. This quantity is known as the baseline and is also measured by the correlator. Given that the Siegert relation [see eqn (2.35)] can be written as

$$\frac{G_I(q,\tau)}{\langle I(q,t)\rangle^2} - 1 = \beta |g^{(1)}(q,\tau)|^2 \qquad (2.40)$$

the experimental $g^{(1)}(q,\tau)$ is determined. (3) The logarithm of $g^{(1)}(q,\tau)$ is plotted against τ (see the inset in Figure 2.24). From a linear fit, the decay rate Γ, and thus D_0, may be obtained from the slope. (4) Finally, the hydrodynamic particle radius can be calculated from the Stokes–Einstein relation [eqn (2.37)].

2.5.3 Neutron Scattering

2.5.3.1 Basic Concepts of Neutron Scattering

Neutron scattering is used to determine the position and motion of atoms in condensed matter. Scattering methods provide statistically averaged information on structure rather than real-space pictures of particular instances. Several advantages may be pointed out: (1) the wavelength is comparable to the interatomic spacing; (2) the kinetic energy is comparable to that of atoms in a solid; (3) bulk properties may be measured and samples can be contained due the high penetration; (4) the weak interaction with matter aids the interpretation of scattering data; and (5) isotopic sensitivity allows contrast variation. On the other hand, some disadvantages should also be mentioned: (1) neutron sources are weak, emitting low signals, and large samples and long observation times are required; (2) several elements (Cd, B, Gd) absorb strongly; and (3) not all energy and momentum transfers are accessed owing to kinematic restrictions.

Neutrons interact with atomic nuclei *via* very short-range ($\sim 10^{-15}$ m) forces and also with unpaired electrons through magnetic dipole interactions. They have both particle-like and wave-like properties, with mass $m_n = 1.675 \times 10^{-27}$ kg, no charge, spin $\frac{1}{2}$, magnetic dipole moment $-1.913\ \mu N$ and nuclear magneton $eh/4\pi m_p = 5.051 \times 10^{-27}$ J T^{-1}. The energy E and wavevector k are related to the velocity v, temperature T and wavelength λ through the well-known relationships

$$E = \frac{m_n v^2}{2} = k_B T = \frac{\left(\dfrac{hk}{2\pi}\right)}{2m_n}$$

and

$$k = \frac{2\pi}{\lambda} = \frac{m_n v}{\left(\dfrac{h}{2\pi}\right)}$$

Table 2.2   Classification of neutrons according to their energies, temperatures and wavelengths.

Category	Energy (meV)	Temperature (K)	Wavelength (nm)
Cold	0.1–10	1–120	0.4–3
Thermal	5–100	60–1000	0.1–0.4
Hot	100–500	1000–6000	0.04–0.1

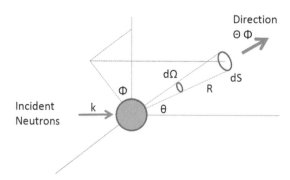

Scheme 2.2   Sketch of a typical neutron scattering experiment.

with E in meV, T in K and λ in nm. Depending on the energy associated with the neutrons, they are classified into three categories (Table 2.2).

As with light, typical neutron scattering experiments consist of an incident beam reaching a sample, which deviates, *i.e.* scatters the initial beam in all directions. However, not all directions are equally probable, the probability being directly related to the structure of the scattering object (sample). It is well known that all particles in the incident beam have an associated wavelength, which limits the observable spatial scale. Thus, light scattering is able to access the nanometer region while neutrons and X-rays are able to observe the ångstrom range. The scattering wave reaches a detector located at a known position, characterized by distances and angles where the signal (wave intensity or counts) is detected on a well-defined surface perpendicular to the signal propagation direction (see Scheme 2.2). The total amplitude at the detector will be

$$\Psi_{\text{scatt}} \approx b \sum_{i=1}^{N} e^{-(i\vec{q} \cdot \vec{r}_i)}$$

where $\vec{q}$ is the wavevector, $\vec{r}_i$ the position of the scatterer i and b the length of the scatterer related to its nature (see below for a further description). The intensity of such radiation measured at the detector is proportional to the square of the scattering function $I \approx |\Psi_{\text{scatt}}|^2$. Note that Ψ_{scatt} plays the same role as $\vec{E}_s(\vec{R},t)$ in light scattering.

The physical information about the sample structure leading to the measured scattering intensity is related to the scattering cross-section σ, which is a hypothetical area describing the likelihood of neutrons (or other radiation) being scattered by a particle. The scattering cross-section is different from the geometric cross-section of a particle as it depends on the wavelength of radiation, the permittivity and the particle shape and size. The total amount of scattering in a sparse medium is determined by the product of the scattering cross-section and the number of particles present. In terms of area, the *total cross-section* σ is the sum of the cross-sections due to absorption, scattering and luminescence: $\sigma = \sigma_A + \sigma_S + \sigma_L$. Typical neutron scattering experiments use an energy window E to $E + dE$ for the incident neutrons, recording the scattering neutron distribution to build the differential scattering cross-section:

$$\frac{d\sigma}{d\Omega} = \frac{\text{number of neutrons scattered per second into } d\Omega}{\Phi d\Omega} \qquad (2.41)$$

where Φ is the incident flux of neutrons on the detector (number of incident neutrons per cm^2 per second) and $d\Omega$ is the solid angle (see Scheme 2.2). For detector positions far enough away from the scattering volume,[103]

$$\frac{d\sigma}{d\Omega} = \sum b_i b_j e^{-i\vec{q}\cdot(\vec{r}_i - \vec{r}_j)} \qquad (2.42)$$

where b_i is the scattering length of the scatterer i, which is characteristic of the nature of the scatterer material. It depends on the nuclear isotope, the spin relative to the neutron and the nuclear eigenstate; for a single nucleus $b_i = \{b\} + \delta b_i$, with $\{b\}$ being a mean value and δb averaging to zero. Using this b_i value and after some calculations,[103,104] eqn (2.42) reads

$$\frac{d\sigma}{d\Omega} = \{b\}^2 \sum e^{-i\vec{q}\cdot(\vec{r}_i - \vec{r}_j)} + \left(\{b^2\} - \{b\}^2\right)N$$

where $\{b\}^2$ is the coherent nuclear scattering length and N is the number of scatterers. The first term corresponds to the coherent scattering and depends on the direction of $\vec{q}$. The second term represents the incoherent scattering, uniform in all directions. In a real experiment, it is very important to increase the coherent term, which contains the structural information, and to reduce the incoherent contribution, which is simply an undesired background. The selection of nuclei with low incoherent scattering capability (low incoherent cross-section) is a keystone when designing a neutron scattering experiment.

Small-angle neutron scattering (SANS) is used to measure objects ranging in size between 1 nm and 1 mm. Complex fluids, alloys, precipitates, polymer structures, biological assemblies, glasses, ceramics, flux lattices, porous media, fractal structures, *etc.*, are typical systems suitable of being studied by SANS. Neutrons can be used with bulk 1–2 mm thick samples. SANS is sensitive to light elements such as H, C and N and also isotopes such as D. The scattering cross-section, $d\sigma/d\Omega$, is also expressed in terms of a function called the structure factor, $S(q)$:[103–105]

$$\frac{d\sigma}{d\Omega} = \{b\}^2 N S(q) \qquad (2.43)$$

$$S(q) = \frac{1}{N}\left\{ \left| \int d\vec{r} \cdot e^{-i\vec{q}\cdot\vec{r}} n_{\text{nucl}}(\vec{r}) \right|^2 \right\} \tag{2.44}$$

where $n_{\text{nucl}}(r)$ is the nuclear density. For an assembly of similar atoms at certain positions, the structure factor $S(q)$ may be expressed as

$$S(q) = \int d\vec{r} \cdot g(\vec{r}) \cdot e^{-i\vec{q}\cdot\vec{r}} \tag{2.45}$$

which is directly related to the position distribution through the atom–atom static pair correlation function $g(r)$:

$$g(r) = \sum \{ \delta(\vec{r} - \vec{r}_j) + \vec{r}_0 \} \tag{2.46}$$

The static pair correlation function gives the probability that there is an atom, i, at distance $\vec{r}_i$ from the origin of a coordinate system at time t, given that there is also a (different) atom at the origin $\vec{r}_0$ of the coordinate system. For simple liquids, $S(q)$ and $g(r)$ show oscillating properties that represent atoms in coordination shells. The function $g(r)$ is expected to be zero for $r <$ particle diameter, and $S(q)$ and $g(r)/r$ both tend to unity at large values of their arguments (see Figure 2.25).

In the case of a dispersion of colloidal particles, the integral in the definition of $S(q)$ is separated into an integral over the positions of the particles and an integral over a single particle [$\rho(r)$ is the scattering length density of particles dispersed in a medium with scattering length density ρ_0].[103–105]

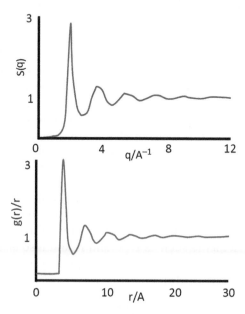

Figure 2.25 Schematic representation of $S(q)$ and $g(r)$ functions.

Thus, SANS can measure particle shapes and inter-particle correlations and eqn (2.43) reads

$$\frac{d\sigma}{d\Omega} = (\rho - \rho_0)^2 |F(\vec{q})|^2 N \int d\vec{r} \cdot G_p(\vec{r}) \cdot e^{i\vec{q}\cdot\vec{r}} \qquad (2.47)$$

where $G_p(r)$ is the particle–particle correlation function (the probability that there is a particle at r if there is one at the origin) and $|F(q)|^2 = P(q)$ is the particle form factor:

$$|F(\vec{q})|^2 = \left| \int d^2\vec{r} \cdot \rho(\vec{r}) \cdot e^{i\vec{q}\cdot\vec{r}} \right|^2 \qquad (2.48)$$

These expressions are similar to those for nuclear scattering except for the addition of a form factor that arises because the scattering is no longer from point-like particles. $G_p(r)$ has no influence for diluted samples and only the particle form factor is relevant in that case. For spherical particles, the particle form factor is determined by the particle shape. For a sphere of radius a, $P(q)$ depends only on the magnitude of $\vec{q}$ [see eqn (2.21)]:

$$F_{\text{sphere}}(q) = 3B \left[\frac{\sin qa - qa \cos qa}{(qa)^3} \right] \qquad (2.49)$$

where B is a normalization constant. Particle size is usually deduced from dilute suspensions in which inter-particle correlations are absent. In practice, instrumental resolution (finite beam coherence) will smear out minima in the form factor. This effect can be accounted for if the spheres are monodisperse. For polydisperse particles, maximum entropy techniques have been used successfully to obtain the particle size distribution.

2.5.3.2 Assessing the Internal Structure of Colloidal Nanoparticles Using Neutron Scattering

This section describes the use of small-angle neutron scattering (SANS) to study the internal structure of particular soft NPs: highly cross-linked temperature-sensitive poly(N-isopropylacrylamide) (PNIPAM) microgels (see Chapter 4). Figure 2.26(a) shows the experimental particle size from light scattering experiments against temperature. The particle size decreases gradually with increase in temperature until a sharp volume transition from swollen to deswollen states occurs, reaching a final collapsed size at a transition temperature $T_c > 307$ K.

SANS is a widely used technique for investigating the structure of polymer networks. Experiments consist of measuring the scattered intensity $I_s(q) = d\sigma/d\Omega$ against the scattering vector q related to the neutron wavelength λ and the scattering angle θ through $q = (4\pi/\lambda)\sin(\theta/2)$, as stated previously. The intensity is directly proportional to the scattering cross-section for SANS and thus to the structure factor $S(q)$ described before. The scattered intensity $I_s(q)$ in absolute intensity scale is given by $I_s(q) = I_{obs}(q) + I_{inc}(q)$, where $I_{obs}(q)$ and $I_{inc}(q)$ are the observed and the

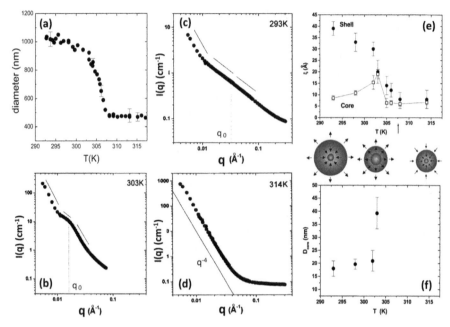

Figure 2.26 (a) Experimental particle size *versus* temperature for PNIPAM micro-gels (from light scattering experiments). (b–d) Double logarithmic plots of SANS intensity for different temperatures. The correlation length is calculated from the curves as indicated in the text. (e) Temperature dependence of the characteristic correlation lengths for the particle core and shell. (f) Estimation of the core size from the q_0 break point (the inverse value represents the characteristic core length). The sketch shows the core and shell behaviour during the phase transition.

Scheme 2.3 Illustrative diagram of the correlation lengths ξ_1 and ξ_2 found in a core–shell particle.

incoherent scattered intensity, respectively (see earlier). This intensity is characterized by a single length scale, namely the correlation length ξ (characteristic size of the polymer concentration fluctuation) or blob size (see Scheme 2.3).

For good contrast between chains and solvent, the neutron scattering intensity results in an Ornstein–Zernicke-type function in the limit $q\xi < 1$:

$$I(q) = \frac{I(q=0)}{(1 + \xi^2 q^2)} \quad (2.50)$$

When the scattering arises from surfaces, the intensity is then written as

$$I(q) \sim q^{-\alpha} \quad (2.51)$$

where $\alpha = 6 - d_s$ and d_s is the fractal surface dimension. This fractal model may be used to characterize the surface roughness. The compactness of that surface is estimated from the value of the fractal dimension d_s [a number ranging between 0 and 2, where 0 is the limit for a completely full-of-holes surface and 2 corresponds to smooth surfaces, the so-called Porod law: $I(q) \sim q^{-4}$]. Neutron scattering allows not only structural information about internal parts (3D information) but also 2D information associated with interfaces to be determined from the same experiment.

The results shown in Figure 2.26 correspond to experiments for PNIPAM microgels that were performed at the D11 spectrometer at the ILL facility (Institut Laue-Langevin, France).[106] A 6 Å average wavelength cold neutron flux with 10% spread was used as the incident beam by setting the mechanical selector at 21126 rpm. In order to cover a large angular range, three instrumental settings were used: the 64×64 cm^2 gas-filled detector with 1 cm^2 resolution was placed 1.2 m away from the sample for large angles, 5.5 m for medium angles and 10.5 m for small angles. The resultant q range was 0.005–0.303 Å^{-1}. The detector efficiency was corrected using H$_2$O as a completely incoherent scatterer standard. Data reduction was performed by correcting the raw scattered counts for dark scattering using a Cd sheet located at the sample position, for cell scattering and D$_2$O scattering, and also for transmission. The contribution of the incoherent scattering due to hydrogen was also corrected by assuming the additivity of the incoherent scattering between PNIPAM and D$_2$O. The microgel particle concentration in D$_2$O was 4 wt%. All samples were contained in 1 mm pathlength Hellma quartz cells, and prior to undertaking the measurements the cells were kept for 15 min in the instrument's temperature control chamber for temperature stabilization. The temperature was controlled with a precision of 0.1 K.

Small-angle neutron scattering results are shown for different temperatures in Figure 2.26(b–d).[106] In each case, the scattering profiles decrease monotonously with increasing wavevector for swollen and collapsed microgel states. At larger q values, *i.e.* $q \geq 0.01$ Å^{-1}, the scattering intensity shows two power laws in q with a break point between both regions at q_0. The presence of two regions is consistent with a core–shell internal structural inhomogeneity where the core dimension is related to the characteristic length coming from the break point q_0. The correlation lengths were calculated using eqn (2.50). Figure 2.26(e) plots the shell and core correlation lengths *versus* temperature.

The first remarkable feature concerns the inhomogeneous character of the particles. At low temperature, the two correlation lengths are very different. This indicates that the nature of the inhomogeneity responsible for the core–shell structure is related to the concentration of cross-linker in the core (see Chapter 4). The shell correlation length is constant above the transition temperature, but it increases drastically from 8 to 40 Å as the temperature decreases owing to particle swelling, as determined already by DLS. The core correlation length also remains constant for temperatures corresponding to the collapsed states. In this region $\zeta_{core} < \zeta_{shell}$, in agreement with the higher core elastic shrinking component, owing to the presence of a higher cross-linker concentration. The characteristic core blob size increases from 8 to 18 Å as the temperature decreases as a consequence of polymer solvency changes and following the growth in shell size. However, the core correlation length decreases again as the temperature is further reduced. This result agrees with the fact that q_0 shifts towards larger values as the temperature decreases, indicating that the core size also decreases. This follows from the fact that the polymer solvency increases as the temperature decreases and the polymer within the shell expands. The whole particle size increases and, simultaneously, the core size decreases since the shell also tries to expand towards the inner particle region, thus pushing the core that shrinks as a consequence of the interaction with the shell. The lowest q-scattering, $q_0 \leq 0.01\text{Å}^{-1}$, follows a decreasing power law, $\sim q^{-\alpha}$ [see Figure 2.26(d)].

The exponent α (see Figure 2.27) crosses from 2, corresponding to extended branches, to 4, characteristic of smooth surfaces (Porod's region). Beyond this point, the exponent α still continues to increase and indicates the presence of rough surfaces. The roughness may be characterized by a 2D fractal dimension d_s, which evolves with the temperature towards smaller values ($\alpha = 6 - d_s$) (see the inset in Figure 2.27). This indicates an increasing surface roughness along with particle compaction with increase in temperature.

In conclusion, it should be pointed out that neutron scattering allows structural information to be accessed at different scales. It also makes it possible to observe and characterize simultaneously the internal structure and the particles surface.

2.5.4   X-Ray Scattering

2.5.4.1   *General Features of X-Rays*

X-rays are electromagnetic radiation with short wavelength, from 0.01 to 10 nm, which correspond to wavelengths below the ultraviolet spectral region. X-rays were first detected by Wilhelm C. Röntgen in Würzburg, Germany, in 1895. He studied the radiation produced by Crookes tubes, which are very similar to the X-ray tubes still in use. In a conventional X-ray tube, electrons are accelerated by a high voltage between a heated cathode and the anode.

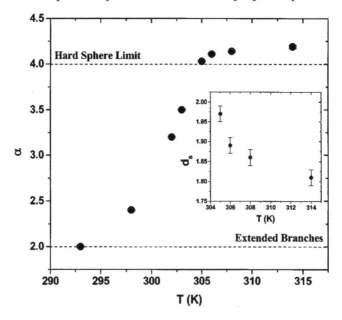

Figure 2.27   Variation of exponent α with temperature. Beyond the hard sphere limit, information about the surface roughness is achieved and described by the value of the fractal dimension.

When the electrons hit the anode, X-rays are produced, giving rise to a combination of two phenomena: fluorescence and Bremsstrahlung. The former occurs when an electron hitting the atom has sufficient energy to expel an electron from the inner electron shells. Electrons from higher energy levels occupy the vacancy, emitting X-ray photons with the precise energy corresponding to the quantum jump between different electronic shells. This effect produces spectral lines. The higher energy lines correspond to jumps down to the electronic K shell and are also termed K. Bremsstrahlung corresponds to the photons emitted by the electrons changing velocity and direction on interacting with the anode atoms. They show continuum spectra with a short wavelength limit corresponding to the maximum energy of the electrons accelerated by the applied voltage.[107] The resulting emission spectra correspond to a wide band with superimposed peaks from the anode fluorescence spectral lines. Because scattering depends on both the scattering angle and wavelength, X-rays have to be filtered. A monochromatic Cu Kα spectral line that corresponds to a wavelength of 0.154 nm can be reasonably filtered using a thin Ni foil. Because of the high temperatures reached by the anode, the X-ray intensity cannot be increased indefinitely by increasing the anode current intensity because the anode would melt. Specially designed anodes such as rotating anodes can reach intensities several times higher than the static variety. Nowadays, the use of synchrotron light sources provides much stronger intensities. In a synchrotron, electrons are obliged to circulate along a closed, nearly circular trajectory. For this purpose, strong magnetic fields are required. The combination of relativistic velocities with direction changes can produce photons in the X-ray region. In a synchrotron light source, X-rays are produced in the bending magnets, but they can also be produced using insertion devices such as wigglers or undulators that force the electron beam to follow a precisely changing magnetic field. Therefore, synchrotron light sources have experimental stations using either one or the other type of X-ray production. The X-rays produced are of very high intensity but, owing to a certain spread of wavelengths, a monochromator is needed. The usual monochromators are based on diffraction by two single crystals. Whereas the traditional tube generators can be used almost everywhere, synchrotron sources are large, immobile installations that are becoming more and more frequent.

Once we have X-rays, we should be able to detect them and, for scattering experiments, we should be able to do this quantitatively and also to measure them as a function of space. Whereas the first X-ray detectors were photographic plates, modern detectors are more efficient having both satisfactory spatial and time resolution.[107] Proportional counters measure the current generated between two electrodes due to the ionization of gas molecules generated by passing X-rays. If the difference in potential is high enough and with the appropriate gas and distances, avalanche ionization is generated from the first ionization, which produces a measurable electric signal. Both time and spatial resolution can be adequate for small-angle X-ray scattering (SAXS). Usually these detectors are one-dimensional and produce some smearing of the signal, particularly at the smallest angles. Charge-coupled devices (CCDs) and complementary metal oxide semiconductor (CMOS) devices coupled with phosphorescent screens allow for two-dimensional detection with adequate time and spatial resolution. These detectors, however, have a high dark current background, which introduces high noise levels in weak signal regions. Solid-state, *i.e.* semiconductor-based, detectors also provide two-dimensional resolution. They can achieve single-photon counting, because of direct conversion of the X-ray photon into an electric signal, and can cover a wide span of intensity with similar efficiency. They are even capable of measuring the direct beam without attenuation.

Owing to their energy, X-rays interact with electrons in matter, and photons with sufficient energy can ionize atoms and molecules. SAXS uses soft X-ray photons[108] and the usual wavelengths are around 0.1 nm. This value, because of its importance in X-ray diffraction, is termed the ångstrom. It is used in the older literature and is now slowly being replaced by SI units. It corresponds to the typical interatomic distances. Regarding their interaction with electrons, two different phenomena can be distinguished: absorption and dispersion. A third possibility, Compton scattering, can be neglected when dealing with SAXS. Similarly, when light is used, the term dispersion means that the incident X-ray, which forces the electrons to perform an oscillatory motion, is re-emitted with the same frequency in all directions. If the sizes of the particles and their features are similar to the radiation wavelength, scattering will be significant at experimentally accessible angles. The typical SAXS setup is similar to that of light and neutron scattering (see Figure 2.18 and Scheme 2.2). The X-rays produced by either a conventional source or a synchrotron are collimated and focused on the sample, which should be contained in a material transparent to the X-rays. The scattered radiation is detected in transmission mode and the non-scattered beam is shadowed out in order to avoid detector overdose. The usual SAXS angle range is from 0.05° up to several degrees, which correspond to q values from 0.04 to 10 nm^{-1}. The distances implied can be obtained from $d = 2\pi/q$, which is one of the formulations of Bragg's law. Taking into account the usually available angles, this limit corresponds to 150 nm as the largest distances that can be measured. We should also mention here the X-ray diffraction (XRD) technique. In this case, the materials subjected to X-ray bombardment are crystalline and, as such, they produce diffraction because of constructive and destructive interferences due to their highly ordered internal structure. In this case, the information available in the sample is higher than that in less ordered materials and the position of the different atoms corresponding to the smallest repeating unit (unit cell) can be obtained with high precision. In the case of SAXS, the interferences are blurred and, consequently, the information obtained from the sample is of lower order.

X-ray absorption depends on the atomic number and the concentration of an atom in the sample. The higher the atomic number, the concentration and the sample thickness, the greater is the absorption. Differences in electron density of different parts of a sample also produce a scattering signal, in the same way as differences in scattering length density for neutrons or refractive index changes for visible light. For X-rays, the differences in electron density define the contrast. The higher the contrast, the higher the concentration and the higher the sample thickness, the stronger is the scattered intensity. These contradictory trends lead to a conjunction where the optimal sample thickness, for a predominant water sample and copper radiation, corresponds to 1 mm. The contrast, that is, what can be seen by X-rays, corresponds to differences in electron density in the sample. Different chemical groups have different electron densities. For instance, water has an electron density of 0.33 e $Å^{-3}$ whereas the typical hydrocarbon electron density is 0.27 e $Å^{-3}$ and the majority of polar head groups (containing O, N, P, S and other high electron density atoms) have electron densities higher than that of water.[109] For a typical phospholipid, the electron density profile across a bilayer looks like that shown in Figure 2.28.

The relative differences of positive–negative electronic density within the medium contribute to the scattered intensity form factor in different ways, in particular producing bands with maxima related to characteristic distances within the particle.

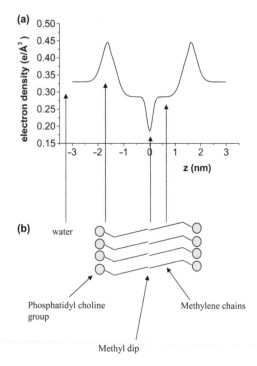

Figure 2.28 (a) Electron density corresponding to a typical phospholipid bilayer. (b) Sketch of a single bilayer.

2.5.4.2 *X-Rays and Soft Nanoparticles*

What can be seen by SAXS? – objects with dimensions from 0.5 up to 500 nm in materials that do not absorb X-rays very efficiently but have sufficient contrast, *i.e.* difference in electron density between the medium and the particles. No prior sample treatment is needed and the samples can contain water and other solvents. SAXS is appropriate for large molecules in solution (proteins, polymers), aggregating molecules in solution (surfactants), lipids in aqueous solutions or dispersions such as liposomes.

In line with what has been shown for light and neutrons, if the samples are diluted enough, single-particle properties can be determined. In the same way, for objects and scattering vectors fulfilling $qr_0 \ll 1$, where, as previously, r_0 means a characteristic length in the system, Guinier's law also applies and a radius of gyration can be obtained from the slope of a plot of $\ln I$ *versus* q^2 [see eqn (2.25)]. The analytical expressions of the particle form factor for spheres is analogous to that found for light and neutrons [see eqns (2.21) and (2.49)]. Also, information concerning the global geometry (size) of the particles can be obtained from plots of $\ln I$ as a function of $\ln q$. A linear behaviour with a slope of -1 is indicative of elongated particles whereas a slope of -2 is indicative of flat particles.[107] If a fractional slope is found over a large enough range, it may be indicative of fractal structures (see Section

2.5.3.2 for neutrons). An important limiting law is that of the Porod regime, when the particles fulfill $qr_0 \gg 1$ and there is a neat interface between them and the medium; a slope of -4 is found and the intensity of this slope is proportional to the total amount of surface and electron density contrast. Based on a hypothesis regarding the sample composition (*i.e.* the contrast) and the concentration of the phases present, a value for the specific surface area of the sample can be obtained from absolute intensity measurements. However, if the integral $\int_0^\infty I(q)q^2 \mathrm{d}q$, known as the invariant, can be evaluated, the specific surface area can be evaluated even without absolute scale calibration. The coincidence between the two methods suggests that the hypotheses regarding the contrast and volume fraction are reasonable.[109]

Analogously to what happens with other types of radiation, the radiation dispersed by the different objects in sufficiently concentrated samples shows interference that is expressed in terms of the structure factor (see previous sections). For spherical particles, it is possible to separate the form and structure factors in the global intensity as a simple product, *i.e.* $I(q) = P(q)S(q)$. For polydisperse and non-spherical particles, some approximations are used to allow for the calculation of the total intensity.[110] As we have seen previously, the form factor $P(q)$ contains the information relating to a single particle and $S(q)$ contains the information relating to the position of the different particles in the sample [eqn (2.47)].

Very ordered samples, such as typical crystals, diffract X-rays. This means that only very well-defined angles give rise to constructive interference. If the order is less than that in a crystal, the phenomenon is called dispersion and the spectra do not correspond to isolated peaks but rather to broader peaks with a considerable non-zero intensity between them. When the order is completely lost, we can talk about glasses and the intermediate states correspond to liquid crystals.

A typical liquid crystalline phase encountered in biological systems is that of the lamellar phases formed by phospholipids (see Chapters 3 and 6). This lamellar phase forms liposomes if appropriately hydrated and sheared. The liposomes can be unilamellar, oligolamellar or multilamellar. Depending on the number of lamellae the scattering pattern will be different.[111] Whereas the unilamellar liposomes or vesicles present a scattering pattern corresponding to the bilayer form factor, in the multilamellar case clear interference peaks are observed [see Figure 2.29(a)]. In the case of the oligolamellar liposomes, the peaks are not very well defined and only bands are observed instead of clear peaks.

If the temperature is low enough, the lamellar phase can become more ordered, close to a crystal state that is known as a lamellar gel. In this case, in addition to the lamellar order, order exists within the bilayer, that is, at much shorter distances and this is reflected in the appearance of peaks in the medium range, which is called wide-angle X-ray scattering (WAXS). The position in the spectra at which a lamellar peak appears corresponds to $q = n(2\pi/d)$, where n is the order of reflection and d is the repetition distance [see Figure 2.29(a)]. This means that the peaks appear at equally spaced distances from the origin of scattering q vector values.

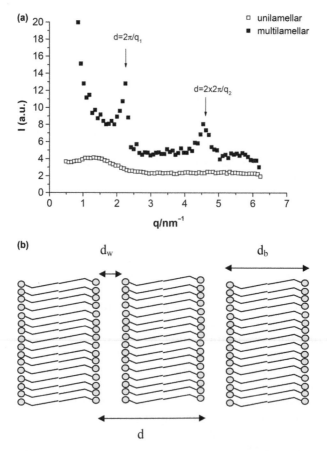

Figure 2.29   (a) Scattered intensity as a function of q corresponding to a single bilayer (open symbols) and to a multilamellar liposome (solid symbols). (b) Sketch of a multilayer.

In Figure 2.29(b), the relevant parameters for a lamellar structure are given; d is the repetition distance (also known as the Bragg distance), which is the distance at which the structural motif repeats by simple translation; d_w corresponds to the thickness of the water layer and d_b to the thickness of the whole bilayer. Further, d_c and d_h, corresponding to the hydrophobic length and the head length (not shown), can be defined. If the system is in a crystal or gel state, there is also repetition in the direction perpendicular to the arrows in the plot and in the direction perpendicular to the page. The repetition distances are considerably shorter than the bilayer repetition distance and the q values at which the reflections show up correspond to the range between 10 and 20 nm^{-1} and concomitant Bragg distances from 0.6 to 0.03 nm. Other types of liquid crystals can also be resolved by using SAXS; hexagonal phases (in which surfactants form cylindrical micelles arranged in hexagonal packing) and cubic phases (with a wide variety of structures from discrete to bicontinuous) are common self-aggregation systems with distances of interest within the X-ray range.[112]

Protein crystallography allows a three-dimensional view of the proteins, necessary for the understanding of their action, to be obtained. However, this crystallographic structure can be critically different from the three-dimensional structure in solution. SAXS and SANS can provide a low resolution of the conformation of proteins in solution.[113] SAXS and SANS provide a picture of the envelope of the protein, *i.e.* the overall dimensions and shape that tell us up to which point the high-resolution structure obtained from crystallography may resemble or differ from that in solution, whether there is dimerization or multimerization, or whether there are some intrinsically disordered domains.

2.6 Fluorescence

2.6.1 General Remarks on Luminescence and Fluorescence

The emission of light from a solution, from the surface of a powdered solid or from membranes is called luminescence and it occurs from molecules excited to higher energy states – some vibrational levels of the excited electronic state – by the absorption of energy. In the phenomenon called fluorescence, after a rapid decay by non-radiational vibrational relaxation to the lowest vibrational level of the excited electronic state, the excited molecule returns to any of the vibrational levels of the ground state by emission of a photon of longer wavelength than the exciting radiation. During the excitation and decay processes, the orientation of the electron spin is preserved. The typical lifetime of fluorescence is around 10 ns. Molecules in excited singlet state S_1 can undergo a spin conversion to the first triplet state T_1 by a process known as intersystem crossing. The decay of a molecule from the T_1 state is called phosphorescence, with a typical lifetime from milliseconds to seconds.

The excitation of substances can occur by external radiation, by chemical reaction (chemiluminescence) or by electrochemical reaction (electrochemiluminescence). Usually, molecules that show fluorescence, called fluorophores, are aromatic in nature and are characterized by their lifetime, *i.e.* the average time that the molecule spends in the excited state prior to returning to the ground state, and the quantum yield, *i.e.* the ratio of the number of emitted to the number of absorbed photons.

A general characteristic of fluorescence is the energy losses between excitation and emission (Stokes shift) due to the rapid decay to the lowest vibrational level of the first excited state before the decay to higher vibrational levels of the ground state. Further Stokes shifts may be due to solvent effects, excited-state reactions, complex formation, and/or energy transfer. Another characteristic of fluorescence is that emission spectra are typically independent of the excitation wavelength (Kasha's rule) owing to the rapid decay of the energy of the fluorophore to the lowest vibrational level of the first excited state before the emission.

The fluorescence intensity can be attenuated by different mechanisms (quenching). When the excited-state fluorophore is deactivated upon contact with other molecules in solution (quencher), it undergoes collisional quenching. There is no chemical reaction and the intensity decrease is given by the Stern–Volmer equation. In addition to collisional quenching, other molecular interactions can also result in quenching, such as excited-state reactions, molecular rearrangements, energy transfer and ground-state complex formation.[114]

Different fluorescence-based techniques are currently in use to characterize soft NPs, including size, shape, surface charge, internal structure and surface microstructure. Below a brief description is given of the latest techniques used and the information they can offer in this field. Fluorescence imaging is restricted to intrinsically fluorescent NPs or fluorescently labelled NPs.[115] In the former case, self-fluorescent NPs allow imaging without the need for external fluorescent probes such as in seed emulsion copolymerization to prepare abietane-based nanogels.[116]

In the latter case, fluorescence bioimaging is usually achieved by conjugation with different fluorescent labels such as dyes, fluorescent proteins, up-converted NPs or quantum dots (QDs).[117] The labelling approach can introduce some problems, such as label instability (leaching), modification of physicochemical properties or photobleaching. All these problems can be minimized by different synthetic strategies related to different architecture design and immobilization strategies such as encapsulation (embedding, entrapment, anchoring), layer-by-layer, hydrophobic or electrostatic interactions and covalent binding, providing versatile nanoscale scaffolds for designing multifunctional NPs. Of all these strategies, covalently linked fluorophores offer higher stability, improve the reproducibility and provide a high shelf and operational lifetime. In order to obtain more reliable fluorescence data, ratiometric measurements are desirable. The ratiometric approach[118] improves the reliability of signal measurement compared with fluorescent NPs based on indicator dyes only, decreasing the influence of the surrounding media or the light source intensity. Accordingly, the fluorescence measurement stems from two different signals, one from the reference dye, which gives the stable reference signal, and the other from a sensing fluorophore.

When using micelles, a ratiometric approach is also developed by using two different fluorescence probes. A hydrophobic probe such as hostasol methacrylate (HMA) is embedded in the micelle core while a hydrophilic probe such as tetramethylrhodamine isothiocyanate (TRITC) is attached on the micelle corona.[119] Fluorescent polymeric assemblies and NPs combined with stimuli-responsive properties can lead to the precise spatial organization of multiple fluorophores. These nanomaterials have been utilized to optimize the photoluminescent properties of covalently or physically attached fluorophores by three different approaches when the polymeric matrix is endowed with stimuli responsiveness. The three main design types are modulation of the Förster or fluorescence resonance energy transfer (FRET) processes, tuning of the fluorescence emission and lighting up fluorescence, where the fluorescence of latent moieties can be generated or switched on through selective chemical reactions or supramolecular recognition. Different types of techniques in which the fluorescence is the analytical signal can be used to elucidate the structure of NPs, such as (a) steady-state fluorescence, (b) resonance energy transfer, (c) fluorescence imaging, (d) fluorescence recovery after photobleaching and (e) fluorescence correlation spectroscopy.

2.6.2   Fluorescence Techniques and Internal Structure of Nanoparticles

2.6.2.1   Steady-State Fluorescence

Steady-state measurements are those performed with constant illumination and observation. The NPs are illuminated with a continuous beam of light and the intensity or emission spectrum is recorded. Because of the nanosecond timescale of fluorescence, most measurements are steady-state measurements. When the sample is first exposed to light, the steady state is reached almost immediately.[114] The surrounding environment of the NP (different materials composing the NPs) could develop differences in fluorescence intensity, involving the possibility of differentiating between different structures. For instance, the structure of stimuli-responsive microgels (Chapter 4) can be elucidated by fluorescence spectroscopy. These microgels undergo a volume phase-transition at a temperature (known as the lower critical solution temperature, LCST) that can be modulated by varying the hydrophilic–hydrophobic balance by increasing the length of the side chain or the degree of cross-linking (Chapter 4). On applying a change in the local environment by changing, *e.g.*, the pH, light, temperature, ionic strength, electric field, pressure or analyte concentration, an environmentally initiated phase transition can be observed. This phase transition involves a change from the swollen state (temperature $<$ LCST) where the more polar solvent–chain interactions dominate towards a collapsed state (temperature $>$ LCST) where the less polar chain–chain interactions are predominant. Hence a change in temperature can cause a phase transition between the hydrophobic and hydrophilic states and fluorescent dyes, involving a change in the fluorescence due to the change in the environment (Figure 2.30). For instance, if a polarity-sensitive (solvatochromic) fluorophore is included in the system, structure-related changes can be measured by fluorescence monitoring. Concretely, the structure and nature of the fluorophore-labelled materials based on acrylamide and on oligo(ethylene glycol) (OEG) bearing thermoresponsive polymers can be determined.[120] Here, a 1,8-naphthalimide derivative (fluorophore) emits with an emission peak centred around 535 nm very weakly below the LCST when incorporated into both poly(*N*-isopropylacrylamide) (PNIPAM) or into a polyacrylate with

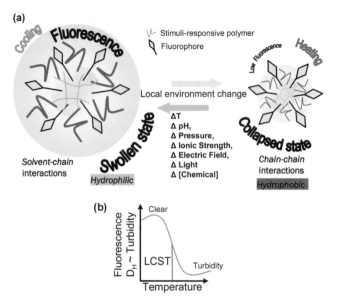

Figure 2.30   General scheme of fluorescence in stimuli-responsive nanogels. Solvent–polymeric chains predominate in the swollen state whereas in the collapsed state chain–chain interactions are predominant. A quenching effect can be observed above the LCST, involving a decrease in the fluorescence, which can be explained by two reasons: (i) a change in the nature of the environment since a more hydrophobic behaviour appears and (ii) due to the shorter distances between fluorophores. (b) General example of fluorescence trend with temperature change. Below the LCST, the nanogel show high fluorescence which decreases when the temperature increases. The turbidity of the sample solution increases on heating, involving a decrease in the hydrodynamic diameter (D_H).

short OEG side chains, whereas it is strongly emissive in polymethacrylates with longer OEG side chains. Nevertheless, on heating above the cloud point, the fluorescence intensity of the labelled PNIPAM is increased whereas no changes are observed with OEG-based polyacrylates and methacrylates. This phenomenon can be explained taking into account that the fluorophore is rather mobile and exposed to water (swollen state, more polar) at lower temperatures. In contrast, on heating above the LCST, polymer-rich and denser microdomains are reached involving a less polar form that develops a change in the steady-state fluorescence properties. This phenomenon is related to the aggregation-induced emission (AIE) process discovered by Tang and co-workers[121] in which light emission coming from weak fluorophores in solution can be rendered highly emissive *via* aggregation or absorption on bulky substrates. Therefore, both swollen and collapsed structures of thermoresponsive polymers (swollen T<LCST<T collapsed) can be elucidated by incorporating fluorescent dyes into responsive polymer matrices where an external stimulus can be converted into fluorescence signals due to changes in the properties of surrounding microenvironments.[122]

In addition, the structure of hollow NPs can also be elucidated *via* steady-state fluorescence. On inducing chemically a hollow core into a polymer nanogel, the degradation of the inner material is evidenced by using phenanthrene-labelled core–shell particles, which shows a significantly attenuated fluorescence (emission peak centred at 373 nm), indicating that the majority of the degraded material has been removed from the particles.[123]

2.6.2.2   Resonance Energy Transfer

The resonance energy transfer (RET) is a process that occurs whenever the emission spectrum of a fluorophore (donor) overlaps with the absorption spectrum of a not necessarily fluorescent molecule (acceptor). The coupling between donor and acceptor is due to a dipole–dipole interaction without intermediate photons. The amount of energy transfer is determined by the distance between donor and acceptor and the extent of spectral overlap.

RET is a non-radiative transfer process where typically a RET acceptor material absorbs energy at the emission wavelength range of the RET donor material involving re-emission of the energy as fluorescence, and this fluorescence signal is measured. RET depends on the distance between the RET acceptor and the RET donor. Therefore, if an NP is functionalized with different kinds of fluorophores located at different shells (outer and inner), the thickness of those shells can be determined since the RET measurement is related to the distance between fluorophores. The major inconvenience is that the RET efficiency is limited to a short range (<100 Å), being dramatically reduced for intermolecular distances beyond the Föster range (R_0, 20–55 Å). Depending on whether RET occurs, the thickness of the different NP shells with chemically linked dyes can be elucidated by measuring the changes in the fluorescence between them. For instance, the features of the nano-scale self-assembly NPs were evidenced by the energy-transfer (ET) efficiency with styrene NPs labelled using a fluorophore (9,10-diphenylanthracene, DPA).[124] Here, the quenching of fluorescence arises through energy transfer (RET) from the dye to the copper–cyclam complexes that form at the NP surface with an operating distance in the 2 nm range.

RET can be significantly increased by placing the donor–acceptor pairs in proximity to metal NPs by a phenomenon called metal-enhanced fluorescence (MEF). MEF depends on the nature of both the metal surface and the fluorophore and the distance between them. Therefore, the structure and the composition of multilayer core–shell NPs can be determined by this technique. For instance, by manipulating a soft polymer spacer (PNIPAM spacer) between silver NPs (used as the core) and the fluorophores (located in a polymeric shell), a tuneable MEF was achieved.[125] When different types of QDs are coated with a thermoresponsive microgel such as PNIPAM, the thermo-induced collapse and swelling of the PNIPAM microgel lead to a tuneable spatial distance between the two types of QDs and result in changes in the RET efficiencies.

Fluorescence quenching, which refers to any process that decreases the fluorescence, may become a useful tool for determining the composition of core–shell structures since metals frequently quench the fluorescence when determining metallic components in soft NPs. In many cases, the right choice of the dye is important. For instance, haematite (iron oxide) NPs absorb up to 630 nm, so quenching can be prevented by using a near-infrared (NIR)-emitting dye. The optimum distance between the emitting dye and the surface of the metal particles depends on their orientation and spectral fitting. This fact is useful in defining fluorescence core-shell structures in hybrid NPs in terms of shell thickness or composition. Finally, the development of fluorescent polymers with polymeric assemblies where the modulation of RET efficiency can be achieved by tuning the spatial distribution of fluorescent donors and acceptors can be useful in the determination of the structure of NPs.[126]

2.6.2.3   Fluorescence Imaging

Fluorescence imaging uses an incident source of light to excite the sample, then the emission is collected through an optical objective connected to a microscope with a detection system. There, the emission (fluorescence) of interest is selected by using different filters. Optical microscopes with high numerical aperture objectives, high-quality optics and high-sensitivity detectors such as avalanche photodiodes (APDs) or charge-coupled devices (CCDs) are used to detect weak fluorescent signals from singly labelled NPs. Generally, the imaging methods can be divided into two groups: wide-field imaging and confocal imaging.[127] The ability to discriminate non-labelled NPs and to identify and quantify almost any type of fluorescent NPs is possible by selecting the optical filters properly.

Fluorescence imaging has become one of the most powerful tools for studying cellular processes, since fluorescent proteins can be attached to NPs. To detect fluorescent NPs, confocal laser scanning microscopy (CLSM) is applied with a resolution of the order of 200 nm, which permits the visualization of the different accumulations of polymeric NPs in time with (sub)cellular resolution in tumours. Hyphenated techniques can lead to a more complete characterization of NPs of different sizes. Detection systems for fluorescence spectroscopy can be combined with imaging techniques to give elemental or structural information. Thus, the combination of a separation technique with a fluorescence imaging microscope provides a simple method for imaging the size of NPs. For example, hydrodynamic chromatography (HDC) coupled with a fluorescence microscope equipped with a CCD camera was used to characterize fluorescent polystyrene NPs of different diameters (26, 44 and 110 nm). In biological imaging, it could be possible to determine, by an endocytosis process, the surface charge of NPs. Thus, with highly fluorescent core–shell hybrid NPs made of a hyperbranched conjugated polymer (HCP), when functionalized with amino groups, the relative geometric mean fluorescence intensities (GMFIs) of HeLa cell-associated HCP@SiO$_2$ (negative surface charge) was lower than those of HCP@SiO$_2$–NH$_2$ (positive surface charge). Another characteristic of the NPs is their shape. The shape of NPs is an important property for biological objects and plays a crucial role in biological functions, involving different properties relating to reactivity and toxicity, and in optical measurements. For example, hydrogel capsules of poly(methacrylic acid) (PMAA), produced as layer-by-layer (LbL) polyacid replicas of cubic inorganic microparticles and capable of keeping cubic geometry[128] at neutral pH, were observed by CLSM. Accordingly, the diameter–temperature dependence of microcapsules composed of a water core and a PNIPAAM shell was monitored by a time–temperature series of confocal microscope images, showing their swelling and release of FITC-dextran upon natural cooling.[129]

2.6.2.4 Fluorescence Recovery After Photobleaching

Fluorescence recovery after photobleaching (FRAP) is an optical technique used to measure diffusion coefficients in a wide range (1×10^{-5} to $<10^{-14}$ cm$^2\cdot$s^{-1}) typically using microscope-based systems. A light-sensitive fluorophore in solution is irradiated with a laser pulse (bleach beam) with a wavelength near the absorption maximum of the sample molecules, inducing their bleaching in a given volume. The recovery of fluorescence due to the exchange by diffusion of bleached molecules with unbleached molecules that originally were outside the irradiated volume permits the diffusion coefficient of the fluorophore or, in most cases, the fluorophore-labelled NP to be calculated.[130] FRAP is a microscopy technique and its main advantage is that the material can be imaged through a microscope while performing FRAP, which makes it possible to obtain an appropriate correlation between structure and diffusion.

FRAP can measure local diffusion in soft materials that possess some fraction of liquid phase in which the probe can be dissolved and transported. The generally accepted procedure for performing FRAP is to introduce a fluorescent probe into the sample, bleach a volume and then monitor the fluorescence recovery.

Owing to the wide range of structures that interact, leading to macroscopic properties, it is of critical importance to be capable of measuring the material both locally and globally in order to define the respective structural properties. Local measurements are vital to determine how specifically a structure influences the material properties whereas global measurements tend to lack information since the influence of different structures is averaged. In contrast, global measurements are crucial to determine the integrated macroscale properties directly related to the heterogeneous microstructures within the material. The length scale of the microstructures is typically 1 μm or even less. Hence FRAP may represent a versatile technique for measuring diffusion locally inside a material.

FRAP is an ideal technique for examining the diffusion of NPs in an un-labelled matrix consisting of colloidal suspensions, permanent or reversible gels, tumours and mucus from different body tissues. Although these experiments may reflect the environment of the particles, some common experimental variables such as matrix concentration, stiffness and temperature may appear.

FRAP has particular advantages when used for analysing dilute solutions of macromolecules, since it requires extremely small sample sizes and no deductions are needed, a critical condition especially for small diffusers that may be affected by dye label attachment.

FRAP shows such high sensitivity in detecting very small materials that it has emerged as a very useful approach for analysing thin films and surfaces of biological relevance. For instance, obtaining a detailed understanding of mobility in lipid layers is a typical biological goal.

2.6.2.5   *Fluorescence Correlation Spectroscopy*

Fluorescence correlation spectroscopy (FCS) measures the absolute value of the fluorescence intensity and the intensity of fluctuations emitted in the confocal volume (focal volume of ~ 10–$15\,\mu L$). Hence FCS can measure very low concentrations of samples (diluted 1000-fold in comparison with DLS). The fluctuations come from the input/output of fluorophores in/from the confocal volume and from physicochemical processes that affect the fluorescent molecules. FCS uses confocal optics (a confocal microscope) to evaluate diffusion coefficients of fluorescent NPs, binding and rate constants of fluorescent molecules involved in chemical reactions or photochemical processes,[131] or NPs with fluorescent coatings whose fluorescence fluctuations depend on the NP shape.[132] FCS can determine size distributions in a very small laser-illuminated volume of much diluted solutions (10^{-8}–$10^{-15}\,mol \cdot L^{-1}$) such that single molecules can be detected in concentration ranges where aggregation is minimized. For example, FCS is useful for characterizing the coating process of a virus model, latex beads with coumarin-labelled N-(2-hydroxypropyl)methacrylamide copolymer containing thiazolidine-2-thione groups.[19] Here, the determination of the free polymer fraction in solution and, consequently, the amount of polymer sufficient for efficient coating of a single latex particle (surface concentration) can be achieved.

Since FCS experiments are restricted to nanomolar concentrations, several complications may be encountered. First, owing to their surface charges, most proteins readily adhere to any surface such as cover-slips or the walls of reaction chambers. This effect is more noticeable the less concentrated the solution is. Another parameter of critical importance is the molecular brightness, μ, which is easily calculated by dividing the average fluorescence count rate by the number of molecules within the illuminated region.

Analysis of higher order autocorrelation, serves to identify subpopulations simple differing in their molecular brightness values. This type of brightness distribution analysis is the method of choice in most biomolecular screenings.[133]

Most FCS applications are related to reactions occurring on a much longer time scale than the short time during which a molecule is within the measurement volume, which can be easily followed by continuous FCS monitoring. One small labelled ligand and a relatively large non-fluorescent counterpart represent an ideal reaction system. In order to detect a significant change in the diffusion time, the mass ratio should have a minimum value of 8, mainly due to the approximate cube root dependence of the diffusion coefficient on molecular mass.

One of the most representative examples is the crucial contribution of FCS to a step in the process of drug discovery. FCS allows the binding affinity of the drug candidate for a specific target receptor to be assessed through monitoring the change in the ligand diffusion time when binding to its receptor. FCS is widely used in drug discovery research, taking advantage of its ability to carry out ligand–receptor binding assays by resolving the bound and unbound fractions of the complex.[134]

2.7   NMR Spectroscopy

2.7.1   Introduction

The progress in NMR spectroscopy in recent years has provided an excellent tool for the characterization and study of the properties of a wide range of materials in different aggregation domains. The complementary application of different NMR spectroscopic techniques has been decisive for advanced applications and the quantitative analysis of the characteristics of homogeneous and multiphase systems, which are of particular importance in the field of polymer materials. Three of the most interesting contributions of NMR spectroscopic techniques and methodologies are outlined below.

The first is the classical high-resolution NMR spectroscopy in solution, with applications to the structural characterization of molecules and macromolecules in solution using high-power equipment (400–600 MHz). The improvement of the data obtained with high chemical shift resolution by the application of refined software of the new equipment offers an excellent tool for the fine chemical structure characterization of complex polymer systems and composites in a suitable solvent. This provides a precise knowledge of the chemical structure and even interactions of specific functional groups of the molecules or macromolecules in solution.

The second is the great contribution of solid-state NMR spectroscopy to the determination of the structure and morphology of supramolecular and macromolecular systems. This is the result of the application of the techniques of cross-polarization (CP) and magic angle spinning (MAS), which provide NMR spectra with good resolution for solid samples, giving a clear idea not only of the structure but also the morphology and physical organization of the systems in a particular application. Very interesting data on the crystallinity and distribution of crystal domains in semicrystalline macromolecular systems and the organization and orientation of molecules and chain segments in polymers or copolymers are examples of the of high level of information offered by solid-state NMR spectroscopy. The technique allows the study and analysis of assembled proteins, NPs, supramolecular structures of polymer and copolymer systems and intra- and intermolecular complexes and even the interactions of organic and inorganic components of complex formulations very common in biotechnology, biomedicine and natural products.

The third is the extensive development and application of NMR imaging (MRI), which provides a detailed insight into the real organization of the molecular and macromolecular systems in different media including the solid state, dispersions in liquids and distributions of specific domains. It represents a major contribution to the precise analysis of compounds and systems not only for clinical diagnostics and exploration but also for a complete knowledge of the morphology and structure of macromolecular systems and composites. The information not only provides the exact radiographic image of an object, system or human body component, but also very precise information about important phenomena that are very frequent in Nature, such as diffusion, reorganization of systems in dynamic equilibrium or self-assembly processes in supramolecular systems.

This section presents an overview of the most recent and advanced contributions of the above three techniques to the precise analysis, characterization and properties of soft macromolecular assemblies and polymer systems and composites.

2.7.2 Basic Principles of NMR

NMR spectra arise from energy transitions of the nuclear spin of certain atoms that has a quantum number of spin different from zero. All isotopes that contain an odd number of protons and/or neutrons have an intrinsic magnetic moment, in other words, a non zero spin, whereas all atoms with even numbers of both have a total spin of zero The nuclei of ^1H, ^{13}C, ^{15}N, ^{17}O, ^{19}F, ^{29}Si and ^{31}P are examples of isotopes sensitive in NMR with a quantum spin number different from zero. The electronic configurations of these atoms result in the appearance of magnetic dipoles that can take different orientations in a magnetic field and each orientation has its own energy. Transitions between the different energy levels of the nuclear spin are responsible for the observed signals in NMR. One important factor is that the nuclei are not isolated and therefore the transition between energy levels is clearly influenced by the chemical environment, which generates the corresponding spectrum as a diagram of intensities and resonance frequencies. After normalization of scale, the frequency is represented by the normalized parameter *chemical shift*, δ, which corresponds to the frequency of resonance of a nucleus expressed in parts per million (ppm) with respect to the frequency of a given reference.

An important characteristic of NMR spectroscopy is that the excited states have a relatively long stability (lifetime in the range of milliseconds), in contrast to other well-known spectroscopic techniques such a fluorescence and infrared spectroscopy. This property allows the transfer of excitation energy from one nucleus to another to be manipulated and offers the possibility of correlating the frequencies of two or more nuclei. This phenomenon generates two-dimensional and even multi-dimensional homonuclear (same kind of nucleus, *e.g.* ^1H–^1H) or heteronuclear (different kinds of nucleus, *e.g.* ^{13}C–^1H) systems, which are of great importance for the identification and assignment of signals to a given chemical group or chemical structure. It is necessary to stress that the correlation spectra are never quantitative, but provide excellent information for the assignment of signals in the corresponding one-dimensional spectra (*e.g.* ^1H or ^{13}C) and the corresponding precise structural information. On the other hand, the influence of the microenvironment of a sensitive nucleus in the corresponding resonance signal is a characteristic of the microstructural arrangement and therefore gives precise information about the chemical structure of the corresponding group or function.[135–138] The facility for the analysis of frequencies associated with the resonance of a given isotope offers additional and very important information about the nature of the functional groups and molecular or macromolecular structures with sufficient clarity. This is a key factor in determining and evaluating the chemical structure and composition of complex polymer systems, including stereochemistry, copolymer composition and distribution of comonomeric sequences and composites.

2.7.3   High-Resolution NMR Spectroscopy

This section considers the most interesting features and contributions of NMR spectroscopy in solution considering different sensitive elements and a variety of strategies for one and two dimensions, and also homo- and hetero-correlation spectra. The characteristic of the technique is that it provides precise information of the chemical structure of molecules and macromolecules, considering not only the compositions but also the stereochemical configuration and the distribution of sequences in copolymer chains. This is also very attractive for the study of complex systems and interactions between different functional groups in a great variety of compounds.

As an example, we can consider the modification of dextran with acetal groups. Dextran, a bacterially derived polysaccharide of glucose, has good biocompatibility and biodegradability and can be easily modified. The modification with acetal groups brings tuneable pH-dependent hydrolysis sensitivity and, depending on the degree of modification, changes the solubility from water to organic solvents.[139] The modification of dextran with 2-methoxypropene under acid catalysis gives an acetal-derivatized dextran that is insoluble in water and soluble in organic solvents, which allows for the preparation of NPs that are stable in water but can be easily hydrolysed with the release of acetone, methanol and water-soluble dextran,

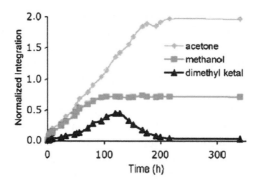

Figure 2.31   Normalized ^{1}H NMR data from the degradation of dextran–acetal NPs
at pH 5.0 and 37 °C, with integration of the signals corresponding to
acetone, methanol and acetal groups.
Reprinted with permission from Ref. 139.

which can be evaluated by ^{1}H NMR spectroscopy. Figure 2.31 shows the
evolution of the release of the three products of the biodegradation process
at 37 °C and pH 5.0. It is clear from the evolution of the NMR spectra and
Figure 2.31 that the initially insoluble NPs release acetone and methanol,
but after a given period the concentration of acetone increases owing to the
presence of cyclic and acyclic acetals. The experiment can be followed in the
NMR equipment directly using the sealed NMR tube and registering the ^{1}H
NMR spectra at different times. The comparison of the integrated intensities
of the signals gives directly the results shown in Figure 2.31 and indicates
that 73% of the hydroxyl groups of the original dextran were modified and
the ratio of cyclic to acyclic acetals was estimated as 1.8 : 1.[139] These values
were calculated using the integration of the acetone and methanol signals
compared with the integration of the anomeric proton signals of dextran.

Homo- and heteronuclear correlation is a useful technique for the identification of signals in a complex spectrum of a 10.6 kDa protein.
Figure 2.32(a) shows the ^{1}H NMR spectrum that corresponds to the proton
signals of the macromolecules, and Figure 2.32(b) shows the ^{13}C HSQC
(heteronuclear single-quantum coherence). The signals distributed in a map
correspond to covalently bonded ^{1}H–^{13}C pairs and correlate strongly with the
secondary structure of the protein. In addition, this 2D heteronuclear correlation spectrum gives an excellent assignment of the proton signals and
allows the identification of the most representative protons and therefore
the evaluation of the corresponding signals for quantitative analysis.[135]

An important parameter in the recording of NMR spectra when polymers
are involved is the temperature, because of the intrinsic viscosity of the
macromolecular systems and the spin–lattice T_1 and spin–spin T_2 relaxation
processes. The stiffness of a macromolecular chain, even in solution, determines the appearance of specific resonance signals and variation of the
corresponding intensity, which logically determines the structural composition of the system. We observed this phenomenon in studying the

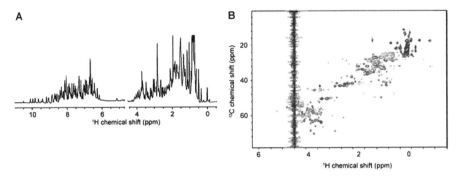

Figure 2.32 (a) ^{1}H NMR and (b) ^{13}C HSQC spectra of CtBP-THAP, a 10.6 kDa protein.
Reprinted with permission from Ref. 135.

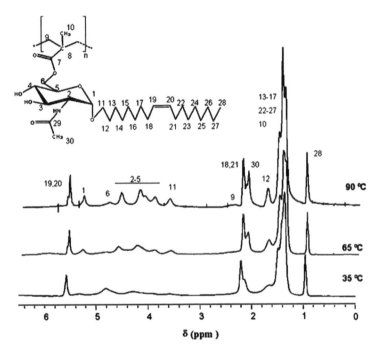

Figure 2.33 ^{1}H NMR spectra of the acrylic polymer POAGMA in deuterated pyridine recorded at different temperatures.

structural characteristics of a poly(acrylic glycoside) derived from glucosamine and a long hydrophobic oleic side group. The polymer chain is very stiff, as demonstrated by the high glass transition temperature, $T_g = 180$ °C, and although the polymer seems to be soluble in organic solvents, as shown in Figure 2.33, the solution of the polymer in deuterated pyridine gives

^{1}H NMR spectra with signals assigned to the glucopyranose protons (3–5 ppm from TMS) that do not appear in the spectrum recorded at 35 °C, but are well resolved at 90 °C with a quantitative coherence with the signals of the same side group at δ 1.0–1.5 ppm.[140] This is a consequence of the inherent stiffness of the polymer chain and gives a clear idea of the resolution of the spectra according to the dynamic mobility of macromolecular segments.

^{2}H NMR spectroscopy is very versatile and gives accurate data on the structure and compositions of a wide range of molecules and macromolecules. The use of different atoms and isotopes brings an additional opportunity to obtain very interesting data on systems not only based on hydrocarbon chains but also nitrogen-, phosphorus-, fluorine- and silicon-containing macromolecules. The application of ^{31}P NMR spectroscopy is very attractive for the study of micelle systems, cell membranes and many other assembled systems that frequently appear in Nature (see Chapter 8). ^{31}P has an isotopic abundance of 100% and a relatively high magnetogyric ratio, with a spin quantum number of $\frac{1}{2}$ making the spectra relatively easy to interpret. These parameters allow the coupling of phosphorus signals with protons and ^{13}C, which gives very interesting information on compounds containing these component atoms, such as phospholipids, cell membranes, stabilized liposomes and micelles.[141] ^{31}P is the only natural isotope of the element and although the nucleus has low sensitivity (6.63% of that of the proton), measurements are easily made even on very dilute solutions by using the standard multi-pulse Fourier transform method. It is common to eliminate coupling with protons [further enhancing the signal size by the nuclear Overhauser effect (NOE)], although the coupling constants and pattern of the multiplets can give additional valuable structural information. Spectra may be obtained on neat liquids, on solutions in common deuterated NMR solvents (especially chloroform, water, benzene, acetonitrile, methylene chloride, *etc.*) and, of growing importance, on solid samples of either pure compounds or compounds attached to solid surfaces. The solvent should always be specified in reporting ^{31}P NMR spectra, since there can be substantial solvent effects on shifts especially where hydrogen bonding can be present. It is the accepted practice to reference chemical shifts to external 85% phosphoric acid at 0 ppm.[142] It is important to consider that the use of deuterated solvents is not necessary except if the proton spectrum should be registered. As occurs with ^{13}C NMR spectroscopy, the application of H–P decoupling simplifies the spectrum considerably and gives more precise information about the different phosphorus atoms of the molecular or macromolecular structure being analysed.

The application of ^{31}P NMR spectroscopy in solution is very interesting for the analysis of complex structures that contain phosphorus. Detergent-based solubilization of membranes is an effective technique for the isolation of lipids and proteins of functional membrane domains and the ^{31}P NMR spectra can be obtained in the presence of the detergent, which makes easier the accurate characterization of the lipid membranes directly from Nature.

This has been demonstrated by several groups, showing that ^{31}P NMR spectroscopy gives quantitative data on the composition and content of phospholipids in membranes[143,144] isolated by the application of different detergents. NMR analysis can be carried out in the presence of the detergent with accurate results and also gives the possibility of determining other components of the membrane such as cholesterol.[145,146]

2.7.4   NMR Spectroscopy in Solids

The study of the structure and dynamics of macromolecules by NMR spectroscopy can be considered as part of a more general methodology for the analysis of materials in the solid state and it is a very efficient tool for establishing the accurate relationship of structure and properties. Polymer behaviour depends on both the molecular structure and the organization of macromolecules in the solid state, which involves relatively large length and time scales. Most polymers are amorphous, where the chains form random coils on a mesoscopic scale. Stereoregularity brings crystallization of ordered lamellar structures separated by non-crystalline regions and multi-component polymer systems are phase separated in most cases to form micelles and cylindrical or lamellar structures, depending on the proportions of the different components.[147]

The application of solid-state NMR spectroscopy has become a very interesting tool since the improvement of the spectral resolution by the applications of two fundamental techniques: cross-polarization (CP) and magic angle spinning (MAS). Detailed information of the principle and application of these techniques can be found in excellent reviews and specialized books.[147–153]

An interesting class of heterogeneous macromolecular systems are core–shell NPs produced by emulsion polymerization or nanoprecipitation of polymer solutions. These dispersions or latexes are of technological importance for applications such as paints, adhesives, coatings, high-impact modifiers of engineering plastics and nano- and microparticles for biotechnological and biomedical applications. According to the morphology and distribution of microdomains and microphases, we can find different structures with a specific phase distribution, as illustrated in Figure 2.34.

The overall size and size distribution of NPs can be determined by TEM or DLS, but solid-state NMR spin diffusion techniques are very sensitive with respect to the morphology of the structures shown in Figure 2.34.[154] Quantitative analysis of the interface structure can be studied by spin diffusion, as shown in Figure 2.35,[147] which compares the spin diffusion behaviour of a core–shell system with complete phase separation and an interface with microdomains with a core–shell system with incomplete phase separation resulting in an interface with a concentration gradient. The intensity of the selected mobile fraction is plotted as a function of the square root of the mixing time, t_{m}. After selection of the mobile component, the spin diffusion curves should level the stoichiometric ratio for the former.

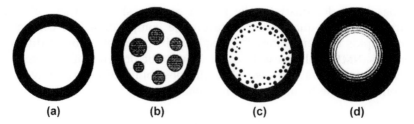

Figure 2.34   Morphology and structure of core–shell latexes or NPs: (a) homogeneous distribution of both phases; (b) microdomains distributed in
the bulk core; (c) heterogeneous nanodomains in the core domain;
(d) onion-like distribution of nano-domains.
Reprinted with permission from Ref. 147.

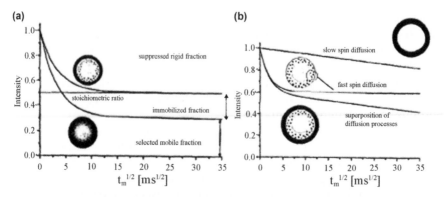

Figure 2.35   Schematic representation of spin diffusion studies of the interface
structure in core–shell NPs. (a) Comparison of microphase separated
and concentration gradient interface. (b) Comparison of ideal structure and interface with microdomains.
Reprinted with permission from Ref. 147.

For the latter, a fraction of the mobile component is immobilized and
cannot pass the filter, which shifts the spin diffusion curve below the stoichiometric ratio.[147] Information on the size and distances between the
microdomains can be deduced from the analysis of the decay of the spin
diffusion curve as shown in Figure 2.35(b). For an ideal core–shell system,
the decay will be very slow, since the spin diffusion has to cover the full size
of the particle, which might be larger than 100 nm; however, in the interface
with microdomains distances of the order of 10 nm have to be covered before the rigid phase is reached. This leads to fast spin diffusion and in a real
system both diffusion processes will be superposed.[147]

   The composition and chemical structures of membrane-forming macromolecules and supramolecular assemblies determine all essential properties
of their interface. Cellular membranes are subject to a hierarchical domain
organization on the mesoscopic length scale.[155,156] In particular, the

attractive interactions between cholesterol and sphingolipids or saturated phospholipids and/or membrane proteins have been identified as the driving force for membrane domain formation.[157] The influence of the hydrocarbon side chain of cholesterol derivatives has recently been studied systematically.[157,158] The measurement of the lateral diffusion of dipalmitoylphosphatidylcholine (DPPC) by solid-state ^{1}H MAS NMR, using the pulsed field gradient (PFG) technique, demonstrated in a single experiment that DPPC and sterol molecules with side substituents of hydrocarbon residues with C_5–C_{10} chain length present the same diffusion coefficients. However, much larger diffusion coefficients were found for androsterol and sterols with side alkyl groups of C_{12} and C_{14}, according to the data obtained from the MAS NMR spectra.[157]

The 100% natural abundance of phosphorus-31 makes the application of ^{31}P solid-state NMR spectroscopy very useful for the analysis of bone tissue and phosphorus-containing systems in the real state. Recent work has shown that solid-state ^{1}H and ^{31}P NMR spectroscopy provide useful data on the composition and characteristics of bone matrix and the MAS and CP techniques have given important information of the bone matrix for frequent pathologies in society, such as osteoporosis and osteomalacia. This approach detected the presence of a protonated phosphate group HPO_4^{2-} having an isotropic chemical shift similar to that in octacalcium phosphate and chemical shift anisotropy similar to that of brushite ($CaHPO_4 \cdot 2H_2O$), which makes a clear difference between the natural bone and synthetic models of bone mineral.[159–162]

While traditional ^{1}H, ^{13}C and ^{15}N solution-state NMR protocols provide a wealth of data on protein structure and dynamics, ^{19}F NMR spectroscopy provides a unique perspective of conformation, topology and dynamics and the changes that ensue under biological conditions. In particular, ^{19}F NMR spectroscopy has provided insight into biologically significant events such as protein folding and unfolding, enzymatic action, protein–protein, protein–lipid and protein–ligand interactions and aggregation and fibrillation in soluble and membrane protein systems. ^{19}F NMR spectroscopy is also often used to complement other biophysical techniques, such as circular dichroism (CD), fluorescence, X-ray crystallography and HCN NMR spectroscopy. In some cases, topological or dynamic information can be obtained from ^{19}F-NMR studies in systems that are too large or unstable for full structural analysis. In the arena of dynamics, the large chemical shift range associated with the fluorine nucleus provides experimental access to additional motional timescales in lineshape and T_2 relaxation studies. With recent and ongoing advances in NMR cryogenic probe technology, multidimensional NMR assignment and labelling approaches, it would seem that ^{19}F NMR spectroscopy may offer even greater possibilities in studies of protein structure and dynamics.[163]

Many integral membrane proteins have a membrane-spanning region that is characterized by the presence of predominantly hydrophobic residues, but is flanked by specific characteristic residues. The helixes that constitute the

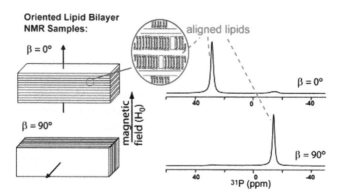

Figure 2.36 ^{31}P NMR spectra of bilayered lipids oriented in a magnetic field. Reprinted with permission from Ref. 165.

membrane-spanning domain of many of these proteins often display a specific tilt relative to the membrane. This can be functionally important and this helix tilting has been found to be involved in the functioning of various membrane proteins. In addition, tilting is one of the ways in which the hydrophobic mismatch between a protein and surrounding lipids can be adjusted. Solid-state NMR methods using macroscopically aligned membranes are able to study their tilt and the behaviour of their tryptophan (Trp) interfacial anchors in a proper, hydrated lipid bilayer environment. The application of ^{2}H NMR spectroscopy to transmembrane helixes containing deuterated Ala residues allowed a direct measurement of their tilt angles,[164] whereas similar measurements with deuterated Trp residues provided insights into these residues.[165] Additional measurements on lipid liposomes allowed us to see the remarkable effect of these peptides on the phospholipid phase transitions, as monitored by ^{31}P NMR spectroscopy. They were found to lead to an increased induction of inverted, non-bilayer structures, which was correlated with the extent of hydrophobic mismatch. Interestingly, the formation of non-bilayer phases is thought to be essential for biological processes that involve membrane fusion and fission. This is represented schematically in Figure 2.36, which shows the two different ^{19}P NMR spectra of bilayered proteins oriented in the direction of the magnetic field B or in an orthogonal direction.[165]

2.7.5   Magnetic Resonance Imaging

Magnetic resonance imaging (MRI) of living organisms (laboratory animals and humans) is a much more popular area of scientific research and practical applications than MRI of non-living objects. It might seem, therefore, that the latter field could benefit from the progress and developments in biomedical MRI research. Unfortunately, however, many of the pulse sequences routinely used in biomedical MRI are often not applicable in porous media studies. In particular, rapid imaging schemes such as multi-echo sequences may not work since in many cases even the first echo in the train may be difficult to detect. For certain porous materials, the short relaxation times of fluids in the pores and the associated problems of reduced signal-to-noise ratio and spatial resolution make the techniques from the arsenal of narrow-line MRI of liquids unsuitable. In fact, in this respect, the MRI of fluids in certain porous materials is closer to the MRI of solids and the drying and sorption processes are often addressed by resorting to the range of broad-line MRI approaches.[166]

One important feature of MRI is the possibility of analysing dynamically drying and swelling processes in heterogeneous systems, gels and films. This offers excellent opportunities to study in practical applications the behaviour of dispersions, latexes or membranes in different media. Drying of aqueous poly(vinyl alcohol) (PVA) layers was studied using both 1D GARField profiling with a spatial resolution of 17 μm and 2D FLASH MRI.[167] The samples were characterized by an initial thickness of 300–1300 μm and different PVA concentrations and were dried at different evaporation rates. The overall thickness of the layers was observed to decrease with drying time. For slow-drying regimes (low evaporation rate, thin films, large mutual diffusivity), the water distribution remained uniform within the layer thickness. In the opposite limit of fast drying, the profiles were not rectangular, with a reduced water concentration near the drying surface. For drying times that were long compared with the characteristic time of amorphous–crystalline transition, the top surface of the drying layers turned into a crystalline skin that was not observable in the profiles, whereas the skin did not develop in the opposite limit. Formation of the skin layer was observed to inhibit measurably the rate of water evaporation. The existence of the skin layer was demonstrated by depositing a fresh wet layer on top of a dry layer. The fresh solution dried without penetrating the more crystalline underlying layer, but permeated the full thickness of a less crystalline layer. Drying of water-based (water-borne) colloidal dispersions has attracted significant attention because of its environmental safety. Aqueous latex dispersions consist of polymer particles stabilized and dispersed in water by surfactants. In such systems, water acts as a solvent and a plasticizer and has an important influence on the film formation process. Drying of latex dispersions can produce polymer films that can be used as paints, glues, coatings, *etc.* It proceeds in three stages. During the first stage, evaporation of water leads to the close contact between the latex particles. Further evaporation leads to an even closer packing of particles due to their deformation. Finally, diffusion of the polymer across the particle boundaries leads to the formation of the final structure, with individual particles that are no longer distinguishable. In such processes, water is present outside the latex particles, in the interfacial surfactant layer and inside the particles.[166]

The protocol described as GARField profiling was used to study the drying process of a latex based on a copolymer of vinyl acetate and ethylene with the reactive monomer and a photoinitiator added for cross-linking.[168] Films

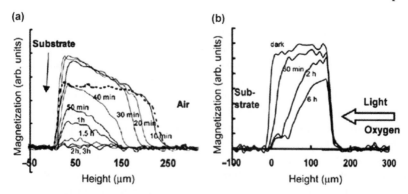

Figure 2.37 (a) MRI profiles of water evaporation in a film of latex only. The dashed line shows a latex–cross-linker mixture imaged 24 h after film formation. (b) Profiles obtained at various times during the cross-linking of a film exposed to light after previously being dried in the dark for 48 h. The experiments were performed at 0.7 T.
Reprinted with permission from Ref. 168.

with an initial thickness of 200–400 μm were prepared for imaging on a glass substrate. During drying, a decrease in film thickness was clearly observed [Figure 2.37(a)]. The signal decreased faster at the film/air interface than at the film/substrate interface. The softened latex may contribute to the detected profile. Indeed, the dried mixture of latex and cross-linker did give a substantial signal, which may have originated from the cross-linker itself and/or the plasticized latex. Upon light-induced cross-linking of the dry latex/cross-linker/photoinitiator system, the signal was observed to decrease as a result of the cross-linking process which started at the film/substrate interface, presumably because of an inhibitory effect of air oxygen on the radical reaction near the film/air interface [Figure 2.37(b)]. As the signal was still observed after 2 days of light exposure, the authors concluded that cross-linking did not make the film harder than the original latex. In another set of experiments, the wet film was imaged while exposed to light immediately after its casting on the glass substrate. In this case, the signal intensity decreased in time because of both water evaporation and latex cross-linking, and the profiles demonstrated the minimum signal intensity at the film centre. This was explained as being the result of the interplay of several processes, including oxygen transport and inhibition of cross-linking, water evaporation and light scattering. For comparison, the cross-linker–photoinitiator mixture (with no water and latex) was imaged while exposed to light. The signal was fairly strong as the cross-linker had a low molecular weight. An induction period with constant signal intensity was followed by a signal decrease at the substrate/film interface and later the gradual displacement of the cross-linking front towards the film/air interface.[166–168]

MRI gives excellent information on the dynamic swelling of 'smart polymers' and sensitive systems. We have applied a microimaging technique to study the changes in the morphology and dimensions of hydrogels based on

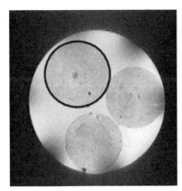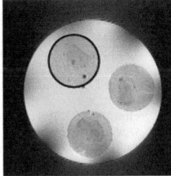

Figure 2.38   NMR images of cross-sections of samples of poly(*N*-isopropylacryl-amide) grafted on to chitosan (PNIPAM-CHI), above (left) and below (right) the VPT. The black circles are drawn for comparison.

thermosensitive macroporous cryogels of poly(*N*-isopropylacrylamide) (PNI-PAM) functionalized with bioactive chitosan (CHI)/bemiparin NPs. Cylin-drical samples of approximately 5 mm diameter and 3 mm height were immersed in vials containing phosphate-buffered saline solution. The vial was depressurized several times inside a desiccator in order to eliminate air bubbles and was maintained under vacuum until used. NMR images were processed using NIH Image (National Institutes of Health, Bethesda, MD, USA) and Bruker Paravision software. Hahn spin–echo images were acquired at echo times (t_e) of 4.349, 10, 20, 40, 80 and 120 ms, at temperatures of 21, 24, 28, 32, 34 and 36.5 °C. Several representative regions of interest (ROIs) were defined avoiding non-homogeneous areas in B1 (artefacts) and the mean signal intensity measured for each sample, medium and background. The average proton density for each sample and the medium was deter-mined by fitting the NMR data to a monoexponential function and ex-trapolating to $t_e = 0$.[169]

Figure 2.38 shows the MRI images of the cross-sections of sample PNI-PAM-CHI below and above the volume phase transition (VPT), revealing a decrease in the sample cross-sectional area as the cryogel was heated from below (left, low t_e value, corresponding to a temperature of 21 °C) to above (right, high t_e value, corresponding to a temperature of 36.5 °C) the VPT. MRI also gives data on the average porosity and the diffusion of water during the hydration or swelling process.

In conclusion, MRI offers interesting potential for the analysis and char-acterization of a great number of polymer matrices, composites and complex hierarchical supramolecular assemblies.

References

1. R. Erni, M. D. Rossell, C. Kisielowski and U. Dahmen, *Phys. Rev. Lett.*, 2009, **102**, 96.

2. L. de Broglie, Recherches sur la théorie des quanta (Research on quantum theory), Thesis, Paris, 1924.

3. E. Ruska and M. Knoll, *Z. Phys.*, 1932, **78**, 318.

4. J. Kuntsche, J. C. Horst and H. Bunjes, *Int. J. Pharm.*, 2011, **417**, 120.

5. W. H. Massover, *Micron*, 2011, **42**, 141.

6. L. C. Sawyer and D. T. Grubb, *Polymer Microscopy*, 2nd edn, Alden Press, Oxford, 1996.

7. R. F. Egerton, *Physical Principles of Electron Microscopy: an Introduction to TEM, SEM and AEM*, Springer, New York, 2005.

8. V. Klang, N. B. Matsko, C. Valenta and F. Hofer, *Micron*, 2012, **43**, 85.

9. A. Desai, T. Vyas and M. Amiji, *J. Pharm. Sci.*, 2008, **97**, 2745.

10. W. H. Massover, *Microsc. Microanal.*, 2008, **14**, 126.

11. S. Ganta and M. Amiji, *Mol. Pharm.*, 2009, **6**, 928.

12. F. A. Araujo, R. G. Kelmann, B. V. Araujo, R. B. Finatto, H. F. Teixeira and L. S. Koester, *Eur. J. Pharm. Sci.*, 2011, **42**, 238.

13. J. Hatanaka, H. Chikamori, H. Sato, S. Uchida, K. Debari, S. Onoue and S. Yamada, *Int. J. Pharm.*, 2010, **396**, 188.

14. S. Brenner and R. W. Horne, *Biochim. Biophys. Acta*, 1959, **34**, 103.

15. K. Burapapadh, M. Kumpugdee-Vollrath, D. Chantasart and P. Sriamornsak, *Carbohydr. Polym.*, 2010, **82**, 384.

16. S. U. Egelhaaf, P. Schutenberger and M. Mueller, *J. Microsc.*, 2000, **200**, 128.

17. M. Mueller, *Encycloped. Hum. Biol.*, 1991, **2**, 721.

18. H. Friedrich, P. M. Frederik, G. de With and N. A. Sommerdijk, *Angew. Chem. Int. Ed.*, 2010, **49**, 7850.

19. M. A. Hayat, *Principles and Techniques of Electron Microscopy: Biological Applications*, Cambridge University Press, Cambridge, 2000.

20. J. R. Harris, *Micron*, 2008, **39**, 168.

21. Z. Li, E. Kesselman, Y. Talmon, M. A. Hillyer and T. P. Lodge, *Science*, 2004, **306**, 98.

22. H. Cui, T. K. Hodgon, E. W. Kaler, L. Abegauz, D. Danino, M. Lubovsky, Y. Talmon and D. J. Pochan, *Soft Matter*, 2007, **3**, 945.

23. M. G. Sceats and S. A. Rice, Amorphous solid water and its relationship to liquid water: a random network model for water. In *Water: a Comprenhensive Treatise*, Vol. 7, ed. F. Franks, Plenum Press, New York, 1982, pp. 83–214.

24. F. Sciortino, P. H. Poole, U. Essman and H. E. Stanley, *Phys. Rev. E*, 1997, **55**, 727.

25. H. Moor, *Philos. Trans. R. Soc. London, Ser. B*, 1971, **261**, 121.

26. J. J. Crassous, M. Ballauff, M. Drechsler, J. Schmidt and Y. Talmon, *Langmuir*, 2006, **22**, 2403.

27. Y. Lu and M. Ballauff, *Prog. Polym. Sci.*, 2011, **36**, 767.

28. M. Adrian, J. Dubochet, J. Lepault and A. W. McDowal, *Nature*, 1984, **308**, 32.

29. M. Adrian, J. Dubochet, S. D. Fuller and J. R. Harris, *Micron*, 1998, **29**, 145.

30. M. Knoll, *Z. Tech. Phys.*, 1935, **16**, 467.
31. J. Xu, Z. Hou, X. Yuan and H. Guo, *Microsc. Res. Technol.*, 2011, **74**, 749.
32. A. M. Donald, C. He, C. P. Royall, M. Sferrazza, N. A. Stelmashenko and B. L. Thiel, *Colloids Surf. A: Physicochem. Eng. Aspects*, 2000, **174**, 37.
33. G. D. Danilatos, *J. Microsc.*, 1990, **162**, 391.
34. R. E. Cameron and A. M. Donald, *J. Microsc.*, 1994, **173**, 227.
35. K. I. Dragnevski and A. M. Donald, *J. Phys. Conf. Ser.*, 2008, **126**, 012077.
36. A. Bogner, G. Thollet, D. Basset, P.-H. Jouneau and C. Gauthier, *Ultramicroscopy*, 2005, **104**, 290.
37. D. B. Willians and C. B. Carter, *Transmission Electron microscopy III. Imaging*, Plenum Press, New York, 1996.
38. J. Goldstein, D. Newbury, D. Joy, C. Lyman, P. Echlin, E. Lifshin, *et al.*, *Scanning Electron Microscopy and X-Ray Microanalysis*, Kluwer Academic/Plenum, New York, 2003.
39. S. Zou, Y. Yang, H. Liu and C. Wang, *Colloids Surf. A: Physicochem. Eng. Aspects*, 2013, **436**, 1.
40. R. Hidalgo-Álvarez, A. Martín, A. Fernández, D. Bastos, F. Martínez and F. J. de las Nieves, *Adv. Colloid Interface Sci.*, 1996, **67**, 1.
41. H. Ohshima, in *Electrical Phenomena at Interfaces: Fundamentals, Measurements and Applications*, ed. H. Ohshsima and K. Furusawa, Marcel Dekker, New York, 2nd edn, 1998, Chapters 1–3, pp. 1–85.
42. J. N. Israelachvili, *Intermolecular and Surface Forces*, Academic Press, London, 1992.
43. P. C. Hiemenz and R. Rajagopalan, *Principles of Colloidal and Surface Chemistry*, Marcel Dekker, New York, 1997.
44. J. A. Molina-Bolívar, F. Galisteo-González and R. Hidalgo-Álvarez, *Colloids Surf. B: Biointerfaces*, 1999, **14**, 3.
45. M. J. Santander-Ortega, J. M. Peula-García, F. M. Goycoolea and J. L. Ortega-Vinuesa, *Colloids Surf. B:Biointerfaces*, 2011, **82**, 571.
46. C. P. Price and D. J. Newman, *Principles and Practice of Immunoassay*, Stockton Press, New York, 1991.
47. J. L. Ortega-Vinuesa and D. Bastos-González, *J. Biomater. Sci. Polym. Edn.*, 2001, **12**, 379.
48. H. J. van den Hul and W. Vanderhoff, *J. Electroanal. Chem.*, 1972, **37**, 161.
49. F. B. Laundry, D. V. Bazile, G. Spenlehauer, M. Veillard and J. Kreuter, *STP Pharma Sci.*, 1996, **6**, 195.
50. N. Csaba, L. González, A. Sánchez and M. J. Alonso, *J. Biomater. Sci. Polym. Edn.*, 2004, **15**, 1137.
51. J. Yao, P. Ravi, K. C. Tam and L. H. Gan, *Polymer*, 2004, **45**, 2781.
52. P. Sánchez-Moreno, J. L. Ortega-Vinuesa, H. Boulaiz, J. A. Marchal and J. M. Peula-García, *Biomacromolecules*, 2013, **14**, 4248.
53. Y. Niu, L. Sun and R. M. Crooks, *Macromolecules*, 2003, **36**, 5725.
54. I. J. Majoros, T. P. Thomas, C. B. Mehta and J. R. Baker Jr, *J. Med. Chem.*, 2005, **48**, 5892.

55. A. Ramadan, F. Lagarce, A. Tessier-Marteau, O. Thomas, P. Legras, L. Macchi, P. Saulnier and J. P. Benoit, *Int. J. Nanomed.*, 2011, **6**, 2941.
56. H. Shirama and T. Suzawa, *Polym. J.*, 1984, **16**, 795.
57. H. Shirama and T. Suzawa, *J. Appl. Polym. Sci.*, 1984, **29**, 3651.
58. T. Hoare and R. Pelton, *Langmuir*, 2004, **20**, 2123.
59. H. Suzuki, B. Wang, R. Yoshida and E. Kokufuta, *Langmuir*, 1999, **15**, 4283.
60. T. Hoare and R. Pelton, *Langmuir*, 2006, **22**, 7342.
61. H. Ohshima, in *Theory of Colloid and Interfacial Electric Phenomena*, ed. H. Ohshima, Interface Science and Technology, Vol. 12, Elsevier, Amsterdam, 1st edn, 2006, Chapter 2, p. 39.
62. J. Ramos, A. Imaz, J. Callejas-Fernández, L. Barbosa-Barros, J. Estelrich, M. Quesada-Pérez and J. Forcada, *Soft Matter*, 2011, 7, 5067.
63. A. V. Delgado, F. González-Caballero, R. J. Hunter, L. K. Koopal and J. Lyklema, *Pure Appl. Chem.*, 2005, **77**, 1753.
64. W. W. Tscharnuter, *Appl. Opt.*, 2001, **40**, 3995.
65. G. Gouy, *J. Phys. Radium*, 1910, **9**, 457.
66. G. Gouy, *C. R. Acad. Sci.*, 1910, **149**, 654.
67. D. L. Chapman, *Philos. Mag.*, 1913, **25**, 475.
68. G. I. Guerrero-García, E. González-Tovar and M. Olvera de la Cruz, *Soft Matter*, 2010, **6**, 2056.
69. R. J. Hunter, in *Zeta Potential in Colloid Science: Principles and Applications*, ed. R. J. Hunter, Academic Press, London, 1981, Chapter 1, p. 1.
70. P. Attard, D. Antelmi and I. Larson, *Langmuir*, 2000, **16**, 1542.
71. S. Madurga, A. Martín-Molina, E. Vilaseca, F. Mas and M. Quesada-Pérez, *J. Chem. Phys.*, 2007, **126**, 234703.
72. R. J. Hunter, in *Zeta Potential in Colloid Science: Principles and Applications*, ed. R. J. Hunter, Academic Press, London, 1981, Chapter 3, p. 59.
73. H. Ohshima, in *Theory of Colloid and Interfacial Electric Phenomena*, ed. H. Ohshima, Interface Science and Technology, Vol. 12, Elsevier, Amsterdam, 1st edn, 2006, Chapter 1, p. 1.
74. M. von Smoluchowski, *Z. Phys. Chem.*, 1918, **92**, 129.
75. E. Hückel, *Phys. Z.*, 1924, **25**, 204.
76. D. C. Henry, *Proc. R. Soc. London, Ser. A*, 1931, **133**, 106.
77. P. H. Wiersema, A. L. Loeb and J. Th. G. Overbeek, *J. Colloid Interface Sci.*, 1966, **22**, 78.
78. R. W. O'Brien and L. R. White, *J. Chem. Soc., Faraday Trans. 2*, 1978, **74**, 1607.
79. M. Lozada-Cassou and E. González-Tovar, *J. Colloid Interface Sci.*, 2001, **239**, 285.
80. P. C. Hiemenz and R. Rajagopalan, in *Principles of Colloid and Surface Chemistry*, ed. P. C. Hiemenz and R. Rajagopalan, Marcel Dekker, New York, 3rd edn, 1997, Chapter 12, p. 534.
81. M. Kosmulski, *Zeta*, http://zeta-potential.sourceforge.net/, 1993.
82. H. Ohshima, *Soft Matter*, 2012, **8**, 3511.

83. H. Ohshima, in *Theory of Colloid and Interfacial Electric Phenomena*, ed. H. Ohshima, Interface Science and Technology, Vol. 12, Elsevier, Amsterdam, 1st edn, 2006, Chapter 9, p. 203.

84. H. Ohshima, K. Makino, T. Kato, K. Fujimoto, T. Kondo and H. Kawaguchi, *J. Colloid Interface Sci.*, 1993, **159**, 512.

85. R. Bos, H. C. van der Mei and H. J. Busscher, *Biophys. Chem.*, 1998, **74**, 251.

86. K. Makino and H. Ohshima, *Sci. Technol. Adv. Mater.*, 2011, **12**, 023001.

87. J. J. Liétor-Santos and A. Fernández-Nieves, *Adv. Colloid Interface Sci.*, 2009, **147–148**, 178.

88. D. Paiva, A. Martín-Molina, I. Cardoso, M. Quesada-Pérez, M. do Carmo Pereira and S. Rocha, *Soft Matter*, 2013, **9**, 401.

89. J. K. G. Dhont, in *An Introduction to Dynamics of Colloids*, ed. D. Möbius and R. Miller, Elsevier, Amsterdam, 1996, Vol. II, Chapter 3, pp. 107–170.

90. M. Kerker, in *The Scattering of Light*, ed. E. M. Loebl, Academic Press, New York, 1969, Chapter 8, pp. 414–486.

91. M. Kerker, in *The Scattering of Light*, ed. E. M. Loebl, Academic Press, New York, 1969, Chapter 3, pp. 27–96.

92. P. C. Hiemenz and R. Rajagopalan, *Principles of Colloid and Surface Chemistry*, Marcel Dekker, New York, 1997, Chapter 5, pp. 193–247.

93. S. Roldán-Vargas, A. Martín-Molina, M. Quesada-Pérez, R. Barnadas-Rodríguez, J. Estelrich and J. Callejas-Fernández, *Phys. Rev. E*, 2007, **75**, 21912.

94. R. Pecora, in *Dynamic Light Scattering: Applications of Photon Correlation Spectroscopy*, ed. R. Pecora, Plenum Press, New York, 1985, Chapter 2, pp. 7–57.

95. B. J. Berne and R. Pecora, *Dynamic Light Scattering: With Applications to Chemistry, Biology and Physics*, Dover Publications, New York, 2000, Chapter 2, pp. 10–23.

96. B. Chu, *Laser Light Scattering: Basic Principles and Practice*, Academic Press, London, 2nd edn, 1991, Chapter 3, pp. 63–92.

97. J. Callejas-Fernandez, M. Tirado-Miranda, M. Quesada-Pérez, G. Odriozola-Prego and A. Schmitt, Photon spectroscopy of colloids, in *Encyclopedia of Surface and Colloid Science*, Marcel Dekker, New York, 2002, pp. 4072–4088.

98. A. J. K. Siegert, *M. I.T. Rad. Lab. Rep.*, 1943, 465.

99. P. C. Hiemenz and R. Rajagopalan, *Principles of Colloid and Surface Chemistry*, Marcel Dekker, New York, 1997, Chapter 2, pp. 62–104.

100. B. J. Berne and R. Pecora, *Dynamic Light Scattering: With Applications to Chemistry, Biology and Physics*, Dover Publications, New York, 2000, Chapter 8, pp. 164–206.

101. B. Chu, *Laser Light Scattering: Basic Principles and Practice*, Academic Press, London, 2nd edn, 1991, Chapter 7, pp. 243–282.

102. D. Koppel, *J. Chem. Phys.*, 1972, **57**, 4814.

103. J. S. Higgins and H. C. Benoit, *Polymers and Neutron Scattering*, Oxford University Press, New York, 1996.

104. J. P. Cohen Addad (ed.), *Physical Properties of Polymer Gels*, Wiley, Chichester, 1996.
105. T. Tanaka, *Experimental Methods in Polymer Science*, Academic Press, New York, 1998.
106. A. Fernández-Barbero, A. Fernández-Nieves, I. Grillo and E. López-Carbarcos, *Phys. Rev. E*, 2002, **66**, 051803.
107. O. Glatter and O. Kratky, *Small Angle X-Ray Scattering*, Academic Press, London, 1982, Chapter 3, pp. 53–118.
108. J. P. Cotton, in *Neutron, X-Ray and Light Scattering*, ed. P. Lindner and Th. Zhemb, North-Holland, Amsterdam, 1991, pp. 3–17.
109. Th. F. Tadros, C. Solans and R. Pons, *Langmuir*, 2002, **18**, 5673.
110. J. S. Pedersen, *Adv. Colloid Interface Sci.*, 1997, **70**, 171.
111. G. Pabst, D. Zweytick, R. Prassl and K. Lohner, *Eur. Biophys. J.*, 2012, **41**, 915.
112. S. Förster, A. Timmann, M. Konrad, C. Schellbach, A. Meyer, S. S. Funari, P. Mulvaney and R. Knott, *J. Phys. Chem. B*, 2005, **109**, 1347.
113. M. V. Petoukhov and D. I. Svergun, *Int. J. Biochem. Cell Biol.*, 2013, **45**, 429.
114. J. R. Lakowicz, *Principles of Fluorescence Spectroscopy*, Springer, New York, 3rd edn, 2006.
115. A. Lapresta-Fernández, P. Cywinski, A. Moro and G. Mohr, *Anal. Bioanal. Chem.*, 2009, **395**, 1821.
116. Y. Chen, P. A. Wilbon, J. Zhou, M. Nagarkatti, C. Wang, F. Chu and C. Tang, *Chem. Commun.*, 2013, **49**, 297.
117. X. Gao, L. Yang, J. A. Petros, F. F. Marshall, J. W. Simons and S. Nie, *Curr. Opin. Biotechnol.*, 2005, **16**, 63.
118. A. Schulz, S. Hornig, T. Liebert, E. Birckner, T. Heinze and G. J. Mohr, *Org. Biomol. Chem.*, 2008, 7, 1884.
119. F. Li, A. H. Westphal, A. T. M. Marcelis, E. J. R. Sudholter, M. A. Cohen Stuart and F. A. M. Leermakers, *Soft Matter*, 2011, 7, 11211.
120. S. Inal, J. D. Kolsch, L. Chiappisi, D. Janietz, M. Gradzielski, A. Laschewsky and D. Neher, *J. Mater. Chem. C*, 2013, **1**, 6603.
121. J. Luo, Z. Xie, J. W. Y. Lam, L. Cheng, H. Chen, C. Qiu, H. S. Kwok, X. Zhan, Y. Liu, D. Zhu and B. Z. Tang, *Chem. Commun.*, 2001, **18**, 1740.
122. C. Li and S. Liu, *Chem. Commun.*, 2012, **48**, 3262.
123. S. Nayak, D. Gan, M. Serpe and L. G. Lyon, *Small*, 2005, **1**, 416.
124. F. Gouanvé, T. Schuster, E. Allard, R. Méallet-Renault and C. Larpent, *Adv. Funct. Mater.*, 2007, **17**, 2746.
125. J. Liu, A. Li, J. Tang, R. Wang, N. Kong and T. P. Davis, *Chem. Commun.*, 2012, **48**, 4680.
126. C. Li, J. Hu and S. Liu, *Soft Matter*, 2012, **8**, 7096.
127. T. Xia, N. Li and X. Fang, *Annu. Rev. Phys. Chem.*, 2013, **64**, 459.
128. V. Kozlovskaya, W. Higgins, J. Chen and E. Kharlampieva, *Chem. Commun.*, 2011, **47**, 8352.
129. B. Kim, H. Soo Lee, J. Kim and S. H. Kim, *Chem. Commun.*, 2013, **49**, 1865.

130. P. S. Russo, J. Qiu, N. Edwin, Y. W. Choi, G. J. Doucet and D. Sohn, Fluorescence photobleaching recovery, in *Soft Matter Characterization*, ed. R. Borsali and R. Pecora, Spinger, New York, 2008, pp. 605–636.

131. N. Fatin-Rouge and J. Buffle, Study of environmental systems by means of fluorescence correlation spectroscopy, in *Environmental Colloids and Particles: Behaviour, Structure and Characterisation*, ed. K. J. Wilkinson and J. R. Lead, Wiley, Chichester, 2006, pp. 507–553.

132. P. Jurkiewicz, C. Konák, V. Subr, M. Hof and P. Stepánek, and K. lbrich, *Macromol. Chem. Phys.*, 2008, **209**, 1447.

133. M. Klumpp, A. Scheel, E. Lopez-Calle, M. Busch, K. J. Murray and A. J. Pope, *J. Biomol. Screen.*, 2001, **6**, 159.

134. A. Van Orden, K. Fogarty and J. Jung, *Appl. Spectrosc.*, 2004, **58**, 122A.

135. A. H. Kwan, M. Mobili, P. R. Gooley, G. F. King and J. P. Mackay, *FEBS J.*, 2011, **278**, 687.

136. J. Keeler, *Understanding NMR Spectroscopy*, Wiley, Chichester, 2nd edn, 2010.

137. J. B. Lambert and E. P. Mazzola, *Nuclear Magnetic Resonance Spectroscopy: an Introduction to Principles, Applications and Experimental Methods*, Pearson Education, Upper Saddle River, NJ, 2004.

138. R. R. Ernst, G. Bodenhausen and A. Wokaun, *Principles of Nuclear Magnetic Resonance in One and Two Dimensions*, Clarendon Press, Oxford, 1990.

139. E. M. Bachelder, T. T. Beaudette, K. E. Broaders, J. Dashe and J. M. Frechet, *J. Am. Chem. Soc.*, 2008, **130**, 10494.

140. M. L. López Donaire, J. Parra-Cáceres, B. Vazquez, I. García, A. Fernández, A. López-Bravo and J. San Roman, *Biomaterials*, 2009, **30**, 1613.

141. M. A. Dubinnyi, D. M. Lesovoy, P. V. Dubovskii, V. V. Chupin and A. S. Arseniev, *Solid State Nucl. Magn. Reson.*, 2006, **29**, 305.

142. L. D. Quin, J. B. Duke and A. J. Williams, *Practical Interpretation of P-31 NMR Spectra and Computer Assisted Structure Verification*, Advanced Chemistry Development, Toronto, 2004.

143. U. Jakop, B. Fuchs, R. Sus, G. Wibbelt, B. Braun, K. Müller and J. Schiller, *Lipids Health Dis.*, 2009, **8**, 49.

144. J. Schiller, M. Müller, B. Fuchs, K. Arnold and D. Huster, *Curr. Anal. Chem.*, 2007, **3**, 283.

145. J. M. Pearce and R. A. Komoroski, *Magn. Reson. Med.*, 2000, **44**, 215.

146. J. Schiller, J. Arnhold, H. J. Glander and K. Arnold, *Chem. Phys. Lipids*, 2000, **106**, 145.

147. H. W. Spiess, *Annu. Rep. NMR Spectrosc.*, 1997, **34**, 1.

148. M. J. Duer, *Solid-State NMR Spectroscopy. Principles and Applications*, Blackwell Science, Oxford, 2002.

149. S. P. Brown and H. W. Spiess, *Chem. Rev.*, 2001, **101**, 4125.

150. A. Samoson, T. Tuhern and Z. Gan, *Solid State NMR*, 2001, **20**, 130.

151. B. Alonso and D. Massiot, *J. Magn. Reson.*, 2003, **163**, 347.

152. M. Ernst, *J. Magn. Reson.*, 2003, **162**, 1.

153. C. Bonhomme, C. Coelho, N. Baccile, C. Gervais, T. Azais and F. Babonneau, *Acc. Chem. Res.*, 2007, **40**, 738.

154. S. Spiegel, K. Landfester, G. Lieser, C. Boeffeld, H. W. Spiess and N. Eidam, *Makromol. Chem.*, 1995, **196**, 985.

155. K. Simons and D. Toomre, *Nat. Rev. Mol. Cell Biol.*, 2000, **1**, 31.

156. S. Munro, *Cell*, 2003, **115**, 377.

157. H. A. Scheidt, T. Meyer, J. Nikolaus, D. J. Baek, I. Haralampiev, L. Thomas, R. Bittman, P. Müller, A. Herrmann and D. Huster, *Angew. Chem. Int. Ed.*, 2013, **52**, 12848.

158. D. J. Back and R. Bittman, *Chem. Phys. Lipids*, 2013, **175–176**, 99.

159. A. C. Seiert, A. C. Wright, S. L. Wehrli, H. H. Ong, C. Li and F. W. Wehrli, *NMR Biomed.*, 2013, **26**, 1158.

160. A. Kaflak, D. Chmielewski, A. Gorecki, A. Slosarcyk and W. Kolodziesjski, *Solid State Nucl. Magn. Reson.*, 2006, **29**, 345.

161. A. Kaflak, A. Samoson and W. Kolodziesjski, *Calcif. Tissue Int.*, 2003, **73**, 476.

162. G. Cho, Y. Wu and J. L. Ackerman, *Science*, 2003, **300**, 1123.

163. J. L. Kitevski-Leblanc and R. Scott Prosser, *Prog. Nucl. Magn. Reson. Spectrosc*, 2012, **62**, 1.

164. P. C. A. Van der Wel, E. Strandberg, J. A. Killian and R. E. Koeppe II, *Biophys. J.*, 2002, **83**, 1479.

165. P. C. A. Van der Wel, N. D. Reed, D. V. Greathouse and R. E. Koeppe II, *Biochemistry*, 2007, **46**, 7514.

166. I. V. Koptyug, *Prog. Nucl. Magn. Reson. Spectrosc.*, 2012, **65**, 1.

167. E. Ciampi and P. J. McDonald, *Macromolecules*, 2003, **36**, 8398.

168. A. C. Hellgren, M. Wallin, P. K. Weissenborn, P. J. McDonald, P. M. Glover and J. L. Keddie, *Prog. Org. Coat.*, 2001, **43**, 85.

169. H. Peniche, F. Reyes, M. R. Aguilar, G. Rodríguez, C. Abradelo, L. García, C. Peniche and J. San Roman, *Macromol. Biosci.*, 2013, **13**, 1556.

CHAPTER 3

The Original Magnetoliposomes: from the Physicochemical Basics to Theranostic Nanomedicine

MARCEL DE CUYPER

Katholieke Universiteit Leuven – Campus Kortrijk, Interdisciplinary Research Centre, University Campus, 8500 Kortrijk, Belgium
Email: marcel.decuyper@kuleuven-kulak.be

3.1 Introduction

In recent years, various nanotechnology platforms in the area of medical biology have attracted remarkable attention and, in particular, the field of magnetic nanoparticles (MNPs) has spurred an exponential growth in both diagnostics and therapy.[1,2] Apart from their useful magnetic properties, the success of MNPs relies mainly on the limited toxicity of the magnetizable core, very often composed of ferrous and ferric oxides, although metals such as cobalt and nickel are also employed. A second key role player influencing toxicity is the coating of the MNPs, which is necessary to stabilize the particles under physiological conditions. Ultimately, it is the MNP envelope that will determine how the particle will interact with its environment, either directly with cells or tissues or indirectly by binding to cell media ingredients or biological fluids, thereby modifying the features of the colloidal structures in an almost unpredictable way.

RSC Nanoscience & Nanotechnology No. 34
Soft Nanoparticles for Biomedical Applications
Edited by José Callejas-Fernández, Joan Estelrich, Manuel Quesada-Pérez and Jacqueline Forcada
© The Royal Society of Chemistry 2014
Published by the Royal Society of Chemistry, www.rsc.org

Different types of coatings have been successfully applied to MNPs to render them suitable for biological applications; these include ions or small organic molecules, synthetic or natural polymers (such as dextrans and proteins)[1,3] and amphipathic molecules such as fatty acids[4] and phospholipids.[5] Further modifications may then be made, such as the addition of specific targeting ligands, dyes or therapeutic agents, and it is these modifications that give the particles their wide-ranging potential in the biomedical arena, for instance, leading to opportunities in magnetic resonance imaging (MRI), as efficient delivery vectors, as mediators of hyperthermia cancer treatment and in drug-targeted therapies.[6–9]

3.2 Magnetoliposomes

3.2.1 Magnetoliposomes: What Are They About?

This chapter focuses mainly on structures built up of Fe_3O_4 (magnetite) and phospholipids, which in the literature are commonly designated *magnetoliposomes*. The term 'magnetoliposome' (ML) was introduced independently by Margolis *et al.*[10] and ourselves.[5,11] The structures described by Margolis *et al.* were produced by simply shaking a suspension of a commercially available, crude phospholipid mixture in the presence of a magnetic fluid (*i.e.* a solution of stabilized nanometre-sized iron oxide grains), but neither a thorough characterization of the starting reagents nor an attempt to describe the structure of the final product was carried out. In contrast, on the basis of both experiments and theoretical calculations (see below), we showed that a single phospholipid bilayer could be directly grafted on to the surface of the iron oxide core. In addition, based on our extensive experience with spontaneous phospholipid transfer processes between artificial membranes such as liposomes or phospholipid vesicles,[12] the mechanism of ML generation could be described in great detail.[13,14] We also demonstrated that the immobilized phospholipid bilayer, just like liposomal membranes, was able to host hydrophobic membrane proteins, giving opportunities to construct magnetically controlled bioreactors using highly purified, hydrophobic membrane enzymes.[15,16] As far as their structural features are concerned, the resulting 'magneto*proteo*liposomes' nicely mimic so-called magnetosomes, which are found in some animals that are believed to rely on the Earth's magnetic field for their guidance. For instance, in magnetotactic bacteria, it has been observed that several magnetosome particles are closely aligned so that they can act as a navigating compass.[17] Although the size of the iron oxide cores in magnetosomes is larger than that in MLs (diameter 50 nm *versus* 15 nm), they are also enwrapped by a strongly tethered phospholipid bilayer containing a variety of membrane proteins. However, as it is a laborious and costly task to raise the microorganisms and to isolate magnetosomes thereof in sufficient quantities, the MLs described above can be considered as a cheap magnetosome surrogate, which is extremely versatile with respect to surface characteristics.

Later, the term 'magnetoliposome' was also used to denote other types of phospholipid–iron oxide constructs with totally different properties, which makes the literature on magnetoliposomes often very confusing. In the following paragraphs, a few other ML constructs will be briefly presented.

Extruded magnetoliposomes consist of large unilamellar vesicles (diameter of the order of a few hundred nanometres) that encapsulate several small nanometre-sized, water-dispersible iron oxide cores in the aqueous cavity.[18,19] With this type of ML, 'superior' iron oxide encapsulation efficiencies of 16% were reported by Domingo *et al.*,[20] but in the original, first-generation MLs this value can be considered to be 100% since all the magnetic spheres are covered with a phospholipid bilayer and thus no internal volume is left.[21] A main advantage of the extruded MLs, however, is that they still have an internal aqueous cavity that can be used to encapsulate water-soluble molecules (*e.g.* chemotherapeutic drug molecules); in the case of the 'original' MLs, such polar molecules have to be linked by chemical means to reactive groups positioned at the surface of the lipid bilayer.

As an alternative to the extrusion method, magnetizable particle encapsulation can also be realized by sonication, inverse phase evaporation, congelation/decongelation or a combination of these techniques.[19,22,23]

Another protocol to produce MLs was originally presented by Mann's group,[24] who claimed the production of uniform particles by precipitation in single-compartment lipid vesicles, acting as *nanoreactors*. In brief, in this technique Fe^{2+} and Fe^{3+} ions/salts are first captured in the inner space of the vesicles during sonication of phospholipids in an aqueous salt solution, the extravesicular reactant cations are removed by dialysis, gel permeation or ion-exchange column chromatography, and the pH of the medium is then raised so that OH^- ions can permeate the vesicle wall and react with the intravesicular Fe^{2+}–Fe^{3+} cations to produce nanosized iron oxide precipitates until all cations are exhausted. Although the authors claimed that the constrained cavities within unilamellar vesicles provide more effective control over the crystal growth and prevention of particle agglomeration than can be obtained by free precipitation methods of the iron ions, in practice, little control can actually be exercised over the size and size distribution of the microstructures and, moreover, only small quantities of iron oxide can be obtained owing to constraints of low reagent concentrations necessitated by this synthetic procedure.[4] Indeed, in order to generate a sufficiently large iron oxide particle that can be quantitatively attracted in a magnetic field, one has to start with a highly concentrated solution of ferrous and ferric salts. Under these conditions, however, the pH is extremely low and this may adversely affect the integrity of both the vesicle structure and the phospholipid building blocks themselves (see, for instance, Ref. 21). In addition, unsaturated phospholipids are reported to be highly susceptible to peroxidation induced by Fe ions.[25,26] Furthermore, the association of Fe^{3+} and, to a somewhat lesser extent, Fe^{2+} ions with the phosphate group in phospholipid molecules is very high,[27] with the result that it is extremely difficult to remove these extravesicular ions completely, *e.g.* by dialysis or ion-exchange chromatography. On increasing the pH, these residual ions precipitate, together with the phospholipid structures, thereby forming irreversible phospholipid–magnetite clusters.

Clearly, we cannot exhaustively cover the complete field of the different types of MLs, so this review will predominantly focus on the production and properties of the 'original' MLs, *i.e.* nanocolloids on which a phospholipid bilayer is intimately associated with the surface of a superparamagnetic iron oxide core, and on studies where these were also used in innovative biomedical applications. In our research endeavours, we have spent a lot of effort to understand and unravel both the physicochemical parameters that play a role in the construction of MLs and the biological pathways involved in cellular processing of MLs. In a broad sense, this approach also permitted us to develop engineered multifunctional nanoparticles, for instance, multimodal MLs which can be monitored by various physical techniques (*e.g.* MRI and fluorescence) or MLs that can be considered as theranostics, *i.e.* which combine both therapy (*e.g.* MLs as drug carriers or mediators for hyperthermia) and imaging.

3.2.2 Magnetizable Nanoparticles

Before discussing the use of magnetic nanoparticles in innovative applications, including those in the biomedical domain, it may be instructive to discuss briefly the behaviour of materials placed in an external magnetic field. In this respect, materials can be categorized into three groups: diamagnetic, paramagnetic and ferro/ferrimagnetic materials.[28] The distinction between them is made on a molecular basis and mainly relies on whether or not (un)paired electrons are present in the substances.

The atoms in *ferro/ferrimagnetic* materials contain several unpaired electrons. When brought into even a weak external magnetic field, the unpaired electron spins will easily align in the direction of the magnetic field (in ferromagnetic substances, however, a minor fraction adopts an anti-parallel orientation). The mutual interaction between the individual electron spins is so strong that even after removal of the external magnetic field the spins remain aligned, exhibiting permanent magnetization. Upon field reversal, the ferromagnetic material will initially oppose the field change but, eventually, most domains will switch their magnetization vectors and the same inverse magnetization is attained. Examples of ferromagnetic structures are metallic iron, nickel and cobalt. Note that many authors use the terms 'ferroand ferrimagnetic' in a loose, practical sense. A solid-state physicist, however, classifies the iron oxide magnetite, which is used in many applications (see below), as ferrimagnetic.

On the other hand, in *paramagnetic* substances, a much weaker interaction between the electronic spins of neighbouring atoms or molecules occurs. Only in the case of very strong magnetic fields is magnetic saturation achieved and, at this point, the induced magnetic moment sometimes equals that obtained with ferromagnets. Examples of paramagnetic materials include organic free radicals and many transition metal complexes, *e.g.* those in which gadolinium ions are tethered.

In contrast, *diamagnetic* materials do not contain unpaired electrons. They are often simply called 'non-magnetic' although they are, in fact, very weakly repelled in an external magnetic field. Most materials are diamagnetic, including water, polymers, glass and many biomolecules such as proteins, nucleic acids, sugars (*e.g.* wood) and (phospho)lipids.

In exploiting the magnetic properties of materials for biomedical purposes, mainly ferromagnetic nanoparticles have attracted the most attention as they are extremely well suited as contrast agents for MRI or as colloidal mediators for cancer hyperthermia. In particular, two very commonly used iron oxides are magnetite (Fe_3O_4) and an oxidized product thereof, maghemite (γ-Fe_2O_3). Particles about 50 nm in diameter act as single monodomain magnetic particles. Above this critical size, the colloidal structure consists of multiple monodomains all having their own magnetization. When exposed to an external magnetic field, a so-called 'domain wall displacement' will occur and, as will be shown below, this has important consequences for their applicability in biomedicine.

3.2.3 Superparamagnetic Fluids

In this chapter, we mainly focus on superparamagnetism. Subdomain iron oxide particles with a diameter of, say, 15 nm are so small that the co-operative phenomenon leading to ferromagnetism is no longer observed. As a result, 'ordinary' magnetic fields are unable to withdraw the particles from solution and so-called high-gradient magnetic fields (HGMs) are needed.[28] On the other hand, a very interesting feature is that the absence of any remnant magnetism circumvents long-lasting, magnetically induced particle clustering and therefore the particles spontaneously resuspend in solution without the need for demagnetization or stirring devices. The MLs that we describe below can be safely classified as demonstrating superparamagnetic properties.

3.2.4 Preparation of Magnetoliposomes

For the production of MLs, we start from Fe_3O_4 cores with a diameter of about 14 nm and which are prepared as a black slurry by wet coprecipitation of Fe^{2+} and Fe^{3+} salts in the presence of ammonia. To improve their stability, amphiphilic laurate ions are attached on the surface, thereby creating a water-adapted magnetic fluid. Apart from a slow oxidation process in which Fe_3O_4 is converted to γ-Fe_2O_3, but which only slightly modifies the magnetic features, Brownian motion provides a solution that is physically stable for several years without showing any tendency for precipitate formation.

MLs are then created by co-incubation and dialysis of this water-compatible magnetic fluid with sonicated small unilamellar vesicles (SUVs) at a

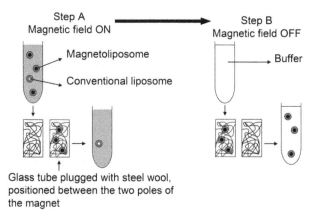

Figure 3.1 Schematic overview of the sequence for separating magnetoliposomes from iron oxide-free vesicles by high-gradient magnetophoresis. Reproduced with permission from Ref. 55. Copyright 2008 IOP Publishing.

temperature that is well above the gel-to-liquid crystalline transition temperature of the lipids involved. For instance, by selecting dimyristoylphospholipids, a working temperature of 37 °C is appropriate. Mechanistically, it has been proved that the process of ML bilayer formation is controlled by spontaneous transfer of phospholipids according to the so-called aqueous transfer model.[12,13] As MLs are in fact 'ultrasmall superparamagnetic iron oxides' (USPIOs), a high-gradient magnetic field (HGM) is needed to separate the particles from the excess of SUVs. The HGM set-up consists of loosely packed, corrosion resistant, micrometre-sized stainless-steel fibres that are placed in a conduit tubing, which in turn is positioned between the two poles of an (electro)magnet. Magnetic filtration then occurs by pumping the incubation mixture through this device by means of a peristaltic pump and collecting the MLs in a buffer stream after switching off the magnetic field (Figure 3.1). More details of the synthesis and separation protocol and also some practical hints can be found elsewhere.[5,28,29]

3.2.5   Cationic Magnetoliposomes

The above-described procedure works very well for the synthesis of anionic and zwitterionic MLs. Experimentally, however, we found that a more complicated pathway has to be followed for the optimal generation of cationic MLs.[30] (As will be shown below, this type of ML is of paramount importance if a high, non-specific ML internalization in biological cells has to be achieved.) Based on information gathered from a qualitative analysis of the electrostatic potential profiles of the intermediate colloidal nanostructures that are involved in the ML formation process, the synthesis protocol has to be adapted and two consecutive steps need to be included: neutral MLs are first prepared with zwitterionic phospholipids, which are then transformed to cationic phospholipids in the presence of SUVs containing a cationic surfactant [for instance, 1,2-dioleoyl-3-trimethylammoniumpropane (DOTAP) or its distearoyl analogue (DSTAP)].[30] During this latter step, spontaneous intermembrane transfer of lipids, residing in the outer leaflet of the bilayer, occurs.[13,29] After reaching an equilibrium state, the desired cationic MLs are withdrawn from the mixture by HGM as mentioned above.

3.2.6 Characterization of Magnetoliposomes

Data derived from phosphate and iron determinations qualitatively prove that phospholipids are really associated with the iron oxide grains. X-ray diffraction patterns further testify that the iron oxide core consists of Fe_3O_4.[21,31] The presence of a bilayered phospholipid coating can be deduced from both theoretical calculations and experimental observations.[5] Given the equality in size (diameter ~ 14 nm) of the Fe_3O_4 cores and of the internal volume of SUVs, it can be reasonably assumed that both structures contain the same amount of phospholipids and that the molar ratio of lipids in the inner/outer leaflet of the highly curved bilayer is about 1:2. Ultimately, based on geometric data on the size of individual phospholipid molecules, the thickness of a bilayer and the density of magnetite, a theoretical ratio of 0.73 mmol of phospholipid per gram of Fe_3O_4 was calculated in the case of an intact bilayer envelope.[5,29]

The theoretical support for the existence of a bilayer configuration, as described in the preceding paragraph, is also supported experimentally.[5] A first piece of evidence comes from transmission electron micrographs (see Chapter 2) of MLs, negatively stained with sodium phosphotungstate (Figure 3.2). The particles have an electron-dense nucleus, surrounded by a sheet of electron-transparent material and then, at the outside, a thin electron-dense zone. This sequence is in accordance with pictures of MLs presented by Gonzales and Krishnan.[31] Consideration of the chemical moieties involved in the complexes suggests that the dark nuclei represent the Fe_3O_4 nanospheres and that the electron light and dark sheets are the hydrophobic interior of the lipid bilayer and the polar headgroups of the outer leaflet phospholipids, respectively.

Another argument in favour of a bilayered architecture of the lipid coat comes from the shapes of the adsorption isotherms, which clearly indicate that first third of the phospholipids (inner layer) are adsorbed on the iron oxide surface in a high-affinity mode, whereas binding of the remaining two-thirds (outer layer) obeys Langmuir adsorption mathematics.[5,13] Also, the bilayered architecture of the lipidic coat can be deduced from experiments in which MLs were treated with Tween 20 (a non-ionic detergent)[5] or with selected organic solvent mixtures.[32] In both cases it was shown that it is difficult to desorb approximately one-third of the amount of phospholipids in the MLs, pointing to strong chemisorption forces between the phospholipid phosphate and the iron at the magnetite surface. This picture is also

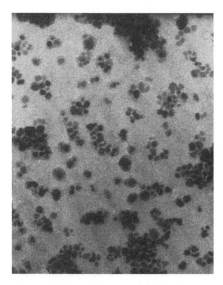

Figure 3.2 Transmission electron micrograph of magnetoliposomes, negatively
stained with sodium phosphotungstate. The diameter of the dark iron
oxide core is 14 nm.

compatible with observed differences in the phase transition temperatures
of the inner and outer leaflet phospholipids (unpublished observations). In
intramembranously mixed dimyristoylphosphatidylcholine (DMPC)–dimyristoylphosphatidylglycerol (DMPG) (molar ratio 9:1) MLs, for instance, about
two-thirds of the phospholipids 'melt' at a temperature corresponding to the
value found for these lipids in liposomes (23 °C),[12] whereas the gel-to-liquid
crystalline phase transition temperature for the remaining one-third, inner
leaflet lipids is located at a significantly higher temperature (31 °C). The
latter value can be understood in terms of reduced repulsion of the polar
headgroups as they are strongly fixed on the solid surface, which in turn
results in a stabilized lipid packing. The existence of two phospholipid
populations can also be deduced from transfer experiments, showing that
the outer layer lipids fully participate in intermembrane transfer events, in
contrast to the inner layer lipids, which are reluctant to move.[14] This behaviour makes MLs most interesting for creating asymmetrically labelled
nanocolloids. For instance, by first incubating the iron oxide cores with a
limited amount of vesicles (corresponding to the amount of phospholipids
needed to build up the inner phospholipid leaflet) containing a fatty acid-
labelled, fluorescent phospholipid (*e.g.* 1% β-BODIPY FL C5-HPC) and
then further incubating the monolayer structures with an excess of vesicles
tagged with another headgroup-labelled phospholipid species [*e.g.* 0.1 mol%
dipalmitoylphosphatidylethanolamine (DPPE)–Lissamine Rhodamine B or
DPPE–Texas Red] dual-tagged nanostructures are produced, allowing one
to monitor separately the behaviour and fate of inner and outer
phospholipids.[7]

3.3   Magnetoliposome Uptake in Cells

In the first experiments in which the interaction of MLs with cells was investigated, murine 3T3 fibroblasts were selected as a standard cell type, because it is a fairly robust cell type and it grows at a relatively fast rate. Among the various physical parameters investigated, we mainly focused on the most important ones: the ML's surface charge, the type and content of the cationic surfactant used and the incubation time. With respect to the incubation time, it is of the utmost importance to follow carefully the kinetics of intracellular iron uptake as the iron content will change in a time-dependent way.

3.3.1   Effect of Surface Charge

Particle internalization was investigated by using neutral DMPC, anionic DMPC–DMPG (90:10 molar ratio) and cationic DMPC–DOTAP (83.34:16.66 molar ratio) MLs.[33] When the cells were incubated for 24 h with anionic or neutral MLs at an Fe concentration of 100 µg mL^{-1}, only a few blue–green dots, located in the cytoplasm and axons, were seen in the optical micrographs of Prussian Blue-stained 3T3 fibroblasts (Figure 3.3). In contrast, when cationic MLs were used, the cells were heavily loaded with the ML colloids. Light microscopic images did not definitely prove whether the MLs were internalized or remained adsorbed at the external side of the cellular membrane. To distinguish between these two possibilities, transmission electron microscopy (TEM) images were taken of 3T3 fibroblasts that were incubated for 24 h with cationic 16.66 mol% DOTAP MLs, and these confirmed that internalization indeed occurred. As expected, it was observed that the MLs were found in lysosomal structures.

3.3.2   Effect of Cationic Surfactant Content

To examine the effect of cationic lipid composition of the ML coating on uptake efficiency, four different types of MLs were prepared that carried 0, 1.66, 6.66 or 16.66 mol% DOTAP in the outer shell of the phospholipid bilayer.[33] The four different ML populations were incubated with 3T3 fibroblasts for various times at an Fe concentration of 100 µg mL^{-1}, then the cells were fixed and stained with Prussian Blue and the extent of ML uptake was visualized by optical microscopy. No significant uptake was seen with neutral MLs (100% DMPC), or with the slightly cationic MLs. Only after 24 h of incubation were a few green–blue pigment spots visible. In contrast, massive uptake was seen on using the 16.66 mol% DOTAP MLs, in particular at longer incubation times (4, 8 and 24 h). Quantitative data on iron uptake per cell revealed a maximum internalization of over 50 pg Fe per cell after 8 h of incubation but about half of the cells were lysed.

To reduce the toxicity brought about by DOTAP, the distearoyl analogue (DSTAP) was investigated as an alternative.[33] The rationale behind this

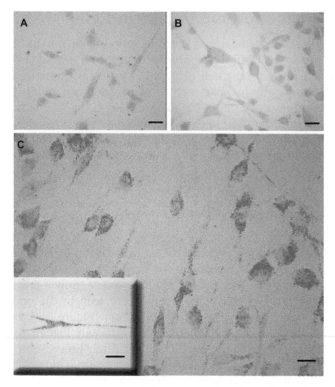

Figure 3.3   Prussian Blue staining of 3T3 fibroblasts that were incubated for 24 h
with (A) neutral DMPC MLs, (B) anionic DMPC–DMPG MLs (90:10
molar ratio) or (C) cationic DMPC–DOTAP MLs (83.34:16:66 molar
ratio). The inset in (C) shows an isolated cell that has a uniform
coloration apart from the cell nucleus. Scale bars: 25 μm.
Reprinted with permission from Ref. 33. Copyright 2007 Wiley-VCH.

choice is based on the knowledge that, in general, amphiphiles containing
saturated fatty acyl chains (such as DSTAP) fit better in a membrane than
those containing non-linear, unsaturated chains (such as DOTAP), which
distort an ideal fatty acyl chain packing within a membrane structure.[13,14]
Hence it is assumed that DSTAP will be more strongly anchored within a
vesicle membrane and, consequently, it will arrive more slowly, for instance,
in the acceptor cytoplasm membrane where it will possibly deregulate the
membrane potential to a lesser extent. Also, since under the experimental
conditions applied (37 °C) DSTAP is well below its gel-to-liquid crystalline
phase transition temperature (65 °C),[34] most probably DSTAP is segregated
in small patches (so-called rafts), thereby significantly retarding the escape
kinetics also.[13,14]

   To verify if this theoretical reasoning tallies with experimental observations, the effect of DSTAP was tested in an approach identical with that
described above for DOTAP. At all cationic surfactant contents tested (0,
1.66, 3.33, 6.66 and 16.66 mol%), the ML uptake was considerably higher

than that obtained by using the corresponding DOTAP-containing MLs (exact numbers at the various DSTAP contents can be found elsewhere[33]). In addition, and most interestingly, it was further found that at the same cationic surfactant content of the donor vesicles and at the same incubation time, the cell viability was less affected with DSTAP than DOTAP.

3.3.3 Incubation of Cationic Magnetoliposomes with Various Cell Types

In addition to 3T3 fibroblasts, uptake of cationic MLs was also followed with N2a neuroblastoma cells, HEK-293 (human embryonic kidney) cells, Gene-Switch TM-CHO (Chinese hamster ovary) cells and HUVECs (human umbilical vein endothelial cells).[34] Using standardized conditions (3.33% DSTAP-containing MLs; incubation time 4 and 24 h; 100 μg Fe mL^{-1}), some variability in uptake efficiency between the different cell types was observed. Fortunately, the particles seem not to be toxic in these experimental circumstances. This strongly suggests that 3.33% DSTAP-containing MLs can be considered as a 'universal' particle that can be employed to label many different cell types efficiently and safely. Interestingly, the variations in iron levels seem to depend on the general transfection efficacies of the cell types used. Internalization was lowest with HUVECs, which are primary cells known to be fairly resistant to transfection, but still reached an impressive 28.63 ± 1.83 pg Fe per cell after 4 h of incubation, which is significantly higher than the uptake levels reported with, for instance, dextran-coated particles in combination with transfection agents.[35,36] The highest uptake $(70.39 \pm 4.52$ pg Fe per cell after 4 h) was found with GeneSwitch TM-CHO cells, which are generally known to be easily transfected.

3.3.4 Uptake Mechanism

Upon incubation of increasing amounts of cationic MLs with fibroblasts, a saturation profile was obtained, *i.e.* the increment in iron uptake decreased as the ML concentration increased. This feature is indicative of unspecific receptor-mediated endocytosis. The actual mechanism by which uptake occurred was unravelled by testing the effect of both specific and unspecific endocytosis inhibitors or combinations thereof and comparing the results with data obtained at 4 °C where active endocytosis no longer occurs.[34] Examples of unspecific inhibitors that were used are cytochalasin D (Cyt D) and Nocodazole (Noco). The specific ones used to determine the actual pathway followed during internalization were monodansylcadaverine (MDC) (which is a competitive inhibitor of transglutaminase, which is essential for the production of clathrin-coated vesicles), filipin III (Fil), which removes cholesterol from the plasma membrane and thus disturbs several internalization pathways,) and genistein (Gen), which is a specific inhibitor of caveolae-mediated endocytosis. Based on these experiments, it could be concluded that ML internalization primarily occurred by clathrin-mediated endocytosis, although macropinocytosis also has to be considered as a minor, secondary, uptake route.

3.3.5 Cytotoxicity

As illustrated above, cytotoxicity is of paramount importance in issues dealing with particle confrontation with biological cells. Theoretically, every parameter that determines the cell's viability and that is influenced by the nanoparticles may endanger a particle's biocompatibility. Just a few of these potentially critical parameters include the size of the particles, the type of coating, the interaction of cell medium components with the particle surface, the chemical composition of the core and the applied dose within a certain time frame. In addition, toxicity may also vary strongly for different cell types and the data obtained may also depend on the experimental method(s) used to detect overall cellular wellbeing.[37–39]

From the experiments with murine 3T3 fibroblasts, murine C17.2 neural progenitor cells and human blood outgrowth endothelial cells (hBOECs) and in which cationic MLs were included, it was observed that high intracellular iron oxide nanoparticle concentrations affect the cell physiology (cell proliferation, migration and maturation) in a concentration-dependent manner *via* alterations in the microtubulin/actin cytoskeleton, changes in the formation and maturation of focal adhesion complexes (FACs), the activity of focal adhesion kinase (FAK) and modified regulation of protein expression levels.[39–41] Overall, the observed concentration-dependent effects may indicate that biological cells have an intrinsic limit on the amount of non-degradable nanomaterial that they can incorporate without any harmful alterations occurring.

Strong evidence has also been delivered that reactive oxygen species (ROS) are involved upon incubation of cells with cationic MLs. A more detailed study of this issue revealed that generation of ROS induces a Ca^{2+} influx due to destabilization of the cell plasma membrane upon transfer of the cationic lipid from the ML bilayer into the plasma membrane.[9,37] Alternatively, at high iron oxide nanoparticle payloads, release of ferric ions may also originate from degradation of the iron oxide core itself in the acidic endolysosomal compartments of the cell.[41] Iron-induced ROS can be easily assessed with desferrioxamine, an iron chelator that prevents their production. It should be stressed that, compared with many other commonly used iron oxide nanoparticles, this effect is far less expressed with MLs, which are covered with a robust bilayered phospholipid shield.[42] The hypothesis of iron-induced ROS production is further backed by the altered TfR1 (transferrin-receptor) expression in the cell. Chen *et al.*[43] reported that the release of free iron from endolysosomes can be a concern for cell homeostasis. In their study, human mesenchymal stem cells (hMSCs) were driven into osteogenic differentiation after labelling them with Resovist (a carboxydextran-coated iron oxide nanoparticle). The results showed that osteogenesis was blocked as a function of the concentration of Resovist used.

In conclusion, any active research on improving the biocompatibility of nanoparticles should focus on the intracellular nanoparticle concentrations achieved, and maximum safe levels for the particles in the cells under investigation should be determined.

3.4   Targeting of Magnetoliposomes

An important issue in using MLs as drug vehicles is that they will leave their cargo at the site(s) envisaged. MLs released into the bloodstream will be distributed within the body, proportionally to the regional blood flow. In practice, however, there are also multiple biological barriers, *e.g.* other organs, cells and intracellular compartments, where they can be neutralized or can cause unpleasant side effects. Therefore, if one wants to implement nanoparticles in therapeutic protocols, *e.g.* in cancer treatment schemes, it is highly desirable that solely tumour cells are exposed to these pernicious particles. Theoretically, this goal can be achieved by specifically targeting the nanoparticles towards the malignant cells or tissues. The main advantages of such a targeting strategy are threefold: (i) the particles can be administered by a method of choice; (ii) the quantity of the MLs can be sharply reduced so that the therapy cost also decreases; and (iii) the concentration in the targeted area can be increased without or with minimal negative side effects in non-target compartments.

In general, targeting strategies can be divided into two main streams: the active system, which will be briefly discussed later, and the passive system.[44] These mechanisms have been fairly well studied for classical liposomes, but not yet in depth for magnetoliposomes. Hence this field has still to be largely explored.

An important *passive* mechanism is the so-called enhanced permeability and retention (EPR) effect. The endothelium of blood vessels in tumours is more permeable than in a normal state. In this condition, the nanosized MLs can leave the vascular blood bed more easily because of the enlarged fenestrations. Another technique, which relies on the ferrimagnetic characteristics of MLs, uses an external magnetic field, so that MLs are pulled to the targeted area *via* magnetic force.[19] This can be done fairly easily in tissues located at the surface or in small animals. However, as magnetic forces generally are short range and small compared with hydrodynamic forces in the body, it remains a major experimental challenge to treat tissues that are located more deeply within the body of larger animals or in the human body. In an attempt to overcome this limitation, the concept of insertion of a ferromagnetic implant (so-called ferromagnetic doughnuts or 'nano-dockers') to create very locally a high-gradient magnetic field near the target site has been successfully tested.[45] In a third approach, so-called pH-sensitive MLs could be exploited. Tumour tissue environments are known to be acidic. If, for instance, an acidic medium-cleavable poly(ethylene glycol) (PEG)–lipid adduct on top of the iron oxide particle is used as a stabilizing substance, the labile linkage will only disintegrate due to proteases, secreted around the tumour tissue, thereby selectively favouring the particle's uptake.[46]

Active targeting, on the other hand, makes use of specific ligand–receptor interactions. In this case, the basic principle is to decorate the ML's surface with a vector for which the cell has a receptor. As specificity is an important characteristic of active targeting, it is clear that cationic MLs, generating a high background uptake, are not well suited here. Also, to improve the ML's blood circulation time, thereby increasing the chance of interaction of the ML with the receptor during its journey through the body, it is strongly advisable to incorporate a limited number (about 5 mol%) of phospholipids in which the polar headgroup has been modified with a water-soluble, functionalized flexible polymer chain, such as PEG.[47,48] The continual conformational change of the polymer at the particle's surface hinders opsonization and recognition of the so-called Stealth MLs by the reticuloendothelial system. For steric reasons, binding of the vector to the terminus of an activatable polymer may further improve the accessibility of the receptor.[49]

To combat numerous malignancies, various vectors such as monoclonal antibodies, lectins, lipoproteins, hormones and peptides have already been successfully evaluated in the past after linking them to magnetic iron oxide particles[1,2] and, therefore, they are assumed to be promising candidates to be linked to MLs also. The folate receptor, for instance, is extensively expressed in tumour cells, hence MLs covered with folate should have great potential to tackle tumour cells. A serious drawback of most coupling protocols, however, is that for each application, a new tailor-made derivative must be synthesized.[50] Fortunately, this cumbersome work can be circumvented by using a universal precursor complex that is able of binding a wide variety of ligands. A phospholipid–PEG–biotin complex is a good candidate for achieving this goal, as it interacts extremely well with the broad range of commercially available (strept)avidinylated proteins ($K_{ass} \approx 10^{15}$; see Ref. 51). The proof-of-concept that the binding strategy indeed works was delivered with alkaline phosphatase chosen as a model protein because it can be readily monitored spectrophotometrically in a very sensitive way.[51] This approach, elaborated with relevant ligands and/or monoclonal antibodies, opens up very promising routes in the development of powerful targeting systems in which tailoring of Stealth MLs to a variety of patients' disease profiles is no longer time consuming.

3.5   Magnetoliposomes as Theranostics

After successful internalization of MLs in the desired cell type(s) or tissue(s), some new avenues may open up in treating malignancies. First, with respect to the diagnosis aspect, the iron oxide core allows localization of MLs to be monitored by MRI. Second, if the MLs are loaded with chemotherapeutic drugs or if the tissue containing the MLs is put in an alternating magnetic field to create heat in a controllable manner, conditions for therapy are fulfilled. Ultimately, within a single particle, the potential to perform diagnosis and therapy is combined, hence MLs can be considered as smart 'thera(g)nostic' nanoparticles.

3.5.1 Diagnosis: Magnetoliposomes as MRI T_2 Contrast Agents

The first *in vivo* applications with the original MLs were performed by Bulte *et al.*,[52] aiming to visualize anatomical structures by MRI. It was shown that MRI of bone marrow following injection of MLs did not show significant differences between MLs and those that were PEGylated. This result, together with the finding that both preparations showed maximum uptake within the first 30 min following injection, indicates that short-circulating MLs can have a high affinity for bone marrow. The bone marrow uptake of ML and ML–PEG in older rats was found to be reduced, corresponding to a conversion of red to yellow bone marrow such as that occurring during a normal ageing process.

In applications in which particle uptake in cells is required, the intrinsic limitation of cells regarding the amount of internalized iron oxide nanoparticles drastically impedes any progress in MRI-mediated tracking of cells. Indeed, relatively high doses of contrast agent are required to achieve good contrast as MRI is intrinsically an insensitive technique. To improve contrast, theoretically larger iron oxide cores could also be used. When magnetic cores are clustered, the magnetic dipoles tend to couple and form one larger magnetic dipole, which significantly reduces T_2 relaxation times. However, larger particles are only poorly internalized by non-phagocytotic cells and are therefore seldom used for cell labelling.[53] Therefore, in an ideal situation, the particles would have to be small (<50 nm) to allow sufficient cellular uptake, but after internalization they should form larger aggregates to promote T_2 contrast. The unique features of the MLs presented in this review permit a close resemblance to this ideal situation (Figure 3.4). In phantom setups,[8] it was indeed found that lysosomal phospholipase A_2 degraded the outer leaflet of the bilayer fairly rapidly, but the inner lipid layer remained intact, most probably owing to an unfavourable orientation of the lipids, leading to monolayer-coated MLs in which the adsorbed phospholipids are directed with their apolar fatty acyl chains towards the aqueous phase. Owing to the hydrophobic effect,[13] the particles become prone to aggregation in aqueous media. The clustering leads to a sixfold increase in hydrodynamic diameter and results in visual aggregation when the dispersion is exposed to an external magnetic field. With respect to MRI, both T_2^* and T_2 relaxation are significantly enhanced (almost threefold). In C17.2 neuronal progenitor cells, for instance, the MLs form large intracellular clusters (over 1 μm in size), which increases their long-term detection compared with commercially available dextran-coated iron oxide cores (Feridex). MLs therefore encompass the combined benefits of small sizes for efficient cell uptake and large intracellular clusters for enhanced, long-term detection.[8] Thus, owing to the amphiphilic nature of the constituting coating molecules and their very defined orientation at the particle's surface, carefully designed MLs may enjoy a privileged position for diagnostic purposes.

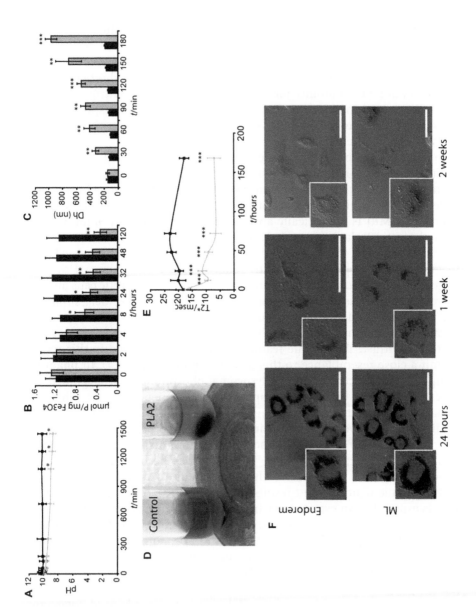

3.5.2 Therapeutic Aspects

3.5.2.1 Magnetoliposomes as Drug Carriers

In addition to their diagnostic value, MLs also have great potential as therapeutic tools, for instance, as a stable delivery system of drug molecules. It must be realized, however, that in contrast to, for instance, extruded MLs, the internal space of the MLs discussed here is completely filled with an iron oxide core. Consequently, their application as drug-delivery systems will be mainly restricted to apolar molecules that are embedded in the phospholipid bilayer, although appending by chemical means *water-soluble* drugs at the exterior of the ML particle should also be feasible.

In general, for *water-insoluble* molecules, a thorough knowledge of their partitioning in the ML's coat is of prime importance for optimizing MLs as efficient and reliable drug carriers. In this respect, it has been shown that both temperature and pH conditions drastically influence the binding strength. For instance, Huth *et al.*[54] demonstrated that thermoresponsive nanocomposites could be prepared by first immobilizing a phospholipid layer on the surface of superparamagnetic Fe_3O_4 nanoparticles, which was then loaded with the hydrophobic dye dansylcadaverine, used as a drug surrogate. It was found that release of the latter molecule was temperature dependent. Above the experimentally determined 'melting' temperature of the phospholipids, the release rate of the 'drug' was about 2.5-fold higher than below this temperature. By carefully selecting the phospholipid composition of the ML coat, this critical temperature can be fixed at, for instance, 39 °C, which is easily reached in cancer tissues.

In addition to the temperature, the pH of the medium is also of importance for the degree of anchoring of an ionizable drug molecule within a

---

Figure 3.4 Clustering of MLs. (A–E) Effect of phospholipase A_2 (PLA_2) activity on MLs as indicated by (A) a decrease in pH of PLA_2-treated samples (grey) compared with control sample (black); (B) ratio of phosphate to magnetite for PLA_2-treated samples (grey) and control samples (black); (C) hydrodynamic diameters of PLA_2-treated samples (grey) and control samples (black); (D) optical micrograph of control particles and PLA_2-treated particles upon exposure to a 1 T magnetic field for 30 min; (E) T2* relaxation times for PLA_2-treated samples (grey) and control samples (black), indicating the enhanced effect of induced clustering of the MLs on MR contrast generation. (F) Representative optical micrographs of C17.2 NPCs labeled with Endorem, a dextran-coated iron oxide nanoparticle (top row) or MLs (bottom row) during 24 h, finally reaching a similar average intracellular iron content. Media were removed and the cells were kept in culture for the duration indicated. Cells were stained for iron oxide using DAB-enhanced Prussian Blue reagent and imaged for 24 h (left column), 1 week (middle column) and 2 weeks (right column) after nanoparticle incubation. Scale bars: 75 nm. The inset is an enlarged view of a single cell displayed in the main image.
Reprinted with permission from Ref. 42. Copyright 2010 Wiley-VCH.

membrane. In this respect, an in-depth study was elaborated with the cationic drug molecule (*R,S*)-propranolol (Ppn), which is used in medicine as a non-selective β-blocker to treat systemic hypertension.[55,56] It is assumed that the hydrophobic naphthalene moiety of the drug molecule is embedded in the acyl-chain interior of the bilayer while the polar headgroup stays near the surface of the membrane.[57] Binding experiments performed with MLs differing in anionic phospholipid content and at different salt concentrations revealed that, apart from a significant contribution of the hydrophobic effect, electrostatics are also important (Figure 3.5). Experimentally, the amount of Ppn sorbed in both weakly anionic DMPC–DMPG (molar ratio 95:5) and pure anionic DMPG MLs was determined after adding different Ppn concentrations in 5 mM TES buffer, pH 7.0, containing 0, 5, 10 or 75 mM KCl. Significantly higher Ppn amounts were captured by the 100 mol% DMPG MLs. These observations were rationalized as follows. Compared with the weakly anionic, mixed DMPC–DMPG MLs, the pure DMPG MLs attract much stronger Ppn as presumably more than 99% of the drug molecule is protonated and thus positively charged at physiological pH. (The pK_a of Ppn in water at 37 °C is 9.14.[58]) Thus, with pure DMPG MLs, the higher local Ppn concentration near the surface inevitably results in more sorption. Also, with both types of anionic MLs studied here, increasing electrolyte concentration leads to less attraction of the protonated Ppn and thus lower surface concentrations and smaller amounts sorbed. Alternatively, the surface

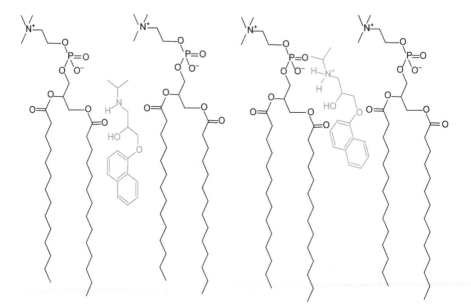

Figure 3.5 Schematic representation of incorporation on non-charged (left) and charged Ppn (right) among DMPC molecules, residing in the ML coat. Reprinted with permission from Ref. 56. Copyright 2008 Taylor and Francis.

concentration of unprotonated Ppn is not affected, either by a difference in surface charge of the ML or by changing the electrolyte concentration.

Although the exact pK_a of propranolol inserted in a negatively charged membrane is difficult to measure experimentally, a detailed picture of the intercalation of the drug at pH conditions below and above the supposed pK_a of Ppn was put forward by Avdeef *et al.*[57] It is assumed that in its neutral form the tertiary amino group of the drug is linked by hydrogen bridges to the glycerol backbone of the phospholipids [the drug being the H-bridge donor and the carbonyl group or (to a lesser extent) the ester oxygen of the phospholipid being the H-bridge acceptor]. At a pH below the pK_a of Ppn, the drug molecule becomes positively charged and, therefore, stronger electrostatic charge–charge interactions between the anionic phosphate group and the cationic amino group come into play. As a consequence, the molecule is pushed up to the aqueous exterior and this is what has been informally designated the 'pH-triggered piston' effect.[57]

Another, very interesting observation with respect to Ppn incorporation into the ML coat is that uptake of the drug is counterbalanced by a proportional expulsion of phospholipid molecules. This behaviour of the ML coat contrasts with that of the bilayer of classical liposomes. The latter are less susceptible to geometric constraints and can easily receive arriving Ppn molecules, but when a certain threshold uptake is exceeded, the vesicle's membrane, ultimately, is much more vulnerable than an ML phospholipid bilayer which is strongly anchored on the surface of the Fe_3O_4 grains.[59–61]

A similar incorporation study was also performed with 10-hydroxycamptothecin (HCPT), a promising anticancer agent, targeting the nuclear topoisomerase I.[62] As the molecule contains a highly hydrolysis-sensitive lactone ring, it absolutely needs an environment shielded from water and here the ML coat can act as an perfect host to preserve the drug's stability and, thus, also its pharmacological activity. Although the exact location of the molecule in the ML coat was not investigated in detail, by using MLs composed of different phospholipid types, we could deduce in an indirect way the importance of hydrogen bridges in positioning of HCPT at the membrane surface.[63] Also, it was observed that HCPT absorption is accompanied by a reduction in the membrane's phospholipid content, indicating that phospholipid molecules have to leave the ML coat to make free space to accommodate arriving HCPT molecules; in this case, one HCPT molecule replaces 2.6 phospholipid molecules.[63]

3.5.2.2 Hyperthermia with Magnetoliposomes

In addition to the use of drug molecules, hyperthermia (HT) is also a well-known approach to treat malignancies, and excellent papers describing the physical fundamentals behind the technique have been published on a regular base.[1,64,65] In short, once magnetizable nanoparticles arrive in a cell or tissue, they can be heated in an oscillating magnetic field. Energy dissipation occurs (i) by rotation of the particle as a whole owing to the torque exerted by the magnetic field on the magnetic moment (Brownian motion) and (ii) when the particle's magnetic moment relaxes to its equilibrium orientation (Néel relaxation). Depending on the intensity of the field and frequency of oscillation, the amount of heat can be carefully fine-tuned. Interestingly, it appears that the optimum diameter of iron oxide-based cores to generate maximum energy, mainly originating from Néel relaxation, is about 10–15 nm,[31,66] which is in the range of the cores we use to produce MLs (14 nm).

A first concept of using MLs in an inductive heating device was presented by Müller-Schulte *et al.*,[67] dealing with the irreversible inactivation of the AIDS virus. It is known that HIV is inactivated by temperatures above 50 °C. To ensure that the heat is transferred solely to the HIV and not to the adjacent tissue, the *in vivo* HIV infection pathway is mimicked, *i.e.* MLs are precoated with CD4 receptor molecules, thus allowing a close attachment of the ML to the HIV *via* its gp120 envelope protein. Using this approach, a means of virus inhibition should be provided which is not impaired by the high mutation rate of the virus.

The proof that HT can be effectively realized *in vivo* with MLs as heating mediators was delivered by Hamaguchi *et al.*,[68] who treated cervical lymph node metastasis of tongue cancer in rabbits. By 24 h after ML injection into the tongue, the highest magnetite concentration was detected in the lymph nodes. Upon applying an alternating magnetic field to the neck region, the temperature in the lymph nodes rose to 44 °C, resulting in a significantly higher percentage of necrosis and apoptosis in comparison with animals that did not receive MLs or where the HT treatment was omitted.

Another example showing the value of MLs to induce HT *in vivo* was presented by Kawai *et al.*,[69] who investigated rat prostate cancer. First, they injected prostate cancer cells into F344 rats. When these cells had grown to 5–6 nm in diameter, they were exposed to an oscillating magnetic field. The problem with other therapies such as HT using radiofrequency irradiation is the increase in body temperature, which exerts a massive burden on the patient. This was solved by the use of MLs. To test this approach, the authors evaluated the temperature at the centre of the tumour and inside the rectum during application of an alternating magnetic field for 30 min. At the centre of the tumour, the temperature reached 45 °C within 5 min and was maintained for 20 min by controlling the magnetic field intensity. In contrast, the temperature inside the rectum remained at 38 °C. This proves that HT with MLs is regional and will not damage healthy tissues. As discussed above, the disadvantage of MLs is that it is difficult to inject them into lesions that are located deep in the body or lesions with hypervascularity such as in hepatic cancer and renal tumours. For these cancer types, it is more appropriate to use HT with radiofrequency irradiation.

Interestingly, the recovery process is not only enhanced by heating the cancer cells but indirectly also by the induction of an immune response.

When a cancer cell is exposed to a temperature above 42 °C, it will die, but the higher the temperature, the lower is the percentage of cells undergoing apoptosis and the higher the percentage of cells undergoing necrosis. In the experiments of Kawai *et al.*,[69] it was clear that HT causes amplification of T lymphocytes and this would lead to destruction of the tumour. The difference between healthy and tumour tissue was that CD4 and CD8 T lymphocytes were detected in the tumour tissue but not in healthy tissue. The function of these T lymphocytes is that CD4 cells cause activation of CD8 cells into cytotoxic T lymphocyte-killing cells.

With regard to the necrosis process, it should be mentioned that not only T lymphocytes appear with HT, but heat shock protein 70 (HSP70) is also present. This protein causes thermotolerance, which means that HSP70 protects the cell from heat damage, and this plays a key role in the modulation of cell growth and differentiation. In contrast to the T lymphocytes, it acts as an inhibitor of apoptosis in prostate cancer. However, products exist that inhibit the production of HSP70, namely quercetin and KNK437. Sahin *et al.*[70] tested the different effects of these two products in prostate cancer cells. They concluded that KNK437 is more efficient than quercetin in hyperthermic therapy for this type of cancer.

In summary, it has been shown that hyperthermia with MLs is a very efficient method to treat cancerous tissues, because of the heat that it produces in an alternating magnetic field and the immune response that it causes. Of particular importance is that this approach opens up possibilities for treating not only primary but also metastatic lesions.

3.6   Conclusion

This chapter illustrates that MLs merge the fields of material science (physics, chemistry), biology and medicine. This interdisciplinary approach allowed us to create a unique magnetizable nanoparticle which provides a particularly useful platform with potentially wide-ranging diagnostic and therapeutic applications. As illustrated above, the use of well-designed and well-characterized MLs has clearly improved our understanding of the interactions of biological cells with iron oxide nanoparticles and has paved the way for elaborating protocols for more efficient and sustained labelling of a wide variety of cell types, leading to minimal cytotoxic effects. Of particular interest is that, in comparison with commonly used iron oxide formulations, MLs are more resistant to intracellular degradation of the iron oxide core and that the bilayered architecture of the ML envelope also permits clustering of the particles when present in lysosomes. Ultimately, these features result in a far better MR contrast remaining for a longer time, which is highly valuable, for example, in cell tracking experiments. Of course, many challenges still remain to be tackled, such as further fine-tuning of stimuli-triggered payload release in ML-based drug-delivery systems and the issue of ML safety *in vivo*.

References

1. S. Laurent, D. Forge, M. Port, A. Roch, C. Robic, L. Vander Elst and R. Muller, *Chem. Rev.*, 2008, **108**, 2064.
2. A. Louie, *Chem. Rev.*, 2010, **110**, 3146.
3. S. J. H. Soenen, M. Hodenius, T. Schmitz-Rode and M. De Cuyper, *J. Magn. Magn. Mater.*, 2008, **320**, 634.
4. L. Shen, P. E. Laibinis and T. A. Hatton, *Langmuir*, 1999, **15**, 447.
5. M. De Cuyper and M. Joniau, *Eur. Biophys. J.*, 1988, **15**, 311.
6. S. J. H. Soenen, M. Hodenius and M. De Cuyper, *Nanomedicine*, 2009, **4**, 177.
7. S. J. H. Soenen, D. Vercauteren, K. Braeckmans, W. Noppe, S. De Smedt and M. De Cuyper, *ChemBioChem*, 2009, **10**, 257.
8. S. J. Soenen, G. Vande Velde, A. Ketkar, U. Himmelreich and M. De Cuyper, *Nanomed. Nanobiotechnol.*, 2011, **3**, 197.
9. S. J. H. Soenen, A. R. Brisson, E. Jonckheere, N. Nuytten, S. Tan, U. Himmelreich and M. De Cuyper, *Biomaterials*, 2011, **32**, 1748.
10. L. B. Margolis, V. A. Namiot and L. M. Kljukin, *Biochim. Biophys. Acta*, 1983, **735**, 193.
11. M. De Cuyper and M. Joniau, *Arch. Int. Physiol. Biochim.*, 1987, **95**, B15.
12. M. De Cuyper, M. Joniau and H. Dangreau, *Biochemistry*, 1983, **22**, 415.
13. M. De Cuyper and M. Joniau, *Langmuir*, 1991, 7, 647.
14. M. De Cuyper and M. Joniau, *Biochim. Biophys. Acta*, 1990, **1027**, 172.
15. M. De Cuyper, B. De Meulenaer, P. Van der Meeren and P. J. Vanderdeelen, *Biocatal. Biotransform.*, 1995, **13**, 77.
16. M. De Cuyper, B. De Meulenaer, P. Van der Meeren and P. J. Vanderdeelen, *Biotechnol. Bioeng.*, 1996, **49**, 654.
17. R. P. Blakemore, *Annu. Rev. Microbiol.*, 1982, **36**, 217.
18. R. Sabaté, R. Barnadas-Rodriguez, J. Callejas-Fernández, R. Hidalgo-Álvarez and J. Estelrich, *Int. J. Pharm.*, 2008, **347**, 156.
19. S. Garcia-Jimeno, E. Escribano, J. Queralt and J. Estelrich, *Nanoscale Res. Lett.*, 2012, 7, 452.
20. J. C. Domingo, M. Mercadal, J. Petriz and M. A. Madariaga, *J. Microencapsul.*, 2001, **18**, 41.
21. M. De Cuyper, P. Müller, H. Lueken and M. Hodenius, *J. Phys.: Condens. Matter*, 2003, **15**, S1425.
22. Y. Chen, A. Bose and G. D. Bothun, *ACS Nano*, 2010, **4**, 3215.
23. D. Frascione, C. Diwoky, G. Almer, P. Opriessnig, C. Vonach, K. Gradauer, G. Leitinger, H. Mangge, R. Stollberger and R. Prassl, *Int. J. Nanomed.*, 2012, 7, 2349.
24. S. Mann and J. P. Hannington, *J. Colloid Interface Sci.*, 1988, **122**, 326.
25. K. P. Ho, L. Li, L. Zhao and Z. M. Qian, *Mol. Cell Biochem.*, 2003, **247**, 219.
26. O. V. Vasiljeva, O. B. Lyubitsky, G. I. Klebanov and Yu. A. Vladimirov, *Membr. Cell Biol.*, 1998, **12**, 223.
27. B. Tadolini, P. Motta and C. A. Rossi, *Biochem. Mol. Biol. Int.*, 1993, **29**, 299.

28. M. De Cuyper, in *Handbook of Nonmedical Applications of Liposomes – from Design to Microrectors*, ed. Y. Barenholz and D. D. Lasic, CRC Press, Boca Raton, FL, 1996, Vol. III, Chapter 18, pp. 325–342.
29. M. De Cuyper and S. J. H. Soenen, in *Methods in Molecular Biology*, ed. V. Weissig, Humana Press, Totawa, NJ, 2010, Vol. 605/1, Chapter 6, pp. 97–111.
30. M. De Cuyper, D. Caluwier, J. Baert, J. Cocquyt and P. Van der Meeren, *Z. Phys. Chem.*, 2006, **220**, 133.
31. M. Gonzales and K. M. Krishnan, *J. Magn. Magn. Mater.*, 2005, **293**, 265.
32. M. De Cuyper and W. Noppe, *J. Colloid Interface Sci.*, 1996, **182**, 478.
33. S. J. H. Soenen, J. Baert and M. De Cuyper, *ChemBioChem*, 2007, **8**, 2067.
34. B. A. Lobo, S. A. Rogers, S. Choosakoonkriang, J. G. Smith, G. Koe and C. R. Middaugh, *J. Pharm. Sci.*, 2002, **91**, 454.
35. J. W. M. Bulte, S. Zhang, P. van Gelderen, V. Herynek, E. K. Jordan, I. D. Ducan and J. Frank, *Proc. Natl. Acad. Sci. U. S. A.*, 1999, **96**, 15256.
36. P. Jendelová, V. Herynek, L. Urdziková, K. Glogarová, J. Kroupová, B. Andersson, V. Bryja, M. Burian, M. Hájek and E. Syková, *J. Neurosci. Res.*, 2004, **76**, 232.
37. S. J. H. Soenen and M. De Cuyper, *Contrast Media Mol. Imaging*, 2009, **4**, 207.
38. S. J. H. Soenen, A. R. Brisson and M. De Cuyper, *Biomaterials*, 2009, **30**, 3691.
39. S. J. Soenen, E. Illyes, D. Vercauteren, K. Braeckmans, Z. Mayer, S. C. De Smedt and M. De Cuyper, *Biomaterials*, 2009, **30**, 6803.
40. S. J. Soenen, N. Nuytten, S. F. De Meyer, S. C. De Smedt and M. De Cuyper, *Small*, 2010, **6**, 832.
41. S. J. Soenen, U. Himmelreich, N. Nuytten and M. De Cuyper, *Biomaterials*, 2011, **32**, 195.
42. S. J. Soenen, U. Himmelreich, N. Nuytten, T. R. Pisanic II, A. Ferrari and M. De Cuyper, *Small*, 2010, **6**, 2136.
43. Y. C. Chen, J. K. Hsiao, H. M. Liu, I. Y. Lai, M. Yao, S. C. Hsu, B.-S. Ko, Y.-C. Chen, C.-S. Yang and D.-M. Huang, *Toxicol. Appl. Pharmacol.*, 2010, **245**, 272.
44. C. C. Berry, *J. Phys. D: Appl. Phys.*, 2009, **42**, 224003.
45. M. O. Avilés, J. O. Mangual, A. D. Ebner and J. A. Ritter, *Int. J. Pharm.*, 2008, **361**, 202.
46. T. J. Harris, G. von Maltzahn, A. M. Derfus, E. Ruoslahti and S. N. Bhatia, *Angew. Chem. Int. Ed.*, 2006, **19**, 3161.
47. T. M. Allen, in *Long Circulating Liposomes: Old Drugs, New Therapeutics*, ed. C. M. Woodle and G. Storm, Springer, Berlin, 1998, Chapter 2, pp. 19–28.
48. N. Nuytten, M. Hakimhashemi, T. Ysenbaert, L. Defour, J. Trekker, S. J. H. Soenen, P. Van der Meeren and M. De Cuyper, *Colloids Surf. B*, 2010, **80**, 227.
49. F. Léon-Tamariz, I. Verbaeys, M. Van Boven, M. De Cuyper, J. Buyse, E. Clynen and M. Cokelaere, *Peptides*, 2007, **28**, 1003.

50. T. Viitala, W. M. Albers, I. Vikholm and J. Peltonen, *Langmuir*, 1998, **14**, 1272.
51. M. Hodenius, M. De Cuyper, L. Desender, D. Müller-Schulte, A. Steigel and H. Lueken, *Chem. Phys. Lipids*, 2002, **120**, 75.
52. J. W. M. Bulte, M. De Cuyper, D. Despres and J. A. Frank, *J. Magn. Reson. Imaging*, 1999, **9**, 329.
53. D. L. Thorek and A. Tsourkas, *Biomaterials*, 2008, **29**, 3583.
54. C. Huth, D. Shi, F. Wang, D. Carrahar, J. Lian, F. Lu, J. Zhang, R. C. Ewing and G. M. Pauletti, *Nano Life*, 2010, **1**, 251.
55. J. Cocquyt, S. J. H. Soenen, P. Saveyn, P. Van der Meeren and M. De Cuyper, *J. Phys.: Condens. Matter*, 2008, **20**, 204102.
56. S. J. H. Soenen, J. Cocquyt, L. Defour, P. Saveyn, P. Van der Meeren and M. De Cuyper, *Mater. Manuf. Processes*, 2008, **23**, 611.
57. A. Avdeef, K. J. Box, J. E. A. Comer, C. Hibbert and K. Y. Tam, *Pharm. Res.*, 1998, **15**, 209.
58. K. Balon, B. U. Riebesehl and B. W. Müller, *J. Pharm. Sci.*, 1999, **88**, 802.
59. B. Božic, V. Kralj-Iglic and S. Svetina, *Phys. Rev. E*, 2006, **73**, 041915-1– 041915-11.
60. S. De Carlo, H. Fiaux and C. A. Marca-Martinet, *J. Liposome Res.*, 2004, **14**, 61.
61. J. A. Rogers, S. Cheng and G. V. Betageri, *Biochem. Pharmacol.*, 1986, **35**, 2259.
62. B. B. Lundberg, *Anti-Cancer Drug Des.*, 1998, **13**, 453.
63. M. A. J. Hodenius, T. Schmitz-Rode, M. Baumann, G. Ivanova, J. E. Wong, T. Mang, F. Haulena, S. J. H. Soenen and M. De Cuyper, *Colloids Surf. A*, 2009, **343**, 20.
64. R. Hergt, R. Hiergeist, M. Zeisberger, D. Schüler, U. Heyen, I. Hilger and W. A. Kaiser, *J. Magn. Magn. Mater.*, 2005, **293**, 80.
65. R. Hergt, S. Dutz, R. Müller and M. Zeisberger, *J.Phys.: Condens. Matter*, 2006, **18**, S2919.
66. F. Gazeau, M. Lévy and C. Wilhelm, *Nanomedicine*, 2008, **3**, 831.
67. D. Müller-Schulte, F. Füssl, H. Lueken and M. De Cuyper, in *Scientific and Clinical Applications of Magnetic Carriers*, ed. U. Häfeli, W. Schütt, J. Teller and M. Zborowski, Plenum Press, New York, 1997, Chapter 39, pp. 517–526.
68. S. Hamaguchi, I. Tohnai, A. Ito, K. Mitsudo, T. Shigetomi, M. Ito, H. Honda, T. Kobayashi and M. Ueda, *Cancer Sci.*, 2003, **94**, 834.
69. N. Kawai, A. Ito, Y. Nakahara, M. Futakuchi, T. Shirai, H. Honda, T. Kobayashi and K. Kohri, *Prostate*, 2005, **64**, 373.
70. E. Sahin, M. Sahin, A. D. Sanlioglu and S. Gümüslü, *Int. J. Hyperthermia*, 2011, **27**, 63.

CHAPTER 4

Nanogels for Drug Delivery: the Key Role of Nanogel–Drug Interactions

JOSE RAMOS,[a] MIGUEL PELAEZ-FERNANDEZ,[b]
JACQUELINE FORCADA[a] AND ARTURO MONCHO-JORDA*[c]

[a] POLYMAT, Bionanoparticles Group, Departamento de Química Aplicada, UFI 11/56, Facultad de Ciencias Químicas, Universidad del País Vasco UPV/EHU, Apdo. 1072, 20080 Donostia-San Sebastián, Spain; [b] School of Physics, Georgia Institute of Technology, Atlanta, GA 30332, USA; [c] Departamento de Física Aplicada, Facultad de Ciencias, Universidad de Granada, 18071 Granada, Spain
*Email: moncho@ugr.es

4.1   Introduction

In the field of soft nanoparticles, micro-/nanogels represent a versatile type of responsive nanomaterials with promising applications. Nanogels are cross-linked colloidal particles with the ability to change their volume drastically in response to changes in their environment, *e.g.* pH, ionic strength, temperature, presence of specific ions and other compounds or light. or through external fields such electric and magnetic fields.[1–3] This enables nanogels to incorporate and release host molecules in a responsive manner, offering opportunities in drug delivery for small molecules and biomacromolecular drugs. In this sense, biomacromolecular drugs, notably proteins and peptides, are becoming increasingly important in drug

RSC Nanoscience & Nanotechnology No. 34
Soft Nanoparticles for Biomedical Applications
Edited by José Callejas-Fernández, Joan Estelrich, Manuel Quesada-Pérez and Jacqueline Forcada
© The Royal Society of Chemistry 2014
Published by the Royal Society of Chemistry, www.rsc.org

development. For such drugs, nanogels are the ideal nanocarriers because they may provide a range of advantages, including conformational stabilization and retained biological activity, protection from chemical and enzymatic degradation, control of drug release rate and reduction of toxicity, immunity and other biological side effects.[4,5] Although nanogels are good candidates for drug delivery, the interaction between the nanogel and the drug is the main cornerstone, controlling not only the uptake of the drug but also its release. Therefore, nanogel networks must be meticulously designed, focusing on the local cross-link density and the functional group distribution within the nanogel.[6,7]

Without any doubt, among the different responsive nanogels, temperature-sensitive nanogels are the most studied type because temperature is an effective stimulus in a number of applications. Nanogels that are able to undergo a volumetric phase change by changing the temperature of the dispersion medium are very interesting in biotechnological applications that require the delivery of an active compound or biomolecule in media in which the main variable to consider is temperature. The design and controlled production of thermo-responsive nanogels have attracted considerable interest owing to their unique feature to swell at low temperatures and being collapsed at high temperatures in aqueous solutions, showing a volume phase transition temperature (VPTT). This unique behaviour makes those nanogels with VPTTs near physiological temperature and based on a bio-compatible polymer very attractive for many potential bio-applications, such as drug delivery.[1] However, owing to the complexity of the drug-delivery systems, other kinds of nanocarriers are required, such as multi-responsive nanogels. This new generation of nanogels combine several stimuli, the most common ones being temperature, pH, ionic strength and the presence of some biomolecules or enzymes.

However, the diffusion and encapsulation of a certain solute inside the nanogel volume is a complex process that depends on many physical and chemical parameters. First, if we are treating charged solutes, the penetration of the solute depends on the electrostatic interaction with the ionic nanogel particle. Second, the diffusion inside the nanogel is also affected by the size of the incoming molecule and the pore size of the polymer network. Indeed, a large-sized solute will experience greater steric repulsion when trying to diffuse inside the nanogel than small solutes, regardless of the electrostatic interaction. On the other hand, a large pore size enhances the penetration of the solute. This means that nanogel particles in the deswollen state are, in principle, more difficult to be penetrated by solutes than swollen particles, provided that the electrostatic interaction and hydrophobic character are the same in both cases. Third, the penetration and attachment of the solute inside the particle can also be strongly affected by the specific interaction between the solute and the polymer, mainly caused by intense short-range hydrophobic or hydrophilic forces that emerge when the water shell surrounding the incoming molecule is lost in the neighbourhood of the polymer chains.

Finally, the stability of a nanogel suspension is also a key parameter when employed in drug delivery. An unstable nanogel suspension could deliver the drug before reaching the target, inducing health problems, or limit the delivery of the drug, making the treatment inefficient. The control of the stability of a nanogel suspension rests on the nanogel–nanogel interactions, which could vary depending on whether the nanogels are ionic or neutral or in the swollen or shrunken states.

This chapter is intended to highlight the importance of the interaction between nanogels and drugs with the objective of achieving a perfect drug-delivery system. Therefore, the first step requires a well-designed synthesis of the nanogel network, followed by an in-depth study of the interaction between the nanogel interior and the desired drug. These points are treated in Section 4.2. The different contributions to the bare nanogel–solute interactions are described in Section 4.3. Finally, Section 4.4 describes the most important nanogel–nanogel effective interactions and their role in colloidal stability.

4.2 Design of Nanogels for Drug Delivery

As noted above, prior to studying the interaction of the nanogel with a specific drug, the design of the entire nanogel structure must be very well planned. The main requirements that one has to bear in mind prior to the nanogel synthesis are the following: (i) particle size between 10 and 200 nm; (ii) biodegradability and, if possible, biocompatibility; (iii) prolonged blood circulation time; and (iv) high drug or enzyme loading and/or entrapment and protection of molecules from the body's immune system. From the synthesis point of view, basically two main parameters must be controlled: the cross-linking density and the distribution and type of chemical functional groups inside the nanogel network. However, from the point of view of the application, other considerations such as the type and size of the drug and its hydrophobicity greatly affect the process. In this sense, the same nanogel can be used for different drugs but with very different results. Hoare and Pelton published a very interesting study on the interaction of water-soluble drugs of different charges and hydrophilicities with carboxylic acid-functionalized poly(*N*-isopropylacrylamide) (PNIPAM) nanogels with different functional group distributions.[8] Their results showed that an increased hydrophobicity of the nanogel at higher temperatures leads to a partitioning of a larger percentage of the hydrophobic drug into the nanogel. However, for cationic drugs, the acid–base interactions between the cationic drug and the anionic nanogel have a significant impact on the efficacy of carboxylic acid-functionalized nanogels as drug-delivery vehicles.

4.2.1 Basics Concepts of the Synthesis of Nanogels

Nanogels have been synthesized by numerous approaches: (1) physical self-assembly of interactive polymers; (2) polymerization of monomers in a homogeneous phase or in a micro- or nanoscale heterogeneous environment; (3) cross-linking of preformed polymers; and (4) template-assisted nanofabrication of nanogel particles.[9] However, focusing attention on thermo-responsive nanogels, emulsion polymerization (also called precipitation polymerization) is the most versatile and common technique used to prepare these nanoparticles.

Different families of thermo-responsive nanogels have been synthesized by means of emulsion polymerization. Of these, it is worth mentioning poly(N-isopropylacrylamide) (PNIPAM)-based nanogels and poly(N-vinylcaprolactam) (PVCL)-based nanogels. Both families show similar thermal behaviour from a macroscopic point of view, but they differ in the phase transition.[10]

Briefly, nanogels synthesized by emulsion polymerization are formed by homogeneous nucleation. The polymerization is carried out at high temperature, always at a temperature higher than the lower critical solution temperature (LCST) of the polymer. Thus, the radicals coming from the initiator (typically sulfate radicals) are generated and they initiate polymerization. After initiation, NIPAM or VCL is attacked by the sulfate radical, and there then follow radical propagation and chain growth. Once the chain reaches a critical length, it collapses upon itself, producing precursor particles. Then the chain collapses because the polymerization temperature is higher than the LCST of the polymer. The precursor particles grow by aggregation with other precursor particles, by being captured by existing particles, by capturing growing radicals and by monomer addition. The charge imparted by the initiator stabilizes the nanogels once they have reached a critical size. This method is extremely versatile from the standpoint of particle size control. More details about the synthesis of thermo-responsive nanogels can be found in several reviews published in recent years.[2,11–16]

4.2.2 Thermo-Responsive Nanogels

Pelton and Chibante[17] reported the first synthesis of cross-linked PNIPAM nanogels in 1986. Since then, a large number of studies have been reported making some improvements, such as modification of the VPTT value or the cross-linking density.[11,12,18–21] Today, PNIPAM-based nanogels can be considered as the model nanogel because in the last 25 years a plethora of studies have been published and various research groups have made tremendous efforts in the development of this type of soft nanoparticles. However, in spite of the large number of articles and patents published on this type of nanogel, the toxicity of PNIPAM prevents their use in biomedical applications. Therefore, in recent years, there very interesting alternatives to PNIPAM have been developed. Among them, poly(N-vinylcaprolactam) (PVCL) has gained rapid success owing to its biocompatibility in the design of biocompatible thermo-sensitive nanogels.[2,3] Both PVCL and PNIPAM are water-soluble, non-ionic polymers and exhibit similar LCSTs. However, they differ in the mechanisms and thermodynamics of the phase transition. PNIPAM shows an almost complete independence of the critical temperature on the polymer chain length, which is a thermo-responsive phase behaviour in water Type II.[22] This means that the cloud point of PNIPAM aqueous solutions is only slightly affected by environmental conditions such as pH or polymer concentration. In contrast to PNIPAM, PVCL exhibits a 'classical' Flory–Huggins thermo-responsive behaviour in water (Type I) and, consequently, its LCST value decreases with both increasing polymer chain length and concentration. Thus, one can easily modify the cloud point of a PVCL-based thermo-responsive system by controlling the polymer molecular weight, with no requirement to use a comonomer.[23] Figure 4.1 shows the thermal response of PNIPAM- and PVCL-based nanogels. As can be seen, the hydrodynamic diameters of a PVCL-based nanogel are larger than those of a PNIPAM-based one [Figure 4.1(a)]. However, the swelling ratio, expressed as the volume ratio with respect to the shrunken nanogel, is similar in both nanogels showing the same VPTT [Figure 4.1(b)].

PVCL has a carboxylic and a cyclic amide hydrophilic group and a hydrophobic carbon–carbon backbone chain. The amide group is directly connected to the hydrophobic carbon–carbon backbone chain, so the hydrolysis of PVCL under strongly acidic conditions, if occurs, will produce a polymeric carboxylic acid and not a small toxic amide compound, which is unwanted for biomedical applications, as in the case of PNIPAM. Therefore, the use of PNIPAM as a biomaterial may be limited because of its higher cytotoxicity and its lower cell viability compared with PVCL,[24] hence its application in drug-delivery systems may be extremely restricted.[25] In addition, the biocompatibility of PVCL-based nanogels was also demonstrated by Imaz and Forcada,[26] confirming their suitability for use in biomedical applications.

Regarding the preparation of PVCL-based nanogels, Gao *et al.*[27] reported the first synthesis in 1999. Since then, the number of publications has been increasing continuously but it has not been possible to surpass the PNIPAM system.[28–32] In addition, the 'Achilles heel' of this system is the hydrolysis

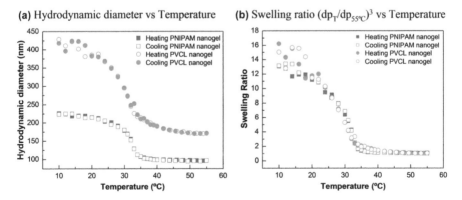

Figure 4.1 Heating and cooling cycles for a PNIPAM-based nanogel (blue) and a PVCL-based nanogel (red).
Reprinted with permission from Ref. 3. Copyright 2012 Royal Society of Chemistry.

that VCL suffers under acidic conditions, making the control of the poly-merization reaction more complex and, in some cases, hindering its func-tionalization. In this regard, Imaz *et al.*[30] proposed the use of sodium bicarbonate buffer to prevent VCL hydrolysis when potassium persulfate (KPS) was used as initiator.

4.2.3 Multi-Responsive Nanogels

Owing to the high demand for new nanocarriers for drug delivery, especially to fight cancer, the use of multi-responsive nanogels has been proposed as one of the best alternatives. The targeted design of the nanogels, especially the in-corporation of different functional groups and the control over their distri-bution inside the nanogel network, is becoming more and more important nowadays. The functionalization of thermo-responsive (PNIPAM- or PVCL-based) nanogels can achieve several objectives. The volume phase transition behaviour of the nanogel particles can be controlled *via* the incorporation of functional groups. Both the absolute value of the volume phase transition and the breadth of the deswelling transition can be influenced through co-polymerization with more hydrophilic or more hydrophobic monomers. Functionalization can also provide reactive sites for post-modification. In addition, other 'smart' environmental triggers can be incorporated into the nanogel particle to provide multivariable control over the particle swelling. These additional triggers may include pH (ionizable comonomers), ionic strength (charged comonomers) or light (photosensitive comonomers). However, by far the most studied nanogels from the point of view of the interaction with drugs are pH- and temperature-sensitive nanogels.

The preparation of pH- and temperature-responsive nanogels by emulsion copolymerization of a thermo-responsive monomer (*e.g.* NIPAM or VCL) with an ionizable comonomer is well reported in the literature. On the one hand, regarding NIPAM copolymerization, several studies can be found detailing the preparation of carboxylic acid-containing PNIPAM nanogels *via* free-radical copolymerization of NIPAM, using a cross-linker and different functional comonomers: acrylic acid (AA), methacrylic acid (MAA), vinyla-cetic acid (VAA), acrylamide (AM), maleic acid (MA), fumaric acid (FA) and allylacetic acid (AAA).[33–42] However, regarding VCL copolymerization, there appears to be only one study on the copolymerization of VCL with AA to obtain carboxylic acid-containing PVCL nanogels.[43] Hence we still have a long way to go in the case of PVCL-based nanogels, but it will surely be tackled in the near future.

4.3 Interactions Between Nanogels and Biomolecules or Drugs

In recent years, experimentalist and theoreticians have directed their at-tention to the investigation of the properties of ionic nanogels.[44] These particles are characterized by a cross-linked polymer network partially composed of charged monomers, distributed inside the nanogel. Examples of ionic nanogels are the previously cited ones obtained by copolymerization of VCL or NIPAM with an ionizable monomer, such as acrylic acid. Moreover, nanogels are permeable materials, so water molecules, ions and many other neutral and charged molecules can flow inside the particle volume, with subsequent changes in the size and net charge of the particle.

Their high permeability and the fact that the nanogel has a multi-responsive internal structure allow the application of this kind or soft particles in many fields of industry and biomedicine. The interior volume of these particles is accessible for incorporation of drugs, they have a large surface area exposed to the solvent and the pore size of the polymer network can be easily tuned in response to external stimuli. In addition, these particles can be prepared to fulfil the requisites of biodegradability and biocompatibility.[45–48] For these seasons, nanogel particles have been proposed to encapsulate active molecules for controlled delivery purposes and diagnostics.[49]

The application of nanogels to the transport of charged or neutral drugs inside the human body involves the attachment of the drug at some point in the network structure and then controlled release at some specific location. For instance, injections of hydrophobic drugs with low solubility (such as amphotericin B and paclitaxel) made with nanogels are less painful than injections of cosolvent-based formulations.[50] Most importantly, toxic drugs encapsulated inside nanogel particles have shown to have reduced toxicity.[47]

In general, the penetration and binding of the drug depend on a complex interplay between electric, steric and specific effects. In fact, the permeation varies widely for different drugs according to their affinity to the polymer network, hydrophobic/hydrophilic character, electrostatic nature and the pore size.[51] Therefore, a good understanding of the interactions between the drug and the polymer matrix is critical for achieving a proper encapsulation and delivery system. Two drugs that are encapsulated through different interactions show significant differences in release behaviour under the same conditions.[52,53] Thus procaine hydrochloride loaded *via* hydrophobic and hydrogen bonding shows higher delivery at high pH, whereas imipramine hydrochloride shows the opposite behaviour due to the electrostatic attraction to the carrier nanogel.

In order to predict the penetrability of ions or other charged or uncharged solutes inside nanogel particles and provide a theoretical background to understand all these experimental results, it is necessary to deduce the effective interaction potential between the nanogel and the incoming solute, $V_{ns}(r)$. This interaction potential depends on many parameters, such as the nanogel charge, the degree of swelling, the cross-linker distribution and the size of the solute. The main purpose of this section is to develop an explicit analytical expression for this interaction potential between a nanogel particle and a single solute molecule or ion (drug, biomolecule, *etc.*). In principle, this interaction is composed of three different terms: electrostatic, steric and that induced by short-range specific forces (such as the hydrophobic interaction). The interaction potential between a single nano-gel particle and a single charged solute can be written to a first approximation in the following additive form

$$V_{ns}\left(r\right) = V_{ns}^{elec}\left(r\right) + V_{ns}^{ster\text{-}spec}\left(r\right) \tag{4.1}$$

where $V_{ns}^{elec}\left(r\right)$ is the electrostatic interaction and $V_{ns}^{ster\text{-}spec}\left(r\right)$ is the interaction caused by the excluded volume repulsion and other existing short-range specific interactions between the solute and the fibres of the internal polymer network. These two contributions to the nanogel–solute interaction energy are calculated in the following subsections. In addition, their role in the permeation of the solute to the interior of the nanogel and the effect of the degree of swelling are also analyzed.

4.3.1 Nanogel–Solute Electrosteric Interaction

The electrostatic interaction is one of the most important contributions. For instance, the electrostatic attraction between the nanogel particle and an oppositely charged solute is usually a prerequisite to guarantee the solute penetration and attachment inside the polymer network.[54] It is well known that nanogel particles show a core–shell structure, having a more concentrated, highly cross-linked core surrounded by a fuzzy corona where the polymer density decays gradually to zero.[55-58] However, to simplify the analytical expression for the nanogel–solute interaction, an approximation for the polymer distribution is chosen here, assuming a homogeneous distribution. If R is the radius of the spherical nanogel, the packing fraction is given by

$$\phi(r) = \begin{cases} \phi & r \leq R \\ 0 & r > R \end{cases} \tag{4.2}$$

where r is the distance to the particle centre; ϕ is not a constant parameter, as it depends on the degree of swelling of the nanogel. By taking the deswollen state as the reference state (with a packing fraction ϕ_0 and particle radius R_0), the packing fraction for any other swollen configuration is

$$\phi = \phi_0 \left(\frac{R_0}{R}\right)^3 \tag{4.3}$$

The polymer volume fraction in the collapsed state is not totally compact, since some amount of solvent molecules remains trapped in the internal volume of the particle. Experimental results indicate that $\phi_0 \approx 0.7$, although this value can fluctuate depending on the internal structure of the cross-linked polymer network inside the particle.

Ionic nanogels contain ionizable comonomers distributed more or less uniformly inside the particle. Therefore, the charge density of the particle follows the same profile as the mass distribution, that is

$$\rho_e(r) = \begin{cases} \dfrac{3Ze}{4\pi R^3} & r \leq R \\ 0 & r > R \end{cases} \qquad (4.4)$$

where e is the elementary charge and Z is the total bare charge of the nanogel particle. The electric field generated by this uniformly charged sphere can be obtained by applying the Gauss law for electrostatics:

$$E(r) = \begin{cases} \dfrac{Zer}{4\pi\varepsilon R^3} & r \leq R \\ \dfrac{Ze}{4\pi\varepsilon r^2} & r > R \end{cases} \qquad (4.5)$$

where ε is the dielectric permittivity of the solvent. Finally, the electrostatic contribution to the interaction potential between the nanogel particle and a single charged solute with charge z is calculated by integration of the electric field[59]:

$$V_{\text{ns}}^{\text{elec}}(r) = \begin{cases} \dfrac{Zze^2}{8\pi\varepsilon R^2}\left(3 - \dfrac{r^2}{R^2}\right) & r \leq R \\ \dfrac{Zze^2}{4\pi\varepsilon r} & r > R \end{cases} \qquad (4.6)$$

This is a soft potential, which shows a Coulombic decay for $r > R$ and tends to a constant value when $r = 0$. A more complicated analytical expression is found when considering core–shell nanogel particles.[60]

The next step is to include the steric repulsive interaction between the nanogel and the solute. Nanogel particles are porous structures that allow the penetration of solvent molecules, ions and other small-sized solutes. This permeability depends strongly on the degree of swelling. In the swollen state, nanogel particles can be regarded as almost perfectly permeable objects, so that the solvent molecules can penetrate freely. However, when the nanogel particle shrinks, the excluded volume fraction of the polymer network increases according to eqn (4.3), so the typical pore size decreases, favouring steric exclusion. This means that the solute is affected by an excluded volume repulsive force that partially inhibits its penetration inside the porous medium. This phenomenon is usually known as the partitioning effect,[61,62] and it is involved in many separation processes in the presence of membranes or gels.[63,64] The steric interaction mainly depends on the polymer packing fraction inside the nanogel particle, the thickness of the polymer fibres and the size of the ions, although there are other secondary parameters, such as the flexibility of the polymer chains and the specific spatial distribution and connectivity of the cross-linking monomers inside the nanogel, that can also modify the steric exclusion between the particle and an incoming solute.

In order to determine the steric repulsion, a simple model based on the equilibrium partitioning effect can be employed. To understand this phenomenon, we consider the system formed by a bulk suspension of uncharged spheres (solute) with diameter σ_s in equilibrium with a fibrous or porous network. The same solvent fills both the porous medium and the bulk phase. It is well known that under these conditions, a concentration difference of the spheres arises between the pore and bulk phases, causing partitioning of the solute. The difference becomes more pronounced when the size of the solute is of the same order of magnitude as that of the pore. The partition coefficient is defined as the ratio between the number density of spheres inside the fibrous network and the number density in the bulk:

$$K = \frac{\rho_{ins}}{\rho_{bulk}} \tag{4.7}$$

The steric nanogel–solute repulsive barrier can be deduced from eqn (4.7)[65]:

$$V_{ns}^{ster}(r) = \begin{cases} -k_B T \ln K & r \leq R \\ 0 & r > R \end{cases} \tag{4.8}$$

where k_B is the Boltzmann constant and T the absolute temperature. The partition coefficient, K, is connected to the internal structure of the cross-linked polymer chains. One of the most appealing models for this structure is that proposed by Ogston,[66] which assumes that the cross-linked polymer network inside the nanogel particle may be represented by a randomly oriented assembly of infinitely long and mutually interpenetrable straight and hard fibres with constant diameter, given by the diameter of the monomer, σ_{mon} (see Figure 4.2).

Using only geometric and probabilistic arguments, Ogston finally obtained the following expression for K, under conditions where the steric exclusion is the dominant effect[66]:

$$K = \exp\left[\left(\frac{\sigma_{mon} + \sigma_s}{\sigma_{mon}}\right)^2 \ln(1 - \phi)\right] \tag{4.9}$$

This model has been found to provide a very good prediction for agarose gels, where the steric repulsion is the dominant interaction.[67] Therefore, the repulsive steric interaction is finally given by[65]

$$V_{ns}^{ster}(r) = \begin{cases} -k_B T(1 + \sigma_s/\sigma_{mon})^2 \ln(1 - \phi) & r \leq R \\ 0 & r > R \end{cases} \tag{4.10}$$

Figure 4.3 shows the bare nanogel–solute interaction potential for different swelling states, for a nanogel bare charge $Z = +100$, with $R_0 = 20$ nm and $\phi_0 = 0.7$. The monomer and solute diameters are assumed to be $\sigma_{mon} = 2\sigma_s = 0.8$ nm. In the collapsed configuration, the steric barrier is about $2.7 k_B T$. This value is large enough to play a significant role, hindering the penetration of the solute inside the nanogel. Nevertheless, the plots shown in Figure 4.3 represent only specific examples of many other multiple

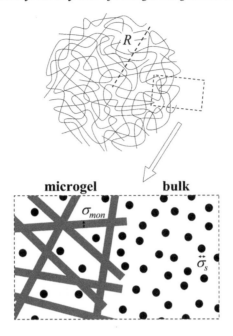

Figure 4.2 Illustration of a nanogel particle. The Ogston model[66] assumes that the polymer fibres inside the nanogel may be represented by infinitely long and mutually interpenetrating rods.

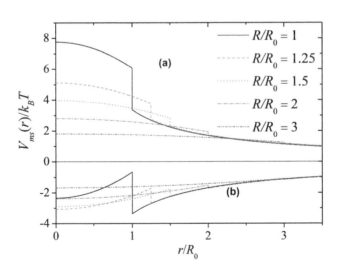

Figure 4.3 Total bare interaction potential between a nanogel particle and a single charged solute for different swelling ratios, R/R_0. The nanogel bare charge is $Z = +100$ and its radius and polymer packing fraction in the deswollen state are $R_0 = 20$ nm and $\phi_0 = 0.7$, respectively. The monomer and solute diameters are $\sigma_{mon} = 2\sigma_s = 0.8$ nm. Plot (a) corresponds to a solute charge of $z = +1$ and (b) to $z = -1$.

possibilities, since the importance of the steric barrier compared with the electrostatic term also depends on the charge of the nanogel particle and on the relative size of the solute compared with the thickness of the fibres. Equation (4.10) also takes into account the fact that large solutes have more difficulty in penetrating inside the nanogel particle, especially in the deswollen state. This effect has been confirmed experimentally. Indeed, the small size of the drug (solute) facilitates its diffusion through the nanogel pores and so increases the drug uptake compared with other larger molecular drugs.[68] Analogously, it has also been observed that the penetration of oppositely charged proteins becomes greatly reduced when the mesh size of the polymer network is smaller than the size of the proteins.[6]

However, this bare interaction between the nanogel and the charged solute is not the only factor to account for. In fact, it is very common to find in the suspension counterions and coions that also contribute to the penetration of the solute inside the nanogel. On the one hand, counterions can screen the bare charge of the nanogel particles, inducing a significant decrease in the range and strength of the effective nanogel–solute interaction. On the other hand, ion–ion and ion–solute finite size correlations can also be very important, especially for high electrolyte concentrations. A simple way to include all these effects is to employ Ornstein–Zernike integral equations.[69] Solving these equations with appropriate closure equations allows the density profiles of the ions and the solute around the nanogel to be calculated.[60,65]

Figure 4.4 depicts the radial distribution function of the solute around the nanogel, defined as $g_{ns}(r) = \rho_{ns}(r)/\rho_{s,bulk}$, where $\rho_{ns}(r)$ is the number density

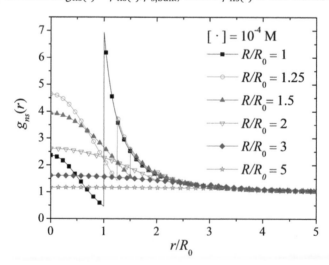

Figure 4.4   Radial distribution function of the solute around the nanogel particle, for several swelling ratios. The nanogel bare charge is $Z = +100$ and its radius and polymer packing fraction in the deswollen state are $R_0 = 20$ nm and $\phi_0 = 0.7$, respectively. The monomer and solute diameters are $\sigma_{mon} = 2\sigma_s = 0.8$ nm and the solute charge is $z = -1$.

of solute molecules located at a distance r from the centre of a nanogel particle and $\rho_{s,bulk}$ is the bulk number density of solute. The calculations are performed in the limit of very low concentrations of solute and nanogel particles, but a finite concentration of 1:1 salt given by 10^{-4} M and assuming a nanogel bare charge of $Z = 100e$ and a solute charge of opposite sign, given by $z = -e$. The size of the solute is assumed to be the same as that of the counter- and coions diameter, $\sigma_s = \sigma_{ions} = 0.4$ nm, and the size of the monomers is $\sigma_{mon} = 0.8$ nm, which corresponds to the average diameter of the PNIPAM chains. The different curves correspond to several values of the degree of swelling. As can be seen, the steric barrier causes a discontinuity in the density profile at the nanogel surface. The radial distribution functions of solute clearly show the competition between steric and electrostatic interactions. For large swelling ratios $(R/R_0 > 3)$, the steric exclusion is very small and the solute is free to penetrate inside the polymer network, driven exclusively by the electrostatic attraction with the nanogel. This leads to a smooth density, which increases due to the effect of the electrostatic attraction as we approach the particle centre.

However, decreasing the swelling ratio below this value gives rise to two competitive phenomena. On the one hand, the decrease in the particle radius causes an increase in the particle bare charge density, which enhances the counterion migration inside the nanogel. This effect induces a progressive growth of the density of the charged solute in the centre of the nanogel. On the other hand, the decrease in the particle size also reduces the size of the pores, so the steric barrier at the nanogel surface becomes more important. This steric interaction compensates the electrostatic interaction, enhancing the condensation of the solute in a more or less thin shell outside the nanogel particle. When the particle shrinkage is very large, the repulsive steric barrier becomes strong enough to screen the electrostatic attraction and the density of counterions inside the nanogel starts to decrease at the time when a large peak arises at the nanogel external surface, in a similar way to that observed in hard colloids. These results clearly show that the penetration of a charged solute or drug inside a nanogel is affected by a rich interplay between steric and electrostatic interactions.[65]

4.3.2   Nanogel–Solute Specific Interaction

In some situations, electrostatic and steric interactions are not sufficient and an additional specific contribution is necessary. For instance, the uptake of water-soluble drugs inside the nanogel particle becomes strongly enhanced by the hydrophobic character of the drug molecules.[68] This kind of interaction is short range, it can be attractive or repulsive and it is mainly generated by changes in the water structure (around the solute and the solvated polymer chain) or by ion adsorption–exclusion effects at interfaces. Although the range of these forces is usually very small (a few water molecule diameters), the strength can be significantly large, giving rise to strong adsorption or desorption of the solute inside the nanogel particle.

This short-range interaction can be included in the calculation of the partition coefficient, introducing a Boltzmann factor.[61] Here, a simple square-well form is used to model the specific interaction between a monomer and the solute:

$$V(r) = \begin{cases} \infty & r < (\sigma_s + \sigma_{mon})/2 \\ V_0 & (\sigma_s + \sigma_{mon})/2 \leq r < (\sigma_s + \sigma_{mon})/2 + \Delta \\ 0 & r > (\sigma_s + \sigma_{mon})/2 + \Delta \end{cases} \quad (4.11)$$

where V_0 is negative for attractive potentials and positive for repulsive potentials. Again, the steric–specific interaction can be written in the form

$$V_{ns}^{ster\text{-}spec}(r) = \begin{cases} -k_B T \ln K & r \leq R \\ 0 & r > R \end{cases} \quad (4.12)$$

but the partition coefficient is now

$$K = e^{-V_0/(k_B T)} \left\{ e^{(1+\sigma_s/\sigma_{mon})^2 \ln(1-\phi)} - e^{[1+(\sigma_s+2\Delta)/\sigma_{mon}]^2 \ln(1-\phi)} \right\}$$
$$+ e^{[1+(\sigma_s+2\Delta)/\sigma_{mon}]^2 \ln(1-\phi)} \quad (4.13)$$

As can be observed, this expression takes both the steric and the specific interactions into account. In the limit $V_0 \to 0$ or $\Delta \to 0$, we recover the steric repulsive barrier [see eqn (4.10)]. However, it should be kept in mind that eqn (4.13) only includes the interaction between the solute and the nearest fibre. Consequently, it represents a good approximation when the interaction range Δ is much smaller than the typical interfibre spacing.

Figure 4.5 shows the density profiles of a charged solute around a nanogel particle, with the same conditions as in Figure 4.4, but assuming a constant swelling ratio of $R/R_0 = 2$ and including a short-range specific interaction between the polymer chain and the solute, with $\Delta = 0.2$ nm. The strength of this additional interaction is varied from $V_0 = +5k_B T$ (strong repulsion) to $V_0 = -2k_B T$ (strong attraction). In order to simplify the calculations, the counterions are removed from the suspension, that is, the electrolyte is formed exclusively by charged solute particles and coions at the same concentration, given as 10^{-3} M.

As the specific interaction goes from repulsion to attraction, the density of solute increases in the nanogel core and decreases in the outer shell of the particle. When $V_0 < 0$, the solute penetration becomes greatly emphasized, so a small decrease in V_0 yields a significant enhancement of the solute adsorption. Although these results only represent a particular example of the many possibilities that may be found in real experiments, they provide clear evidence for the importance of these short-range specific interactions (hydrophobic, hydrophilic, *etc.*) on the encapsulation of drugs inside nanogel particles.

In the previous treatments, the size of the nanogel was assumed to be independent on the solute and electrolyte concentration. However, it should be noted that binding is sometimes accompanied by a significant deswelling

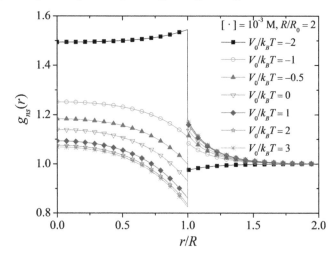

Figure 4.5 Radial distribution function of the solute around the nanogel particle, for several specific interactions with the polymer network, from repulsive to attractive. The nanogel bare charge is $Z = +100$ and its radius and polymer packing fraction in the deswollen state are $R_0 = 20$ nm and $\phi_0 = 0.7$, respectively. The monomer and solute diameters are $\sigma_{mon} = 2\sigma_s = 0.8$ nm and the solute charge is $z = -1$.

of the nanogels.[70,71] This deswelling is mainly driven by two effects. First, the neutralization of the net charge of the nanogel (caused by the screening of oppositely charged solute) reduces the Donnan equilibrium and the repulsion between the internal charges of the particle. Both factors favour the particle deswelling. Second, the hydrophobic character of the solute can reduce the mixing enthalpy between the polymer–solute complexes and the solvent.[72]

4.4 Stability of Nanogel Suspensions: Modelling the Nanogel–Nanogel Interactions

The stability of a nanogel suspension rests on the nanogel suspension Helmholtz free energy $(A = U - TS)$, where U is the internal energy, S the entropy and T the absolute temperature of the system. The internal energy is defined as the statistical average of the total energy of the system, $U = \langle K + \Phi \rangle$, where K is the kinetic energy of the particles and Φ the total potential energy. As in other kinds of colloidal suspensions, Φ can be assumed to be given by a sum over pair interactions, $\Phi = \sum_{i<j} V_{nn}^{i,j}(r_{ij})$, where $V_{nn}^{i,j}(r_{ij})$ is the interaction potential between the nanogels i and j (r_{ij} is the distance between the centres of mass of the two nanogels). $V_{nn}(r)$ is one of the most important magnitudes controlling the stability of the colloidal system. It can be purely attractive (unstable suspension), repulsive (stable suspension) or a combination of both.

The nanogel–nanogel interaction is affected by the swollen/deswollen ratio of the particle. For $R/R_0 = 1$ (shrunken nanogel), $V_{nn}(r)$ could be modelled with the classical DLVO theory[73]: a combination of the London–van der Waals short-range attraction and a long-range electrostatic repulsion. Both contributions lead to an electrostatic repulsive barrier at intermediate distances with an infinite attraction at contact. In that case, a balance between the nanogel-borne charge in the deswollen state and the salt concentration of the suspension determines the stability or instability of the suspension. When $R/R_0 > 1$ (swollen state), the nanogel–nanogel interaction picture becomes more complex, depending on many parameters, such as the chemical composition of the nanogel, solvent properties, ambient conditions and/or degree of cross-linking inside. Next we briefly revise some of the more common contributions to the total nanogel–nanogel interactions, paying special attention to the swollen state.

4.4.1 London–van der Waals Interaction

Two bodies can feel an effective attraction between each other known as the London–van der Waals attraction.[74] The origin of this interaction comes from the oscillating dipolar moments of the molecules forming the object. The simplest expression for this attractive force is found when the distance between body surfaces, h, becomes smaller than the body dimension, R. In the case of a spherical body it is given by

$$V_{vdW}(r) = -f(A)\frac{R}{12h} \tag{4.14}$$

where R is the sphere radius and $f(A)$ is a function depending on the Hamaker constant A.[75] In general, $f(A)$ can be expressed for a nanogel suspension as

$$f(A) = \left(A_{nanogel}^{\frac{1}{2}} - A_{solvent}^{\frac{1}{2}}\right)^2 \tag{4.15}$$

where $A_{nanogel}$ and $A_{solvent}$ are the Hamaker constants of the nanogel and solvent, respectively. In the deswollen state, $A_{nanogel}$ is equal to that for the polymer, $A_{polymer}$. However, if we consider a swollen nanogel, the solvent inside the particle cannot be neglected and must be considered in the effective value of $A_{nanogel}$. Assuming that a swollen nanogel is a homogeneous mixture of two components (polymer and solvent), the following equation for $A_{nanogel}$ can be used[76]:

$$A_{nanogel} = \left[\phi A_{polymer}^{\frac{1}{2}} + (1 - \phi)A_{solvent}^{\frac{1}{2}}\right]^2 \tag{4.16}$$

where ϕ is the polymer volume fraction in the nanogel volume. Hence the London–van der Waals attraction between two swollen nanogels can be written as

$$V_{nn,vdW}(r) = -\left(A_{polymer}^{\frac{1}{2}} - A_{solvent}^{\frac{1}{2}}\right)^2 \frac{\phi^2 R}{12(r - 2R)} \tag{4.17}$$

where $h = r - 2R$ is the surface separation between two spherical particles. This equation indicates that the smaller is ϕ, the weaker is the van der Waals attraction. The polymer packing fraction on the nanogel is connected with its swollen state through eqn (4.3). Therefore, it is a characteristic feature of nanogels that the London–van der Waals attraction can be tuned by varying the swelling ratio. For example, aqueous suspensions of PNIPAM microgel particles cross-linked with N,N-methylenebisacrylamide (BIS) swell at temperatures higher than 32 °C. By using a lower BIS concentration during the synthesis, the maximum achievable swelling ratio $(R/R0)_{max}$ becomes larger. In that situation, the van der Waals attraction will be weaker, as eqn (4.17) shows for $\phi \to 0$. Hence ultrasoft nanogels (nanogels with a low degree of cross-linking) do not undergo this attractive interaction and are stable.

4.4.2   Depletion Interaction

In general, the presence of a different kind of particle in a colloidal suspension could induce an effective attraction between colloids, and this should be considered when studying, for instance, nanogels dispersed in blood for drug-delivery purposes. If the depletant particles are ideal polymer chains and the colloids are hard spheres, the induced effective nanogel–nanogel attraction takes the following analytical form[77]:

$$V_{nn,dep}(r) = \begin{cases} -\dfrac{4\pi}{3}\Pi(R+\delta)^3\left[1 - \dfrac{3}{4}\dfrac{r}{R+\delta} + \dfrac{1}{16}\left(\dfrac{r}{R+\delta}\right)^3\right] & 2R < r < 2(R+\delta) \\ 0 & r > 2(R+\delta) \end{cases}$$

(4.18)

where Π is the polymer osmotic pressure and δ is the depletion layer thickness around the colloid where the polymer chains will not penetrate since this implies that the polymer acquires an entropically unfavourable deformation from its equilibrium conformation. In dilute conditions, δ is equal to the polymer radius of gyration, R_g. This expression can also be used for mixtures of polymer chains and deswollen nanogels assuming that the latter can be modelled as hard spheres. In contrast, the picture changes when the nanogel is swollen. Then, the nanogel becomes softer, favouring the interpenetration of the polymer chains inside the nanogel depletion layer. This was first analyzed by Vincent and co-workers using colloids with a layer of chemically grafted polymer chains.[78,79] Their experiments indicated an induced attraction softer than that predicted by eqn (4.18). The explanation lies in a reduction of the depletion layer, which is promoted by the penetrability of the polymer chain in the particle corona.

The depletion attraction in the presence of swollen nanogels is therefore more complicated to model. In a nanogel–polymer mixture, two important factors have to be analyzed: the relative size between the polymer radius of gyration and the mesh size of the nanogel polymeric network, L, and the interplay among the three Flory parameters, X_{sp} (solvent–polymer), X_{np} (nanogel–polymer) and X_{ns} (nanogel–solvent). Depending on the combination of these parameters, the physical picture could change drastically, so the simple induced attraction provided by eqn (4.18) could be not applicable. A formal approach to this kind of mixtures was carried out by Flory[80] and a detailed description of the different parameters scenarios may be found in the literature.[81] For the particular case where $2R_g > L$ and $X_{np} > 0$, the polymer chains are able to interpenetrate the periphery of the nanogel, generating a weak depletion attraction, as mentioned above. For that case, the stiffer the nanogel, the stronger the depletion attraction is, as was shown experimentally.[82] A careful analysis of the effect of adding polymer on the swollen nanogels was performed by mixing 2-vinylpyridine (2VP) nanogels with non-adsorbing dextran (MW 70 000) at different concentrations.[83] When the osmotic pressure of the dextran solution, π, is higher than the nanogel osmotic pressure, the outside–inside nanogel unbalanced osmotic pressure deswells the particles, inducing flocculation at high dextran concentrations.

In contrast to the London–van der Waals attraction, the flocculation induced by the depletion phenomenon is reversible, since the attractive well is finite at contact and depends on the polymer concentration of the bulk. Nevertheless, if the colloid is not sterically stabilized, the depletion attraction can bring colloids sufficiently close that the London–van der Waals attraction makes the aggregation irreversible. In swollen nanogels, the reversibility of the flocculation after polymer depletion will depend on how strong the van der Waals attraction is after the polymer-induced deswelling of the nanogel.

4.4.3   Elastic Repulsion

Elastic repulsion between two objects acts when the surface-to-surface distance h becomes negative, that is, when the two particles overlap. Then, a force or mechanical work has to be employed to compress the elastic object at such overlapping distances. The work exerted by two elastic spheres of radius R at centre-to-centre distances $r < 2R$ was modelled by Hertz,[84] obtaining the following interaction potential[85]:

$$V_{nn,el}(r) = \begin{cases} \dfrac{\sqrt{2R}}{5D}(2R - r)^{\frac{5}{2}} & r \le 2R \\ 0 & r > 2R \end{cases} \qquad (4.19)$$

where the parameter D depends on the Poisson ratio (ν) and on the bulk modulus of the material, K. It is given by

$$D = \frac{1}{2K}\left(\frac{1 - \nu^2}{1 - 2\nu}\right) \qquad (4.20)$$

Swollen nanogels are a good example of soft colloids. The softness of the swollen nanogel allows such overlapping distances to be reached by compressing or partially deswelling the particle. In general, the more cross-linked the nanogel, the larger is the nanogel bulk modulus and so the stronger is the elastic repulsion for $r < 2R$.[86] In that case, the range of the repulsion will be also reduced since R becomes smaller for a higher cross-linker concentration.[87] Although the latter statement is true in general, the dependence of $V_{nn,el}(r)$ on r mostly depends on the specific cross-linker distribution in the polymer network.

A more versatile potential to model the repulsion between nanogels is given by the so-called soft-repulsive interaction:

$$V_{nn,el}(r) = \varepsilon \left(\frac{R}{r}\right)^{n} \tag{4.21}$$

where ε is the repulsive potential at the nanogel–nanogel contact and n is the softness parameter, which depends on the bulk modulus of the polymer network. The connection between the softness parameter and the nanogel cross-linker distribution or the nanogel bulk modulus is a problem that has not yet been solved. Therefore, n is used as a fitting parameter. One method employed for that purpose is to find the value of n that reproduces the experimental liquid-to-crystal phase transition of the nanogel suspension.[88] Nevertheless, this method for obtaining n is limited to the nanogel compression at the liquid-to-crystal phase transition packing fractions. Larger or smaller compressions (larger or smaller packing fractions) than this will change n since the nanogel cross-linker distribution is dependent on the nanogel compression. Therefore, in the absence of a softness parameter dependent on the nanogel concentration, the model of the elastic repulsion could be imprecise overall when studying dense nanogel suspensions. Nevertheless, this contribution to the nanogel suspension stability could be masked by the electrostatic repulsion when the polymer on the nanogel is ionized. In the following section, we present recent findings in the modelling of ionic nanogel electrostatic interactions.

4.4.4 Electrostatic Interaction

When the polymer network bears a certain amount of charged groups, the interactions among the ionic nanogels, the counterions and the coions must necessarily be included. However, the large size asymmetry between the nanogel particles and the ionic species allows the ionic degrees of freedom to be traced to obtain a nanogel–nanogel effective electrostatic pair potential.[89] This interaction is the combination of the direct Coulombic repulsion between two charged nanogels and the interaction induced by the counterion and coion density profiles inside and around the particles. The role of the ion-induced interaction is of great importance, as it is responsible for the screening of the long-range Coulombic forces between colloids in salty suspensions. For the particular case of nanogel particles, the effective electrostatic interaction is a soft potential for overlapping distances $(r < 2R)$. The exact functional form of this interaction is complicated, as it depends on the spatial distribution of charged monomers inside the nanogel and on the degree of swelling.[59,60,65,90] An analytical expression for this was given by Denton[59] for the particular case of weakly and homogeneously charged, perfectly permeable nanogel particles. However, the situation becomes simpler for non-overlapping distances $(r > 2R)$. It has been proved in many colloidal systems that the effective electrostatic interaction for non-overlapping distances may be written as the following Yukawa potential:

$$V_{\mathrm{nn,elec}}(r) = \frac{Z_{\mathrm{eff}}^{2} L_{\mathrm{B}}}{(1 + \kappa R)^{2}} \frac{\exp[-\kappa(r - 2R)]}{r} \qquad r > 2R \qquad (4.22)$$

where $L_{\mathrm{B}} = e^{2}/4\pi\varepsilon k_{\mathrm{B}}T$ is the Bjerrum length and κ is the inverse Debye length, defined as

$$\kappa = \left(4\pi L_{\mathrm{B}} \sum_{i} z_{i}^{2} \rho_{i,\mathrm{bulk}} \right)^{\frac{1}{2}} \qquad (4.23)$$

where z_{i} is the valence of each ion in the suspension and $\rho_{i,\mathrm{bulk}}$ its bulk number density. Z_{eff} is the effective charge of the nanogel, which is a renormalized charge that includes the effect of the ionic clouds around the colloids. In particular, this functional form for the electrostatic effective interaction has been found to hold in salty suspensions of ionic nanogel suspensions for both uniformly charged and core–shell ionic nanogel suspensions in the swollen or deswollen state.[59,60,65] Therefore, although the ions can penetrate inside the nanogel network, the screened electrostatic interaction can be regarded as the one between hard colloids bearing an effective charge Z_{eff}.

For very weakly charged nanogel particles, the effective charge is given by the net charge of the particle, Z_{net}, defined as the charge after taking into account the counterions adsorbed within the nanogel.[65] Explicit expressions for Z_{net} of uniform and core–shell nanogels may be found in the literature.[91] However, for moderate or highly charged nanogels, the strong electrostatic coupling between particles and ions gives rise to an additional screening, caused by the accumulation of counterions in the immediate vicinity of the nanogel surface.[65] As a result, the effective charge becomes smaller than the net charge, $Z_{\mathrm{eff}} < Z_{\mathrm{net}}$.

4.5 Conclusion and Perspectives

This chapter focused on the great importance of the interaction between nanogels and drugs. In fact, the preparation of nanogels must be well designed in order to achieve new nanogel families with the purpose of optimizing the encapsulation and delivery of drugs and biomolecules under specific conditions, such as the physiological conditions.

In this respect, materials scientists and chemists should join biologists to strengthen this field, synthesizing more complex nanostructures with biological applications.

The effects that the electrostatic, steric and other specific interactions have on the penetration of drugs inside porous cross-linked nanogel particles have been studied. For this purpose, a theory based on the partition coefficient has been used to coarse-grain the internal degrees of freedom of the polymer network. The results clearly indicate that the diffusion and binding of drugs, biomolecules or any other kind of solute is the consequence of a complex interplay between all these interactions. It should be emphasized that this model assumes that the polymer chains inside the nanogel are rigid. However, the polymer segments between two cross-linker monomers are flexible and can be deformed to some extent around a neighbouring ion of another incoming solute. In other words, the solute permeability can be higher than that predicted using eqns (4.10), (4.12) and (4.13). Another important factor affecting the penetration of a charged solute is the cross-linker density and the distribution of functional groups in the nanogel. Experimental results obtained with PNIPAM show that a uniform distribution of charged groups inside the particle is able to bind more drug than surface-localized nanogels, whereas a distribution of bare charged groups in the external shell of the particle leads to condensation of the drug on the nanogel surface that hinders the diffusion of additional drug.[68,92]

Finally, the stability of the nanogel suspension has been studied in terms of the nanogel–nanogel pair interaction. The final pair potential is obtained as the addition of the attractive London–van der Waals and depletion forces, with the repulsive elastic and electrostatic contributions. These interactions depend on the specific properties of the nanogel particles, such as the charge distribution, swelling ratio and elastic behaviour of the internal cross-linked polymer chains. However, this picture based on two-body pair potentials becomes useless at large particle packing fractions, where the overlap between the electric double layers and the deformation of the polymer network lead to the appearance of many-body interactions. In this respect, it would be interesting to develop density-dependent nanogel–nanogel pair potentials and use them to reproduce the phase diagram of nanogel suspensions.

References

1. J. Ramos, J. Forcada and R. Hidalgo-Alvarez, *Chem. Rev.*, 2014, **114**, 367.
2. J. Ramos, A. Imaz, J. Callejas-Fernández, L. Barbosa-Barros, J. Estelrich, M. Quesada-Pérez and J. Forcada, *Soft Matter*, 2011, 7, 5067.
3. J. Ramos, A. Imaz and J. Forcada, *Polym. Chem.*, 2012, **3**, 852.
4. S. Frokjaer and D. E. Otzen, *Nat. Rev. Drug Discov.*, 2005, **4**, 298.
5. M. Malmsten, *Soft Matter*, 2006, **2**, 760.
6. G. M. Eichenbaum, P. F. Kiser, A. V. Dobrynin, S. A. Simon and D. Needham, *Macromolecules*, 1999, **32**, 4867.

7. G. M. Eichenbaum, P. F. Kiser, D. Shah, S. A. Simon and D. Needham, *Macromolecules*, 1999, **32**, 8996.
8. T. Hoare and R. Pelton, *Langmuir*, 2008, **24**, 1005.
9. A. V. Kavanov and S. V. Vinogradov, *Angew. Chem. Int. Ed.*, 2009, **48**, 5418.
10. J. Ramos, A. Imaz and J. Forcada, *Polym. Chem.*, 2012, **3**, 852.
11. B. R. Saunders and B. Vincent, *Adv. Colloid Interface Sci.*, 1999, **80**, 1.
12. R. Pelton, *Adv. Colloid Interface Sci.*, 2000, **85**, 1.
13. S. Nayak and L. A. Lyon, *Angew. Chem. Int. Ed.*, 2005, **44**, 7686.
14. A. Pich and W. Richtering, *Adv. Polym. Sci.*, 2010, **234**, 1.
15. B. R. Saunders, N. Laajam, E. Daly, S. Teow, X. Hu and R. Stepto, *Adv. Colloid Interface Sci.*, 2009, **147–148**, 251.
16. N. M. B. Smeets and T. Hoare, *J. Polym. Sci., Part A: Polym. Chem.*, 2013, **51**, 3027.
17. R. H. Pelton and P. Chibante, *Colloids Surf.*, 1986, **20**, 247.
18. M. J. Snowden and B. Vincent, *J. Chem. Soc., Chem. Commun.*, 1992, 1103.
19. D. Duracher, A. Elaissari and C. Pichot, *J. Polym. Sci., Part A: Polym. Chem.*, 1999, **37**, 1823.
20. W. Leobandung, H. Ichikawa, Y. Fukumori and N. A. Peppas, *J. Appl. Polym. Sci.*, 2003, **87**, 1678.
21. G. Huang, J. Gao, Z. Hu, J. V. St. John, B. C. Ponder and D. Moro, *J. Controll. Release*, 2004, **94**, 303.
22. S. Zhou and B. Bhu, *J. Phys. Chem. B*, 1998, **102**, 1364.
23. M. Beija, J. D. Marty and M. Destarac, *Chem. Commun.*, 2011, **47**, 2826.
24. H. Vihola, A. Laukkanen, L. Valtola, H. Tenhu and J. Hirvonen, *Biomaterials*, 2005, **26**, 3055.
25. L. Jun, W. Bochu and W. Yazhou, *Int. J. Pharmacol.*, 2006, **2**, 513.
26. A. Imaz and J. Forcada, *J. Polym. Sci., Part A: Polym. Chem.*, 2010, **48**, 1173.
27. Y. Gao, S. C. F. Au-Yeng and C. Wu, *Macromolecules*, 1999, **32**, 3674.
28. A. Imaz and J. Forcada, *J. Polym. Sci., Part A: Polym. Chem.*, 2008, **46**, 2510.
29. A. Imaz and J. Forcada, *J. Polym. Sci., Part A: Polym. Chem.*, 2008, **46**, 2766.
30. A. Imaz, J. I. Miranda, J. Ramos and J. Forcada, *Eur. Polym. J.*, 2008, **44**, 4002.
31. A. Imaz and J. Forcada, *Eur. Polym. J.*, 2009, **11**, 3164.
32. A. Imaz and J. Forcada, *Macromol. Symp.*, 2009, **281**, 85.
33. K. Kratz, T. Hellweg and W. Eimer, *Colloids Surf. A*, 2000, **170**, 137.
34. J. D. Debord and L. A. Lyon, *Langmuir*, 2003, **19**, 7662.
35. T. Hoare and R. Pelton, *Macromolecules*, 2004, **37**, 2544.
36. M. Bradley, J. Ramos and B. Vincent, *Langmuir*, 2005, **21**, 1209.
37. S. Zhou and B. Bhu, *J. Phys. Chem. B*, 1998, **102**, 1364.
38. T. Hoare and R. Pelton, *Langmuir*, 2004, **20**, 2123.
39. T. Hoare and R. Pelton, *Langmuir*, 2006, **22**, 7342.
40. W. Qin, Y. Zhao, Y. Yang, H. Xu and X. Yang, *Colloid Polym. Sci.*, 2007, **285**, 515.
41. M. Das, N. Sanson, D. Fava and E. Kumacheva, *Langmuir*, 2007, **23**, 196.
42. M. Karg, I. Pastoriza-Santos, B. Rodriguez-Gonzalez, R. von Klitzing, S. Wellert and T. Hellweg, *Langmuir*, 2008, **24**, 6300–6304.

43. A. Imaz and J. Forcada, *J. Polym. Sci., Part A Polym. Chem.*, 2011, **49**, 3218.
44. D. M. Heyes and A. C. Branka, *Soft Matter*, 2009, **5**, 2681.
45. A. Kabanov and S. Vinogradov, *Multifunctional Pharmaceutical Nano-carriers*, Springer, New York, 2008, pp. 67–80.
46. J. Oh, R. Drumright, D. Siegwart and K. Matyjaszewski, *Prog. Polym. Sci.*, 2008, **33**, 448.
47. S. V. Vinogradov, *Curr. Pharm. Des.*, 2006, **12**, 4703.
48. M. Yallapu, M. Reddy and V. Labbaserwar, *Biomedical Applications of Nanotechnology*, Wiley, Hoboken, NJ, 2007, pp. 131–171.
49. N. Murthy, M. Xu, S. Schuck, J. Kunisawa, N. Shastri and J. M. J. Frechet, *Proc. Natl. Acad. Sci. U. S. A.*, 2003, **100**, 4995.
50. J. M. Lee, K. M. Park, S. J. Lim, M. K. Lee and C. K. Kim, *J. Pharm. Pharmacol.*, 2002, **54**, 43.
51. S. V. Vinogradov, *Structure and Functional Properties of Colloidal Systems, Surfactant Science Series*, Vol. 146, Taylor & Francis, New York, 2010, pp. 367–386.
52. J. P. K. Tan and K. C. Tam, *Structure and Functional Properties of Colloidal Systems, Surfactant Science Series*, Vol. 146 , Taylor & Francis, New York, 2010, pp. 387–411.
53. J. P. K. Tan, C. H. Goh and K. C. Tam, *Eur. J. Pharm. Sci.*, 2007, **32**, 340.
54. H. Bysell and M. Malmsten, *Langmuir*, 2009, **25**, 522.
55. A. Fernández-Barbero, A. Fernández-Nieves, I. Grillo and E. López-Cabarcos, *Phys. Rev. E*, 2002, **66**, 051803.
56. M. Stieger, W. Richtering, J. S. Pedersen and P. Lindner, *J. Chem. Phys.*, 2004, **120**, 6197.
57. S. Meyer and W. Richtering, *Macromolecules*, 2005, **38**, 1517.
58. I. Berndt, J. S. Pedersen, P. Lindner and W. Richtering, *Langmuir*, 2006, **22**, 459.
59. A. R. Denton, *Phys. Rev. E*, 2003, **67**, 011804. Erratum: *Phys. Rev. E*, 2003, **68**, 049904
60. A. Moncho-Jordá, J. A. Anta and J. Callejas-Fernández, *J. Chem. Phys.*, 2003, **138**, 134902.
61. E. M. Johnson and W. M. Deen, *J. Colloid Interface Sci.*, 1996, **178**, 749.
62. M. J. Lazzara, D. Blankschtein and W. M. Deen, *J. Colloid Interface Sci.*, 2000, **226**, 112.
63. J. L. Anderson and J. A. Quinn, *Biophys. J.*, 1974, **14**, 130.
64. R. Langer and J. Folkman, *Nature*, 1976, **263**, 797.
65. A. Moncho-Jordá, *J. Chem. Phys.*, 2013, **139**, 064906.
66. A. G. Ogston, *Trans. Faraday Soc.*, 1958, **54**, 1754.
67. E. M. Johnson, D. A. Berk, R. K. Jain and W. M. Deen, *Biophys. J.*, 1995, **68**, 1561.
68. T. Hoare and R. Pelton, *Langmuir*, 2008, **24**, 1005.
69. J. P. Hansen and I. R. McDonald, *Theory of Simple Liquids*, Academic Press, New York, 3rd edn, 2006.
70. C. Johansson, J. Gernandt, M. Bradley, B. Vincent and P. Hansson, *J. Colloid Interface Sci.*, 2010, **347**, 241.

71. C. Johansson, P. Hansson and M. Malmsten, *J. Colloid Interface Sci.*, 2007, **316**, 350.

72. T. Hoare and R. Pelton, *J. Phys. Chem. B*, 2007, **111**, 11895.

73. J. N. Israelachvili, *Intermolecular and Surface Forces*, Academic Press, London, 2007.

74. F. London, *Trans. Faraday Soc.*, 1937, **33**, 8.

75. H. C. Hamaker, *Physica*, 1937, **4**, 1058.

76. H. M. Crowther and B. Vincent, *Colloid Polym. Sci.*, 1998, **276**, 46.

77. S. Asakura and F. Oosawa, *J. Chem. Phys.*, 1954, **22**, 1255.

78. B. Vincent, J. Edwards, S. Emmett and A. Jones, *J. Colloid Interface Sci.*, 1986, **17**, 261.

79. A. Jones and B. Vincent, *Colloids Surf.*, 1989, **42**, 113.

80. P. J. Flory, *Principles of Polymer Chemistry*, Cornell University Press, Ithaca, NY, 1953, Chapter X.

81. A. Fernandez-Nieves, H. Wyss, J. Mattsson and D. A. Weitz, *Microgel Suspensions*, Wiley-VCH, Weinheim, 2011, Chapter 5.

82. M. Rasmusson, A. Routh and B. Vincent, *Langmuir*, 2004, **20**, 3536.

83. A. Fernandez-Nieves, A. Fernandez-Barbero, B. Vincent and F. J. de las Nieves, *Prog. Colloid Polym. Sci.*, 2000, **115**, 1273.

84. J. L. Johnson, Normal contact of elastic solids – Hertz theory, in *Contact Mechanics*, Cambridge University Press, Cambridge, 1985, Chapter 4.

85. J. Riest, P. Mohanty, P. Schurtenberger and C. N. Likos, *Z. Phys. Chem.*, 2012, **226**, 711.

86. J. J. Lietor-Santos, B. Sierra-Martin and A. Fernandez-Nieves, *Phys. Rev. E*, 2011, **84**, 060402.

87. B. Sierra-Martin and A. Fernandez-Nieves, *Soft Matter*, 2012, **8**, 4141.

88. H. Senff and W. Richtering, *J. Chem. Phys.*, 1999, **111**, 1705.

89. M. Dijkstra, R. van Roij and R. Evans, *Phys. Rev. Lett.*, 1998, **81**, 2268.

90. D. Gottwald, C. N. Likos, G. Kahl and H. Löwen, *Phys. Rev. Lett.*, 2004, **92**, 068301.

91. D. Gottwald, C. N. Likos, G. Kahl and H. Löwen, *J. Chem. Phys.*, 2005, **122**, 074903.

92. R. Yoshida, K. Sakai, T. Ukano, Y. Sakurai, Y. H. Bae and S. W. Kim, *J. Biomater. Sci. Polym. Ed.*, 1991, **3**, 155.

CHAPTER 5

Polymeric Micelles

P. TABOADA,[a] S. BARBOSA,[a] A. CONCHEIRO[b] AND
C. ALVAREZ-LORENZO*[b]

[a] Departamento de Física de la Materia Condensada, Facultad de Física, Universidad de Santiago de Compostela, 15782 Santiago de Compostela, Spain; [b] Departamento de Farmacia y Tecnología Farmacéutica, Facultad de Farmacia, Universidad de Santiago de Compostela, 15782 Santiago de Compostela, Spain
*Email: carmen.alvarez.lorenzo@usc.es

5.1 Introduction

Polymeric micelles are self-assemblies of amphiphilic polymers that resemble to a certain extent the aggregates (micelles) of classical surfactants. Compared with surfactants, amphiphilic polymers have higher molecular weight and can combine a variety of chemical functionalities in both the hydrophobic and hydrophilic regions, which allows the preparation of micelles with diverse structures and performances. In this regard, polymers in which the regions of different polarity are well differentiated are more prone to self-assemble at low concentration to lead to thermodynamically and kinetically stable polymeric micelles. That is the reason for the predominant role that block copolymers play in this field, which has prompted the development of a variety of synthesis routes and of characterization techniques. Moreover, the peculiar hydrophobic core–hydrophilic shell architecture of block copolymer assemblies in an aqueous medium has extended the applications of polymeric micelles to a large variety of fields, in particular addressing relevant needs in the biomedical field. Polymeric

RSC Nanoscience & Nanotechnology No. 34
Soft Nanoparticles for Biomedical Applications
Edited by José Callejas-Fernández, Joan Estelrich, Manuel Quesada-Pérez and Jacqueline Forcada
© The Royal Society of Chemistry 2014
Published by the Royal Society of Chemistry, www.rsc.org

micelles are particularly suitable for improving the solubility and stability of diverse molecules *via* encapsulation in the hydrophobic core or at the core–shell interface depending on the molecule polarity. Such capability to encapsulate molecules in conjunction with its ability to accumulate in specific tissues or cells is being largely explored in the targeting of active substances. In this context, block copolymers that can sense external stimuli or internal variables are being shown to be particularly suitable for preparing micelles that can precisely regulate the site and rate of drug delivery in the body. This chapter begins with a brief description of the procedures for synthesizing and characterizing block copolymers in order to introduce the reader to the mechanisms of micelle formation. Then, micellization of neutral, ionic and double hydrophilic block copolymers is covered in detail together with the techniques useful for characterizing the micellization process and the structure of the micelles. The last section is devoted to the pharmaceutical applications of micelles in drug solubilization, targeting, stimuli-responsive release and biological response modulation.

5.2   Block Copolymer Synthesis and Characterization

To generate linear diblock AB (where A and B denote the two different block components or monomers) and triblock ABA structures, several strategies can be followed (Figure 5.1): (i) sequential addition of monomers (route I) from monofunctional (both AB and ABA structures) or difunctional (only ABA) initiators, this being the traditional and easiest synthetic route for most block copolymers; (ii) coupling of two preformed (co)polymer segments with antagonist functional end-groups (route II); (iii) combination of different modes of polymerization for the preparation of a well-defined block co-polymer not accessible from a single polymerization mechanism (route III); and (iv) one-pot initiation from dual bifunctional initiators for AB block copolymer synthesis (route IV). These routes can be implemented by using any of the polymerization techniques described below, which are chosen taking into account the structure of the aimed-for block copolymer and monomer, the desired molecular weight range and monodispersity of each block and the purity of the end product. This last point involves several extra requirements: (a) initiation should not be slower than propagation; (b) absence of transfer and termination reactions; (c) deactivation equilibrium must be dynamic and exchange between active and dormant states must be much faster than propagation, so that a chain undergoes a large number of activation–deactivation cycles during the whole course of polymerization; and (d) adventitious initiation should be minimized.

5.2.1   Polymerization Methods

Sequential anionic polymerization is a well-established method for the synthesis of tailored block copolymers in which the active centre is anionic in character (*e.g.* carbanion, oxyanion).[1,2] Monomers are sequentially

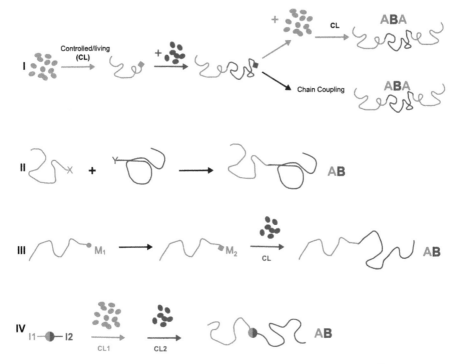

Figure 5.1 Synthetic routes for diblock and triblock copolymers: (I) sequential addition of monomer in controlled/living polymerization; (II) coupling of linear chains with terminal antagonist chemical end-groups (X and Y); (III) switching of polymerization mechanism; and (IV) dual functional initiators for simultaneous copolymerization.

polymerized in order of increasing electronegativity, starting with the monomer that forms the higher reactive propagating centre.[3–5] The anionic active species are usually highly reactive, hence the preparation of well-defined polymers requires high-purity starting monomers, solvents and reagents and high-vacuum conditions to exclude air and moisture. Recent advances have been directed mainly towards the synthesis of block co-polymers with exotic architectures, such as mixed-arm stars,[6] H-shaped,[7] ring-shaped (cyclic)[8] and pom-pom[9] copolymers. Acidic monomers, such as acrylic and methacrylic acid, require protection/stabilization of the carbanionic species before polymerization.[10–13] Ethylene oxide (EO) is one of the most interesting monomers polymerizable by sequential anionic polymerization to obtain amphiphilic block copolymers such as poly-styrene–poly(ethylene oxide) (PS–PEO) or polybutadiene–poly(ethylene oxide) (PB–PEO), or linear triblock copolymers based on hydrophobic poly(propylene oxide) (PPO) and hydrophilic PEO blocks, known as poloxamers or Pluronics and their star-shaped counterparts, denoted poloxamines or Tetronics (Figure 5.2).[14–17]

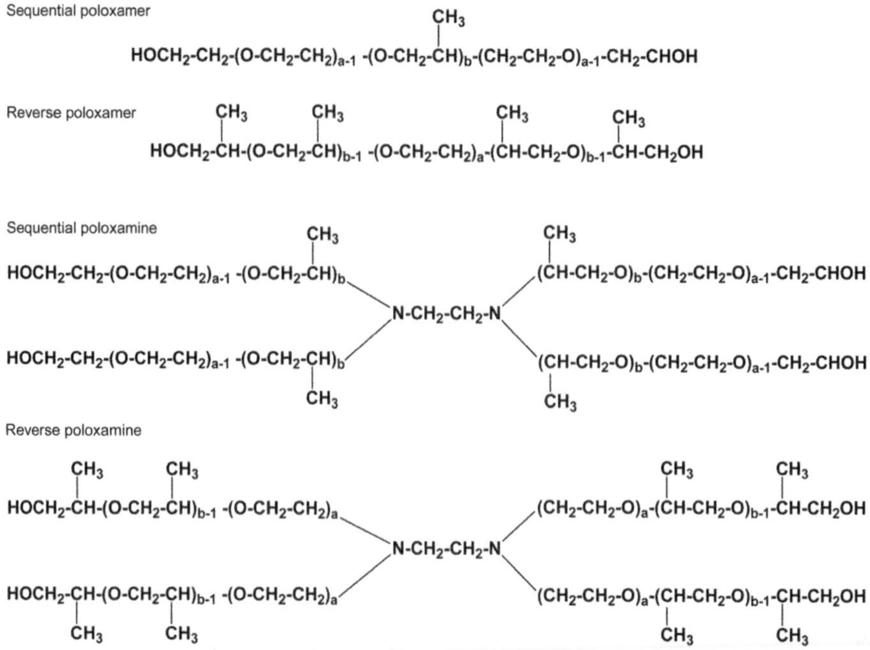

Sequential poloxamer

$$CH_3$$
$$HOCH_2\text{-}CH_2\text{-}(O\text{-}CH_2\text{-}CH_2)_{a\text{-}1}\text{-}(O\text{-}CH_2\text{-}CH)_b\text{-}(CH_2\text{-}CH_2\text{-}O)_{a\text{-}1}\text{-}CH_2\text{-}CHOH$$

Reverse poloxamer

$$CH_3 \qquad CH_3 \qquad\qquad\qquad CH_3 \qquad\quad CH_3$$
$$HOCH_2\text{-}CH\text{-}(O\text{-}CH_2\text{-}CH)_{b\text{-}1}\text{-}(O\text{-}CH_2\text{-}CH_2)_a\text{-}(CH\text{-}CH_2\text{-}O)_{b\text{-}1}\text{-}CH\text{-}CH_2OH$$

Sequential poloxamine

$$CH_3 \qquad\qquad\qquad\qquad\qquad CH_3$$
$$HOCH_2\text{-}CH_2\text{-}(O\text{-}CH_2\text{-}CH_2)_{a\text{-}1}\text{-}(O\text{-}CH_2\text{-}CH)_b \qquad (CH\text{-}CH_2\text{-}O)_b\text{-}(CH_2\text{-}CH_2\text{-}O)_{a\text{-}1}\text{-}CH_2\text{-}CHOH$$
$$N\text{-}CH_2\text{-}CH_2\text{-}N$$
$$HOCH_2\text{-}CH_2\text{-}(O\text{-}CH_2\text{-}CH_2)_{a\text{-}1}\text{-}(O\text{-}CH_2\text{-}CH)_b \qquad (CH\text{-}CH_2\text{-}O)_b\text{-}(CH_2\text{-}CH_2\text{-}O)_{a\text{-}1}\text{-}CH_2\text{-}CHOH$$
$$CH_3 \qquad\qquad\qquad\qquad\qquad CH_3$$

Reverse poloxamine

$$CH_3 \qquad CH_3 \qquad\qquad\qquad\qquad\qquad CH_3 \qquad\quad CH_3$$
$$HOCH_2\text{-}CH\text{-}(O\text{-}CH_2\text{-}CH)_{b\text{-}1}\text{-}(O\text{-}CH_2\text{-}CH_2)_a \qquad (CH_2\text{-}CH_2\text{-}O)_a\text{-}(CH\text{-}CH_2\text{-}O)_{b\text{-}1}\text{-}CH\text{-}CH_2OH$$
$$N\text{-}CH_2\text{-}CH_2\text{-}N$$
$$HOCH_2\text{-}CH\text{-}(O\text{-}CH_2\text{-}CH)_{b\text{-}1}\text{-}(O\text{-}CH_2\text{-}CH_2)_a \qquad (CH_2\text{-}CH_2\text{-}O)_a\text{-}(CH\text{-}CH_2\text{-}O)_{b\text{-}1}\text{-}CH\text{-}CH_2OH$$
$$CH_3 \qquad CH_3 \qquad\qquad\qquad\qquad\qquad CH_3 \qquad\quad CH_3$$

Figure 5.2 Structures of sequential and reverse poloxamers (Pluronic, Lutrol) and poloxamines (Tetronic).

Sequential group transfer polymerization (GTP) permits the anionic polymerization of (meth)acrylates[18] and several other monomers containing electron-deficient double bonds, *e.g.* acrylonitrile, methacrylonitrile, *N,N*-dimethylacrylamide and 2-methylene-4-butyrolactone. The initiator, generally a silyl ketene acetal, is reversibly activated by a nucleophilic or an electrophilic catalyst, which attacks the electron-deficient Si centre in the dormant polymer (cleaving the Si–O bond) in order to generate a polymeric ester enolate, which is the active propagating species in GTP.[19] The polymerization is terminated by protic compounds, but not by oxygen. Hence it must be conducted under anhydrous conditions. *Sequential cationic polymerization* is suitable for monomers bearing electron-donating groups such as styrenics, vinyl ethers, 2-alkyloxazolines or tetrahydrofuran. The initiators are organic halides and lead to homopolymers capped with a halogen atom (Cl, Br, I), which can be used as macroinitiators for the copolymerization with other monomers.[20–22] *Non-radical metal-catalysed polymerization*, also known as coordination polymerization, is a chain polymerization in which both the monomer and the active centre are coordinated to the metal catalyst prior to the incorporation of the monomer in the polymer chain. Ethylene, α-olefins and conjugated dienes are the most important monomers polymerized by this route.[23–26] *Ring-opening polymerization* (ROP) is suitable for a wide variety of cyclic compounds containing at least a heteroatom or an unsaturation centre in the ring. The heteroatom–carbon bond is opened through a nucleophilic reaction,[27-32] whereas the double bond is opened through an olefin metathesis reaction [ring-opening metathesis polymerization (ROMP)].[33-36] Diblock copolymers can be also formed by direct *coupling* of two 'living' polymeric chains[37] or by reacting two different polymeric chains functionalized at their respective ends with suitable end-groups.[38] This methodology can be expanded to triblock or multiblock copolymers when difunctional polymer precursors or coupling agents are used.[5]

Controlled/living radical polymerizations (C/LRP). Free radical polymerization is suitable for a wide range of monomers, operating temperatures and media.[39-41] However, it leads to copolymers with high polydispersity indexes due to chain breakings by irreversible terminations and the so-called cage effect. C/LRP, based on the concept of reversible chain termination,[42] can overcome this drawback. Polymerization of the first monomer (the most reactive one) is discontinued before its total consumption in order to preserve its end functionality intact.[43] There are three main approaches: atomic transfer radical polymerization (ATRP), nitroxide-mediated polymerization (NMP) and reversible addition–fragmentation chain transfer polymerization (RAFT) (Figure 5.3).

ATRP involves the reversible homolytic cleavage of a carbon–halogen (RX) bond by a redox reaction between the organic halide and a metal (*e.g.* Cu^I, Fe^{II}, Pd^I) halide (in the presence of a ligand, *e.g.* bipyridine), which yields the initiating radical and the oxidized metal complex.[44,45] As the polymer chain end still contains a halogen group, this can be used to initiate the polymerization of a second monomer for the preparation of block copolymers.[46] To synthesize block copolymers sequentially, the sequence of monomers should be in decreasing order of ATRP equilibrium constants for monomers of different reactivity, for example, acrylonitrile > methacrylates > acrylates > (meth)acrylamides; otherwise, a complete changeover from one block to another will not occur owing to slower and incomplete initiation.[47,48] Despite its evident advantages, ATRP has also several drawbacks, such as the tedious metal removal steps and the impossibility of using monomers that poison the catalyst.[49] NMP involves the reversible coupling reaction of the nitroxide radical with a carbon-centred growing radical, at relatively high temperature.[46,50] The advantage of this nitroxy-mediated technique is that the first block could be isolated and characterized before starting the polymerization of the second block. RAFT polymerization had its origin in the use of thiocarbonylthio compounds, of general structure Z–C(=S)SR, as reversible chain transfer agents (CTAs).[51,52] These compounds react though their C=S double bonds with propagating oligomers, leading to a transient radical whose fragments reversibly produce a new thiocarbonylthio RAFT agent and an expelled new radical capable of reinitiating the polymerization. One of the advantages of RAFT polymerization is its applicability to a wide range of monomers in both organic and aqueous solvents.[53-60]

When the different monomers of a block copolymer cannot undergo the same polymerization reaction, a preformed polymer can be used as a

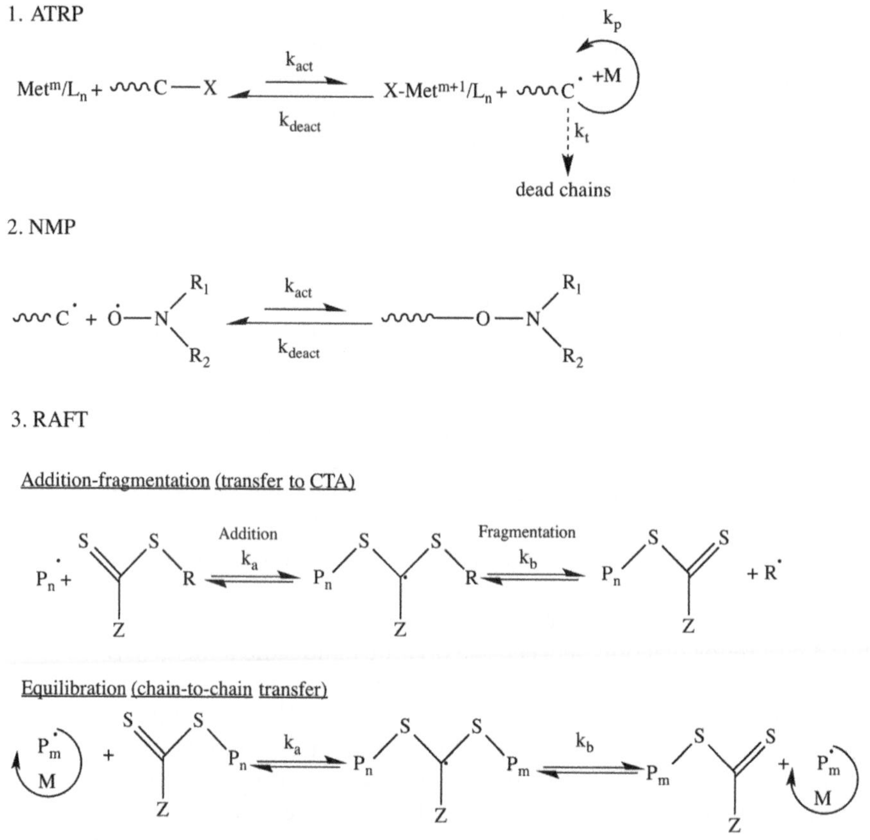

1. ATRP

2. NMP

3. RAFT

Addition-fragmentation (transfer to CTA)

Equilibration (chain-to-chain transfer)

Figure 5.3 Main events in atomic transfer radical polymerization (ATRP), nitroxide-mediated polymerization (NMP) and reversible addition-fragmentation chain transfer polymerization (RAFT).

macroinitiator for the growth of the second block by another mechanism.[61] The change in the polymerization mechanism can be accomplished by means of an *in situ* conversion of the active centre or through isolation of the first block and chemical modification of the active centre.[62–66]

5.2.2 Block Copolymers with Complex Architectures

Advances in synthetic methodologies permit the preparation of block co-polymers with complex architectures (linear, star, heteroarm, dendritic) where polymer segments of different types (rigid and/or 'soft' blocks, hydrophobic and/or hydrophilic, amorphous and/or crystallizable or iso-tropic/nematic blocks) and architectures are combined within the same molecule.[67–69] Moreover, the development of controlled/living radical poly-merizations, the design of multifunctional initiators and the control of coupling reactions allow the manipulation of the chemical nature and the overall compositions of the final copolymers, leading to novel families of

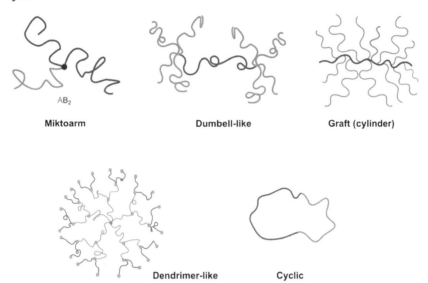

Miktoarm   Dumbell-like   Graft (cylinder)

Dendrimer-like   Cyclic

Figure 5.4 Some representative branched block copolymer structures.

block copolymers comprising dendritic segments, rod-like conformations based on synthetic polypeptides or rigid conducting blocks,[70–75] and permitting the introduction of specific functions (*e.g.* H-bondings or electrical charges) within the blocks that can direct the self-assembly process and control the resulting ordered nanostructures.[76]

Multifunctional initiators combined with sequential CLP (the so-called 'core-first approach') or coupling 'living' linear copolymeric chains on to a multifunctional comonomer ('arm-first' approach) may lead to star-shaped block copolymers.[77] Alternatively, a precursor bearing multiple complementary functions to that of the living chains can be used to create the arm-first star copolymer. Also, graft copolymers can be obtained through 'grafting onto' ('living' linear chains obtained by CLP are deactivated onto a preformed macromolecular backbone with functional side groups), 'grafting from' (a macromolecular precursor serves as a multifunctional initiator for CLP, from which the chains are grown) and 'macronomer' or 'grafting through' (a preformed polymer contains a group at one end of the chain that can be polymerized) approaches. Other complex structures can be generated by combining several branching reactions and CLP methods (Figure 5.4).

5.2.3   Polymer Characterization

Different empirical and statistical models have been developed to characterize the molecular weight distribution of block copolymers. Mathematical models such as the most probable distribution and the Poisson distribution and empirical models such as Schulz, Beall, Tung and Lasing–Kraemer distributions can be used to analyse molecular weight data. The use of one distribution or another relies on the characteristics of the synthetic procedure, that is, the reactivity of the monomers or the types and recurrence of transfer and termination reactions, which finally determine the probability of obtaining a polymer chain within a specific averaged molecular weight range (statistical molecular weight distribution models have been described elsewhere).[78,79] Provided that a molecular weight distribution exists, different types of average molecular weight can be defined, such as number-average molecular weight, M_n, weight-average molecular weight, M_w, z-average molecular weight, M_z, and viscous-average molecular weight, M_v:

$$M_n = \frac{\sum N_i M_i}{\sum N_i} = \sum n_i M_i \tag{5.1}$$

$$M_w = \frac{\sum N_i M_i^2}{\sum N_i M_i} = \sum w_i M_i \tag{5.2}$$

$$M_z = \frac{\sum N_i M_i^3}{\sum N_i M_i^2} \tag{5.3}$$

$$M_v = \left(\frac{\sum N_i M_i^{1+a}}{\sum N_i M_i}\right)^{\frac{1}{a}} \tag{5.4}$$

where N_i, M_i, n_i and w_i represent the number, the molecular weight, the molar fraction and the weight fraction, respectively, of polymer molecules with polymerization degree i (i is a positive integer), and a is an empirical parameter that depends on the polymer–solvent pair and the temperature. M_n can be determined through end-group analysis, measuring colligative properties of polymer solutions or by means of nuclear magnetic resonance (NMR) spectroscopy. M_w is usually evaluated by applying scattering techniques (*e.g.* light, small-angle X ray and neutron scattering methods) and mass spectrometric techniques such as MALDI-TOF. M_z can be determined by applying sedimentation equilibrium methods, and M_v can be obtained by means of viscosity measurements.[80,81]

Importantly, before the experimental characterization of a copolymer, it is necessary to reduce/eliminate the presence of impurities such as homopolymers, short-chain copolymers or copolymers of different structure. This can be achieved by fractionating the block copolymers according to their molecular weight or chemical composition through batch or column elution fractionation.[82] The first method is based on the partitioning of polymer molecules between two liquid phases; the polymer is first dissolved in a good solvent for all copolymer units to give a dilute solution and then, upon addition of a non-solvent and/or by lowering the temperature, the polymer solution is separated into polymer-rich and polymer-poor phases. The partitioning of the polymer into the two phases depends on molecular weight and chemical composition. The same principle governs the column elution fractionation technique, but in this the polymer-rich phase is selectively deposited on the packing material in the column as a thin layer. Copolymer fractions can be collected by elution with the solvent.

5.3 Micelle Formation and Characterization

Block copolymers in a solvent selective for one block usually tend to form micelles at a certain concentration and/or temperature. The concentration at which micellization begins at a certain temperature is termed the critical micelle concentration (CMC), and the temperature at which block copolymer micelles first appear at constant concentration is defined as the critical micelle temperature (CMT). Both the CMC and CMT depend on the nature and length of the copolymer blocks, type of solvent, presence of additives and temperature in much the same way as small amphiphilic molecules do. The micellization of block copolymers has been studied for decades[64,83–85] both theoretically and experimentally owing to the potential applications of the micellar structures in a wide range of technological areas such as pharmaceutics, cosmetics, emulsification, environmental remediation, lubrification and coatings. Nevertheless, the micellar self-assembly of block copolymers still presents many unresolved questions and considerable effort is being expended in new studies on the self-assembly of different classes of block copolymers (*e.g.* polyethers, polyelectrolytes and conducting polymers) in aqueous and non-aqueous solvents, supercritical fluids, and in the absence of solvent with the help of a wide range of experimental tools.[84,86,87] Here, some basic concepts and practical information (mostly based on our own experience) regarding polymeric micelle formation and characterization are outlined.

5.3.1 Thermodynamics of Micelle Formation and Critical Micelle Concentration

There are two possible models to describe the association of block copolymer chains into micelles.[88–89] For the so-called open association model, there exists a continuous distribution of micelles containing different numbers of chains, with an associated continuous series of equilibrium constants; this model of association does not lead to a well-defined CMC. Conversely, in the closed association model, the CMC does not correspond to a thermodynamic exactly defined quantity, but denotes the concentration at which a sufficient number of micelles is formed to be detected by a given method (Figure 5.5).

The equilibrium constant for equilibrium between copolymer molecules (M) and micelles (M_N) can be written as follows:

$$K_c = \frac{[M_N]_{eq}^{\frac{1}{N}}}{[M]_{eq}} \qquad (5.5)$$

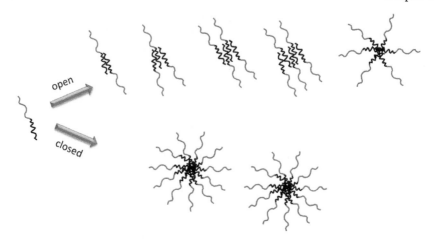

Figure 5.5   Association of block copolymer unimers according to the open and closed association models.

If the association number is sufficiently large, then the equilibrium constant can be approximated by

$$K_c = \frac{1}{[M]_{eq}} \tag{5.6}$$

where $[M]_{eq}$ is the CMC at a given temperature. Accordingly, the standard Gibbs energy of micellization is

$$\Delta_{mic}G^\circ = -RT\ln K_c = RT\ln(CMC) \tag{5.7}$$

assuming that copolymer chains are in their standard state of ideally dilute solution at unit concentration going to copolymer chains in micelles in their ideally dilute state.

The standard enthalpy of micellization is

$$\Delta_{mic}H^\circ = \frac{RTd\ln K_c}{d\left(\frac{1}{T}\right)} = \frac{RTd\ln(CMC)}{d\left(\frac{1}{T}\right)} \tag{5.8}$$

Hence, assuming that $\Delta_{mic}H^\circ$ is approximately constant within a certain temperature range, eqn (5.8) can be integrated to yield the so-called van't Hoff equation:

$$\ln(CMC) = \frac{\Delta_{mic}H^\circ}{RT} + \text{constant} \tag{5.9}$$

Hence the logarithm of the CMC can be plotted against the inverse of temperature to extract information on the micellization enthalpy.[90] Nevertheless, the data should be interpreted with caution considering the differences that exist between $\Delta_{mic}H^\circ$ values derived from eqn (5.9) and those measured experimentally. Such differences arise from the distribution of chain lengths including the small proportion of short chains generated during transfer reactions: the enthalpy of micellization derived from the van't Hoff equation relates to the most readily micellizable species, whereas the experimentally determined enthalpy relates to the whole sample.[91]

The energetics of micelle formation in organic solvents are generally different from those in aqueous media: block copolymer micelle formation in organic solvents is an exothermic (enthalpy-driven) process, as observed for polystyrene–polyethylene–polypropylene block copolymers.[92,93] This process stems from the replacement of polymer–solvent interactions by polymer–polymer and solvent–solvent interactions and the loss in combinatorial entropy, because the copolymer chains are less swollen in the micelles than in the unassociated state and the block junctions are placed at the core/shell interface. By contrast, in aqueous solvents micelle formation is often an endothermic (entropy-driven) process owing to the hydrophobic interactions between non-exposed blocks and solvent and the changes in the water structure in the vicinity of the copolymeric chains – a phenomenon more pronounced in PEO-containing block copolymers where micelles form at elevated temperatures.[94,95]

5.3.2 Micellization of Block Copolymers

Block copolymer micelles differ from micelles formed by small amphiphilic substances in terms of size (polymeric micelles being larger) and degree of segregation between the blocks that form the micelle core and those that form the shell (the surfactant tail and head group are strongly segregated), but most functional properties are common. Compared with the micelles formed by common low molecular weight surfactants, polymeric micelles have (i) lower CMC, (ii) greater thermodynamic and kinetic stability to withstand dilution as a result of the slow exchange kinetics between single chains and micellar aggregates,[96] and (iii) enhanced drug solubilization and stabilization capabilities as a consequence of larger cores and shells.[97,98]

Two general procedures can be applied for the preparation of block co-polymer micelles. In one, a solid sample of the copolymer is directly dissolved in a selective solvent; in the other, the copolymer is dissolved molecularly in a common solvent, *i.e.* one that is 'good' for both blocks, and then conditions such as temperature or composition of the solvent are changed in a way that leads to the formation of micelles. This is commonly achieved by adding gradually a selective precipitant of one of the blocks, eventually followed by elimination of the common solvent, for instance through dialysis.[99]

One very striking difference when comparing block copolymer aggregates with those formed by small surfactants relates to the kinetics of macro-molecule exchange between micelle and solution. In some cases, when the insoluble block presents a high glass transition temperature, T_g (as for the commonly used polystyrene), glassy cores are formed causing an extremely low chain mobility, sometimes making it difficult to attain equilibrium conditions and hence preventing the application of some thermodynamic models and scaling approaches.[100] A representative example was reported by

Tuzar *et al.*[101] for di- and triblock poly(methacrylic acid)–polystyrene (PMAA–PS) copolymers in water–dioxane mixtures. When prepared in dioxane-rich mixtures, a dynamic equilibrium between micelles and unimers was found, but as the aggregates were transferred to a water-rich environment *via* dialysis, 'frozen' micelles behaving like autonomous particles were detected. Subsequent sections mainly deal with micellization in an aqueous medium owing to its particular relevance in a number of application areas, including the biomedical field.

5.3.2.1 Micellization of Neutral Block Copolymers

Micellization of neutral block copolymers in an aqueous medium is mainly driven by changes in solvation with increases in temperature, especially for those copolymers bearing hydrophilic blocks of PEO, poly(2-oxazoline),[102–104] poly(N-isopropylacrylamide)[105–107] or poly(vinyl ether).[108,109]

In the particular case of poloxamers, both PEO and PPO are water soluble at temperatures below 15 °C. As the temperature is raised, the blocks become increasingly hydrophobic; PPO is more sensitive and triggers micelle formation. Further increase in temperature makes the solvent quality for the PEO block decrease, which in turn causes PEO dehydration (shell shrinking) and eventually leads to clouding and phase separation due to micelle aggregation. Micellization of this type of copolymer is entropy driven; the usual large unfavourable enthalpy of micellization comes from the hydrophobic effect. Exceptions to this rule are copolymers bearing very long[90] or extremely hydrophobic (*e.g.* polystyrene-based) blocks,[110,111] whose enthalpy of micellization is negligible as a consequence of the hydrophobic block being tightly coiled in its dispersed unimer state, so that its hydrophobic interactions with water are small and their CMC is independent of temperature.[112]

Micellization of PEO-based copolymers largely depends on composition, architecture and length of the blocks and also on temperature and the presence of additives such as cosolvents or added salts. The effects of these variables, which have been addressed in detail in a number of reviews,[84,90,96,113,114] are briefly discussed below.

As the solvent becomes poorer, the CMC decreases and the association number increases. The corresponding micelle sizes are not very temperature dependent as a result of a compensation between the increase in association number and the decrease in the expansion of the hydrophilic block shell in the poorer solvent at high temperature.[115] This effect is particularly large for copolymers with PPO blocks, for which $\Delta_{mic}H^\circ{}_{app}$ (which reflects the dependence of CMC on temperature) exceeds 200 kJ mol^{-1} and less relevant for copolymers with PS, poly(butylene oxide) (PBO) or poly(styrene oxide) (PSO) blocks, for which $\Delta_{mic}H^\circ$ can be very low owing to more efficient packing of hydrophobic monomers inside the micelle, which makes them more insensitive to temperature changes.[116]

Micellization of triblock and cyclic copolymers is entropically disfavoured compared with that of diblocks owing to looping either in the shell or in the core, which involves that two junctions being located at the core–shell boundary compared with just one for a diblock copolymer. This leads to larger CMC values and lower micellar aggregation numbers and sizes for triblock and cyclic block copolymers than for structurally related diblocks.[90,114] Micellization is favoured for cyclic copolymers compared with triblock copolymers; the former showing lower CMC values and larger micelles.[117] Also, the CMC of direct ABA triblock copolymers is lower than that of their reverse (BAB) counterparts, as predicted in simulation studies[118] and confirmed experimentally.[119–122] The effect of copolymer architecture is also manifested in the bridging of chains between micelles that occurs with telechelic block copolymers composed of PEO and two short hydrophobic blocks at the ends, as in the case of PBO–PEO–PBO block copolymers.[117,123] Micellization of tapered block copolymers has also been investigated for PEO–PBO,[124] PBO–PEO–PBO[125] and PEO–PPO.[126] The micellar properties of these copolymers are fairly similar to those of their true block counterparts, but tapered statistical PBO–PEO–PBO showed, for example, lower cloud points.

Micellization is also largely controlled by the length of the hydrophobic block and only slightly modified by the length of the hydrophilic block.[114] An increase in the number of hydrophilic EO units leads to a small increase in CMC and CMT.[113] Conversely, an exponential decrease in CMC accompanied by important increases in both micellar size and aggregation number is observed upon changes in the hydrophobic block length,[113] as predicted from scaling relations for a given family of block copolymers.[114,127–130] From these relations, a surprising decrease in association number with increasing hydrophilic block length (in particular for PEO-bearing copolymers) and an increase in micelle size compatible with a looser corona can be inferred. Looping of the triblock copolymers in the core halves the maximum length of the core chains compared with diblocks and thereby the maximum possible radius of the core. This geometric restriction implies a reduction of $2^3 = 8$ times the association number of triblock copolymers compared with their diblock counterparts.

The enthalpy of micellization decreases to zero as the hydrophobic block length increases owing to the shielding of the hydrophobic blocks from water in the unimer state, giving rise to the formation of the so-called unimolecular micelles, which is in contrast to the interaction enthalpies of the units of short blocks that are relatively extended in the molecular state.[90] For spherical micelles, the average length of the hydrophobic block limits the attainable core volume and hence the attainable association number. This situation is reached at moderate temperatures for copolymers with lengthy hydrophobic blocks and so relatively high association number.

For PEO-bearing copolymers, the relative hydrophobicity of the hydrophobic blocks was quantified as PPO : PLL : PC : PBO : PVL : PSO : PS : PCL : PG = 1 : 4 : 5 : 6 : 10 : 12 : 12 : 12 : 14, where PLL refers to poly(D,L-lactide), PC to polymethylene, PVL to poly(γ-valerolactone), PS to polystyrene and PCL to poly(ε-caprolactone). The dependence of the CMC on the hydrophobic block length increases as the hydrophobicity increases.[90,131,132] In addition, the

CMT and CMC decrease with increasing total molecular weight of the copolymer, if comparisons are made for a constant hydrophobic block/ hydrophilic block ratio. In the case of PEO-bearing copolymers, the lower the relative PEO content, the greater is the influence of the total molar mass.[113]

Cosolvents, salts or other additives also have an important impact on micellization of non-ionic block copolymers and their phase behaviour.[133] For example, alkanols and 1,4-dioxane increase the CMC of Pluronic[134–136] and PSO-based block copolymers,[137,138] indicating that organic solvents behave as better solvents than water. Also, the addition of electrolytes having anions and cations with different sizes and polarizabilities has a significant influence on micellization.[139,140] Salts affect the solvent quality of water, acting as structure makers or breakers, either increasing or decreasing the cloud point of the block copolymer solution. The influence of the ions generally follows the Hofmeister series.[141,142] For example, addition of NaCl to Pluronic solutions gave rise to a decrease in Θ temperature, enabling a sphere-to-cylinder micelle transition in a more accessible temperature range.[142,143] Also, salts lead to a decrease in the excluded volume of PEO–PBO micelles,[144] in contrast to observations with Pluronic block copolymers under similar conditions, for which the micelle dimensions remained almost unchanged.[145]

5.3.2.2   Micellization of Ionic Block Copolymers

As in the case of non-ionic block copolymers, micellization occurs in a solvent that is selective for one of the blocks. In general, solutions of ionic block copolymers display trends similar to those observed for non-ionic block copolymers. The behaviour of ionic copolymers is additionally governed by the repulsive electrostatic interactions between charged polymer units, introducing new factors governing the structure and properties of the micellar system such as pH, addition of salts, polar interactions with other components or the presence of a water-miscible cosolvent. The sensitivity to all these factors makes ionic copolymers interesting candidates for preparing stimuli-responsive micelles. Interestingly, the existence of repulsive electrostatic interactions between micelles may result in liquid crystalline behaviour or gel formation at sufficiently high concentrations.[100]

Depending on the solvent selectivity, the aggregates formed by these copolymers are classified as block ionomers or block polyelectrolytes.[146] Block ionomers occur in organic solvents, where the ionic moieties form the core of the aggregate and are surrounded by the apolar blocks, in a way similar to the formation of reverse micelles of low molecular weight surfactants. Conversely, in water or mixtures of water and polar organic solvents, the ionic copolymer arranges in micellar structures with the non-ionic blocks at the core and the ionic blocks at the shell. In any case, the copolymer may contain different blocks with ionic or ionizable groups including poly(4-vinyl-pyridiniumalkyl halides) (P4VP), poly(methyl acrylate) (PMA), poly(methyl methacrylate) (PMMA), poly(styrene sulfonate), polypeptide blocks such as polyarginine, polylysine and poly(aspartic acid) or polydendritic charged blocks.

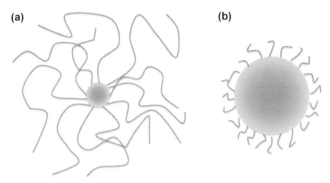

Figure 5.6 Representation of (a) a star and (b) a crew-cut block copolymer micelle.

Ionic copolymers can be also classified as a function of the relative lengths of the soluble ionic and insoluble blocks and the subsequent arrangement of the polymeric chains in solution (Figure 5.6). Star ionic block copolymers are composed of insoluble blocks that are relatively short, leading to the formation of micelles with compact cores and large, extended shells that resemble a star shape; by contrast, the so-called crew-cut copolymers are formed by relatively long insoluble blocks that form micelles with large cores surrounded by a relatively thin shell of hydrophilic corona chains. Star micelles can be prepared *via* direct dissolution of an asymmetric block copolymer; in contrast, the large weight fraction of the insoluble block makes this method impracticable for the preparation of crew-cut micelles. In this latter case, the copolymer is initially dissolved in a good solvent for both blocks and then water is slowly added in such a way that the quality of the mixed solvent for the hydrophobic block gradually decreases until micelle formation.[147]

Owing to the greater incompatibility between the ionic and non-ionic (usually hydrophobic) blocks, micelles start to form at very low concentrations and, consequently, have high stability. In general, an increase in temperature shifts the single chain–micelle equilibrium towards single chains, hence increasing the CMC, but this behaviour is additionally modulated by the salt content and the length of the hydrophobic block. As occurs for non-ionic copolymers, the CMC decreases and the association number increases as the hydrophobic block enlarges. Regarding the influence of the ionic block on CMC, contradictory experimental data have been reported. The dependence of CMC on the ionic block length was found to be parabolic for polystyrene–poly(sodium acrylate) (PS–PANa) block copolymers owing to the concurrency of two opposite effects. Shorter ionic blocks enhance the solubility and, in turn, CMC and association numbers decrease; as the ionic block length increases, the solution ionic strength rises, which leads to a reduction of the solubility of longer ionic blocks, thus favouring micellization (lowering the CMC).[148] This dependence of micellar parameters on ionic block length becomes less important as the hydrophobic block is enlarged. For example, CMC values of polyisobutylene–poly(methacrylic acid) (PIB–PMAA) copolymers were found to be insensitive to changes in the length of the PMAA block.[149]

The CMC decreases and the association number increases with increase in salt concentration until a threshold value after which these effects level off as a consequence of changes in chain conformation of the ionic block.[148] The molecular weight and the architecture of the block polyelectrolyte also influence the aggregation behaviour: diblock copolymers readily form micelles and possess larger association numbers than their triblock counterparts of comparable mass and composition, as occurs for non-ionic copolymers. Micellar sizes are depend largely on the hydrophobic block length and, to a minor extent, on the hydrophilic block length, as observed for polystyrene–poly-(methacrylic acid) (PS–PMAA) copolymers.[150,151] Also, micellar sizes increase with increase in the degree of ionization of the hydrophilic block owing to unscreened electrostatic repulsions between ionized moieties, which result in stretching of shell chains; for instance, micelles of PIB–PMAA in water increased their hydrodynamic radii from 50 to 100 nm as the pH increased from 4 to 8, as a result of increased dissociation of the acid PMAA block.[149]

5.3.2.3   *Micellization of Double Hydrophilic Block Copolymers*

A special class of block copolymers is those composed of only hydrophilic blocks, the so-called hydrophilic–hydrophilic copolymers or double hydrophilic copolymers, such as PEO-poly(2-vinylpirydine) (PEO-P2VP), PEO–poly(*N*-isopropylacrylamide) (PEO–PNIPAM) and PEO–poly(*N*,*N*-dimethyl-aminoethyl methacrylate) (PEO–PDMAEMA).[152–155] These copolymers can bear only non-ionic, non-ionic–ionic and ionic–ionic blocks in their structure. If there exists one anionic block and one cationic block, the resulting copolymer is termed a polyampholite copolymer. In aqueous solution, double hydrophilic copolymers behave as unimers similar to classical polymers or polyelectrolytes. The amphiphilic characteristics, such as surface activity and micelle formation, only appear under the influence of a given external stimulus, mainly temperature (especially for fully non-ionic ones), pH or ionic strength changes. Micellization can further be induced by complex formation of one of the blocks either by electrostatic interaction with oppositely charged polymers, by hydrophobic interaction with surfactants or by insolubilization in the presence of metal derivatives to form polyion complexes.[156]

5.3.3   Characterization of the Micellization Process and the Micelle Structure

Numerous methods can be applied to study block copolymer micellar systems in an effort to shed light on different aspects of the micellization process and also of the structure of the resulting micelles in aqueous solution, as described below. Different techniques inform about different features of the system and on some parameters responsible for the observed behaviour. Additionally, each method is influenced in a different way by changes in micelle–unimer equilibrium and has sensitivity in different concentration ranges. Sometimes even the process of measurement itself may induce changes in the thermodynamics or the structural parameters of the system, complicating the interpretation of results.

5.3.3.1 Surface Tension

A familiar method for the determination of the CMC of a block copolymer is measurement of the surface tension (γ) of aqueous solutions over a wide range of concentration. The method detects completion of the Gibbs monolayer at the air/water interface and is a secondary indicator of the onset of micellization. A sharp break in $\gamma(c)$ at the CMC and a constant value of γ thereafter are typical of block copolymers with narrow block-length distributions. In some cases, two breaks in the curve of surface tension against concentration are observed, the first break corresponding to chain rearrangements in the air/water interface and the second to the 'true' CMC, as observed for some Pluronic copolymers (Figure 5.7).[157] Surface tension can be measured using du Noüy ring and Wilhelmy plate techniques.[158]

5.3.3.2 Viscometry

The hydrodynamic properties of block copolymer micellar systems can be evaluated from the value of the intrinsic viscosity $[\eta]$ estimated as follows:

$$\frac{\eta_{sp}}{C} = [\eta] + k_{H}[\eta]^{2}C \qquad (5.10)$$

where η_{sp} and k_{H} represent the specific viscosity and the Huggins coefficient, respectively. By studying the concentration dependence of viscosity, the CMC and changes in micellar size and mass can be determined. For example, the intrinsic viscosity decreases and the Huggins coefficient increases as micelles become smaller.

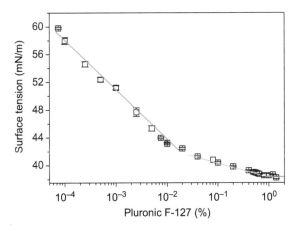

Figure 5.7 Surface tension values of Pluronic F-127 in water at 20 °C. Reproduced from Ref. 157 with permission of Springer.

5.3.3.3   Fluorescence Spectroscopy

Fluorescence probes can be used to provide insight into probe mobility and the local microenvironment of block copolymer micelles. For example, the most extended case is the monitoring of the intensity change of the first and third vibronic peaks of pyrene when the dye transfers from the aqueous phase to the micelle core upon block copolymer micellization.[159,160] Steady-state, time-resolved fluorescence and depolarization fluorescence can be used to characterize the micelle formation and equilibrium with unimers, and also the micellar association number and the micropolarity.[161] Owing to their sensitivity on the local scale in contrast with the global scale sensitivity of scattering methods, the use of spectroscopic techniques allows the determination of very small CMC values that are inaccessible by scattering techniques, probing the internal viscosity of the micellar core and solubilization of low molecular weight substances into micelles. Micelle–unimer equilibrium or chain exchange between micelles can also be studied using specially labelled block copolymers with fluorophore groups. Fluorescent donor–acceptor pairs can also be used to detect the formation of micelles and the properties of the interface between the micellar core and the shell by Förster or fluorescence radiation energy transfer (FRET). More detailed information can be found in Chapter 2.

5.3.3.4   Scattering Techniques

Scattering techniques have been proven to be among the most powerful approaches for investigation of micellar structures and properties. By analysing the interactions between the particles and the probing radiation (light, X-rays or neutrons), physical properties such as micellar size, shape and internal structure can be investigated.

5.3.3.4.1   Static Light Scattering.   Static light scattering (SLS) provides a convenient means of determining the CMT. Measurement of the temperature dependence of the excess intensity ($I_{ex} = I - I_s$) of scattering from solution (I) over that from pure solvent (I_s) allows the CMT to be estimated as the temperature at which the intensity curve leaves the baseline set at low temperatures for unassociated molecules (unimers).[90] The CMC can be also located at the concentration where the function Kc/I_{ex} departs from its value for unimers. See eqns (2.26)–(2.30) in Chapter 2.

The average molecular weight of the micelles can be determined using SLS if measurements are performed well above the CMC, where the micellization equilibrium is shifted towards micelles and the amount of unimers in solution is very small and therefore their contribution to the light-scattering intensity is negligible, by means of the following expression:

$$\frac{Kc}{\Delta R_\theta} = \frac{1}{MP(\theta)} + B_2 c + B_2 c^2 + \cdots \qquad (5.11)$$

where c is the polymer concentration, K is an optical constant:

$$K = \frac{4\pi^2 n^2 \left(\frac{dn}{dc}\right)^2}{\lambda^4 N_A}$$

n being the refractive index of the solvent, dn/dc the specific refractive index increment, λ the wavelength of light and N_A Avogadro's number, $P(\theta)$ is the micellar form factor, which describes the angular dependence of the scattered light, B_i are the virial coefficients, M is the micellar molecular mass and ΔR_θ is the excess Rayleigh ratio (Chapter 2). Commonly, block copolymer micelles in aqueous medium are small compared with the wavelength of light and act as point scatterers, hence the intensity does not depend on scattering angle θ. Conversely, for large micelles, intensity is sensitive to θ. This is usually tested by measuring the dissymmetry ratio, Z, which is defined as the ratio of the intensity scattered from the solution at $45°$ to that at $135°$ ($Z = I_{45°}/I_{135°}$), as a function of the concentration and the temperature. In any of the two former situations, extrapolation to zero angle gives values of micellar M_w corrected by intramolecular interference, whereas the angular dependence gives values of z-average radius of gyration, R_g, for large block copolymer micelles.

The dependence of scattering intensity on concentration provides information about the interactions of the micelles in solution *via* the second virial coefficient B_2 and so the excluded volume of one micelle to another. Fitting of experimental data to the first two terms on the right-hand side of eqn (5.11) [with $P(\theta) = 1$, the so-called Debye equation] is often unsuccessful provided that they are markedly curved (Figure 5.8). This indicates a strong intermicellar interaction, so that the data can be alternatively fitted using the Carnahan–Starling equation[162,163] (which is equivalent to using the virial expansion for the structure factor of effective hard spheres taken to its seventh term) to obtain the following expression:

$$\frac{Kc}{R_\theta} = \frac{1}{SM_w} \tag{5.12}$$

where

$$\frac{1}{S} = \left[(1 + 2\phi)^2 - \phi^2 \left(4\phi - \phi^2\right)\right](1 - \phi)^{-4} \tag{5.13}$$

S being the interparticle structure factor and ϕ the volume fraction of equivalent uniform spheres, which may be used to compute the swollen volume once the dry volume calculated from the intercept is known.

5.3.3.4.2 Dynamic Light Scattering.

As stated in Chapter 2, dynamic light scattering (DLS), also known as quasi-elastic light scattering (QELS) or photocorrelation spectroscopy (PCS), provides direct evidence of the existence of micelles. This technique measures the temporal fluctuations of

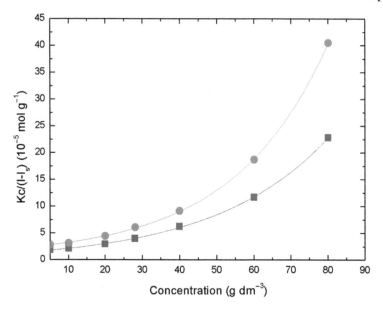

Figure 5.8 Debye plots for aqueous solutions of block copolymers (●) $EO_{137}SO_{18}EO_{137}$ and (■) $EO_{10}SO_{10}EO_{10}$ in aqueous solution at 25 °C. Adapted with permission from Ref. 181. Copyright 2008 American Chemical Society.

the light scattered intensity. Hence, given a reliable method of data analysis (*e.g.* CONTIN),[164] DLS provides information on the intensity distribution of the apparent translational diffusion coefficient (D_{app}) for the micelles and, through the Stokes–Einstein equation:

$$r_{h,app} = \frac{k_B T}{6\pi\eta D_{app}} \qquad (5.14)$$

where k_B is the Boltzmann constant, T the absolute temperature and η the solvent viscosity, the corresponding distribution of apparent hydrodynamic radius ($r_{h,app}$), *i.e.* the apparent radius of the hydrodynamically equivalent hard sphere, can be estimated. Extrapolation to zero concentration allows the true hydrodynamic radius to be obtained. The weight fractions of unimer and micelle species can be estimated through the intensity contribution of each species coming from the bimodal size distribution together with the association number. Information on frictional and thermodynamic interactions in micellar systems can also be obtained from the concentration dependence of the diffusion coefficient D_{app}:

$$k_d = \left| \frac{\partial \left(\dfrac{D_{app}}{D_0} \right)}{\partial c} \right|_{c=0} = 2B_2 M_w - k_f - 2V \qquad (5.15)$$

where V is the partial specific volume (generally small compared with the other terms in the right-hand side of the equation), k_d is the concentration coefficient and k_f is the frictional coefficient. The use of depolarized light in DLS experiments can also allow information to be gained on the rotational diffusion coefficient of non-spherical micelles.

5.3.3.4.3 Small-Angle X-Ray and Neutron Scattering. Small-angle scattering techniques are powerful tools to determine micellar properties, including overall micelle size, individual core and shell dimensions, core/shell and shell/solvent interface sharpness and core swelling by the selective solvent by means of model fitting of the experimental scattering curves.[165,166] In dilute conditions, intra-micellar scattering can be measured, *i.e.* the so-called form factor. In semi-dilute and concentrated conditions, inter-micellar scattering becomes predominant and positions of the found reflections from microstructural periodicities provide information on morphology. Small-angle X-ray scattering (SAXS) is appropriate when the electron density differences between the solvent and the different parts of the soluble species are sufficient to diffract X-rays; in this way, the individual dimensions of the core and the shell can be determined, usually through the assumption of a model. For small-angle neutron scattering (SANS), selective isotope labelling of part of the block copolymer structure permits changes in scattering contrast that can be used to obtain local information on the chain configuration and the intra-micellar structure. For example, making the scattering cross-section of the core or the shell invisible, the dimensions of the visual part of the micelles can be determined.[167] Nevertheless, SANS is used less often than SAXS, mostly owing to technical difficulties regarding the availability of instrumentation and isotopically labelled samples.

5.3.3.5 Nuclear Magnetic Resonance Spectroscopy

Nuclear magnetic resonance (NMR) spectroscopy is useful for characterizing the local environment of block copolymers in solution by detecting the chemical shifts of certain atoms such as ^{1}H and ^{13}C before and after micellization (see Chapter 2). Submicellar properties such as chain distribution and chain mobility are readily accessible by this technique.[168] For example, the mobility of insoluble segments is very much reduced when the core of the micelle is formed, resulting in reduced intensity of the corresponding NMR peaks. Whereas DLS gives the mutual-diffusion coefficient of micelles in solution, the self-diffusion coefficient is available from pulsed-gradient spin–echo NMR (PGSE-NMR). The concentration dependence of the two diffusion coefficients differ,[169] as do the averaged diffusion coefficients obtained from DLS and NMR methods,[170] but given a narrow distribution of micelle size they should be identical in the limit of infinite dilution. PGSE-NMR is also effective for characterizing micellar shape, structure and intermicellar interactions.[171]

5.3.3.6 Gel Permeation Chromatography

Gel permeation chromatography (GPC) provides information on the unimer–micelle equilibrium and also the size distribution of the micelles formed.

In the course of the analysis, large species (micelles) are continuously separated from smaller species (unimers), which may disturb micellization equilibria and lead to erroneous results. This problem can be overcome using sample concentrations well above the CMC to avoid dissociation of the micelles as the solution is diluted in its passage through the GPC column. In this respect, a useful adaptation is eluent GPC (EGPC). In this method, a copolymer solution is used as eluent and the eluent is probed by injecting a solution of the same copolymer at different concentrations.[172] EGPC gives the distribution of hydrodynamic volume, and hence the apparent hydrodynamic radius,[173] of the species in the eluent using universal calibration, similar to that available from DLS but mass weighted rather than intensity weighted.

5.3.3.7   Microscopy

Transmission electron microscopy (TEM) allows the visualization of the size, shape and internal structure of block copolymer micelles.[174] The technique relies on the production of phase contrast between the core and the shell by selectively staining one of the micelle parts. For sample preparation in TEM, a drop of a dilute solution of micelles is spread on a carbon film and the solvent is allowed to evaporate. Then the dry isolated micelles can be stained and observed under a microscope. The dehydration process undergone by micelles during the sample preparation leads them to be in a collapsed state, which implies some limitations in the correct determination of the micellar size, but valuable semiquantitative information can be obtained. Cryo-TEM can overcome this limitation: the micellar solution is rapidly frozen, *via* immersion in liquid nitrogen, to 'trap' structures formed in solution by means of vitrification of the sample and avoiding crystallization in the solvent. The sample is subsequently stained and observed under appropriate low-temperature conditions.[175] Section 2.2 in Chapter 2 contains more detailed information about these imaging techniques.

Atomic force microscopy (AFM) permits the imaging of block copolymer micelles on the surface of a substrate. The technique relies on the interaction force between a sharp tip and the substrate. The deflection of a cantilever to which the tip is attached due to the force that it experiences as it approaches the surface is measured using a reflected laser beam or the interference pattern of a light beam from an optical fibre. The experiment is usually performed with the so-called 'tapping mode,' where the tip oscillates in proximity to the sample surface to avoid damage to the sample. The sample or the tip is then moved so that the tip rasters over the surface to build up an image. This image contains information on surface topography and phase contrast of the micellar sample.[176]

5.3.3.8   Other Methods

Vapour pressure and osmometric techniques are standard methods for the determination of the number-average molecular weight and CMC values.

Since micellization in aqueous systems is usually an endothermic event, differential scanning calorimetry (DSC) can be used to detect its onset (so defining the CMT) as a consequence of the pseudo-phase undergone, which causes changes in the differential power supplied to the sample. This technique can be used also to detect the CMC and the enthalpy of micellization. The change of sound velocity upon micellization is small, but the specific velocity (the ratio of the sound velocity change upon micellization relative to the sound velocity in water) has been also used to detect the CMC.[177] Similarly, isoperibol microcalorimetry provides information about CMC and enthalpy of micellization.[95] An alternative method is to measure the ultrasound absorption caused by the influence of changes in pressure and temperature on the molecule–micelle equilibrium. The ultrasound absorption coefficient (α) at frequency f is measured; the quantity α/f^2 relates to the relaxation time (or distribution of relaxation times) of the perturbation, allowing the estimation of CMC or CMT depending on the chosen conditions.[178] Sedimentation velocity analysis can give valuable information on the size and size distribution of block copolymer micelles.[179] In this method, the velocity at which each solute species is displaced, due to the influence of a strong centrifugal force, is measured. The sedimentation velocity depends on the molecular weight, size and density of the species itself and on the friction forces developed on the solute by the medium. This method has been employed for the study of micelle–unimer equilibrium in a variety of block copolymer colloidal systems.

5.3.4 Micelle Structure

Two of the main questions that predominate in the study of micelle formation by block copolymers are what the structure of an isolated micelle is and which configuration of a block copolymer chain is incorporated in a micelle. The generally accepted idea is that most micelles possess a relatively compact core consisting of the insoluble blocks and a shell consisting of the soluble blocks (Figure 5.9). The chains in the core can be considered to be in a state resembling the homopolymer melt if the solvent is very poor for them (*i.e.* highly selective), although some swelling of the core by solvent molecules cannot be ruled out, especially in cases of low selectivity.[99] The shell is formed by the soluble blocks, which are in a well-solvated state but considerably stretched because of geometric constraints, especially in the region near the core. Although the micelles are generally spherical with a narrow size distribution, changes in shape and size distribution may occur under certain conditions. Micelles are stable in time at fixed conditions and their characteristics depend on the thermodynamic quality of the solvent and on temperature. For this reason, it is impossible to study the same system under different conditions, *e.g.* in a different solvent, at a different temperature or at different concentrations. To circumvent this problem, cross-linking of the micellar core or corona by physical (photo- or electron irradiation-induced cross-linking) or chemical means helps to stabilize the micellar structure.[180]

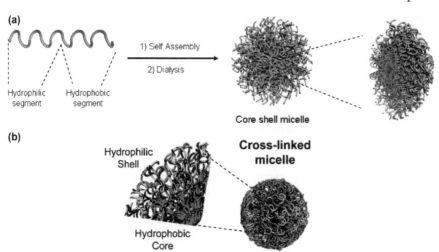

Figure 5.9 (a) Aqueous solution self-assembly of an amphiphilic block copolymer into a spherical micelle and (b) scheme showing a cross-linked block copolymer micelle.
Reproduced from Ref. 180 with permission of The Royal Society of Chemistry.

Micellization of block copolymers in the form of cylindrical micelles has also been observed[167,181–183] (Figure 5.10). This occurs when the radius of the micelle core exceeds the stretched length of the hydrophobic block. This change in the overall structure may be due to the high content of the block copolymer in the insoluble part, to the block copolymer architecture and to the selectivity of the solvent. Also, a transition from spherical to cylindrical shape can be observed for both neutral and ionic block copolymers on changing the solution temperature, the solvent quality or the concentration of additives such as salts or homopolymers.[184–187] Transition from spherical to cylindrical shape has been observed for Pluronic P85 and F88 micelles on raising the solution temperature as a result of the increase in the association number.[90,186] The problem of morphological transitions in block copolymers, and especially in block polyelectrolytes, was studied extensively by Eisenberg and co-workers.[147] For instance, PS–PAA diblock copolymers prepared with constant PS block length (PS_{200}) and decreasing PAA block lengths (PAA_{21} to PAA_4) led to micellar shapes ranging from spherical crew-cut to rod-like micelles and to vesicular aggregates for the lowest PAA contents. A similar structural transition has been observed upon modification of the hydrophobic/hydrophilic block ratio in PEO–PSO–PEO block copolymers with short PEO block lengths,[181] and for poly(2-hydroxypropyl methacrylate)–poly(glycerol monomethacrylate) (PHPMA–PGMA) block copolymers.[188] In block copolymers bearing long hydrophobic blocks (such as BO blocks), highly hydrophobic monomers (*e.g.* PS, PG or PSO) or long asymmetric structures, the insoluble block has a collapsed conformation

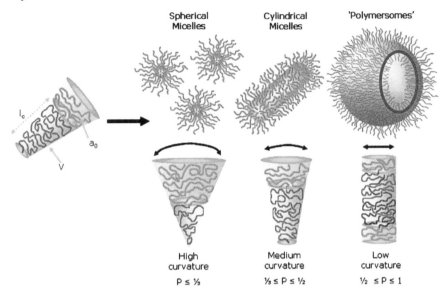

Figure 5.10   Various self-assembled structures from amphiphilic block copolymers in aqueous solution. The type of structure formed depends on the inherent curvature of the copolymer chain estimated through the molecular packing parameter *p*.
Reproduced from Ref. 183 with permission of Wiley.

and is protected from the hostile solvent by the soluble block, which gives rise to structures called unimolecular micelles.[114,189,190]

   Triblock copolymers with short hydrophobic end blocks (so-called telechelics), such as HEUR (hydrophobic ethoxylated urethanes), PBO–PEO–PBO or polyacrylamide–PEO–polyacrylamide (PAAm–PEO–PAAm) copolymers, can self-assemble into flower-like micelles, in which the midblocks are looped. Bridging of the micelles occurs when the middle block spans the space between different micelles.[191,192] Other, more exotic, micellar shapes have also been found. For example, the formation of spherical core–shell–corona micelles ('onion-type micelles') was observed for triblock ABC copolymers such as PS–P2VP–PEO,[193] PPO–PDMAEMA–PEO,[194] polybutadiene–poly-styrene–poly(ethylene oxide) (PB–PS–PEO)[195] or alkynyl–poly[oligo(ethylene glycol) monomethyl ether methacrylate]–poly[2-(dimethylamino)ethyl metha-crylate]–poly[2-(2-diethylamino)ethyl methacrylate] (alkynyl–POEGMA–PDMA-PDEA).[196] Poly(fluorinated butadiene)–polystyrene–poly(ethylene oxide), poly(2-methylvinylpyridinium iodide)–PEO (P2MVP–PEO) and PAA–PAAm block copolymers formed core–shell disk-shaped micelles,[197,198] whereas pH-sensitive vesicular aggregates have been obtained with a polybutadiene-poly(sodium glutamate) (PB–PGANa) copolymer due to the stiffness asym-metry introduced by the rigid polypeptide block, and with PS–PAA diblocks with short PAA blocks (Figure 5.10).[147]

Segmented micelles were observed for a mixed-arm star block terpolymer architecture composed of PEO, poly(perfluoropropylene oxide) (PFPO) and poly(ethylethylene) (PEE) blocks. The connection of the water-soluble PEO and the two hydrophobic but immiscible components at a common junction leads to molecular frustration when dispersed in aqueous medium and the formation of the compartmentalized micelles.[199] Micelles with two different faces, *i.e.* Janus micelles, were obtained from an ABC triblock copolymer of PS-PB-PMMA with a PB cross-linked block at the interface between PS and PMMA fringes, having a typical PB sphere–PS/PMMA lamellar mesomorphic structure in the solid state,[200] or from the co-assembly of complementary block copolymer chains.[198,201]

Experimental observations on micelle structure have been accompanied by a great number of theories and models that can be classified into two major groups: scaling and self-consistent mean field theories. These models were developed to describe the micellization process and the structure of the micelles, relating fundamental properties of block copolymer micelles at equilibrium with the molecular characteristics of the block copolymers and the solvent quality. Interested readers are referred to comprehensive reviews.[67,96,202–204]

5.3.5 Safety Evaluation

Before testing *in vivo*, novel polymers used for preparing micelles should be carefully purified and then tested in cell cultures at concentrations below and above the CMC to gain an insight into their potential bio- and immune compatibility. Then, simplified *in vivo* studies are carried out to confirm the safety. There are several guidelines of the International Conference on Harmonization (ICH) that could be helpful for testing the safety of novel polymers in non-clinical studies, for example, guidelines ICH S1A, S2B, S3A, S5A, S7A and M3. Most regulatory issues refer to the use of the novel polymers as pharmaceutical excipients and depending on the intended administration route and the frequency and duration of the use, different protocols should be followed.[205,206]

In brief, the first safety-evaluation steps are devoted to determining by means of computational methods and *in vitro* tests the potential genotoxicity, cytotoxicity and metabolism and the ability of the polymer to be absorbed across biological membranes. Quantitative structure–activity relationship (QSAR) models enable the structure of the novel compound to be compared with those of chemicals previously identified as toxic, particularly carcinogenic or mutagenic. There are several commercially available human cell lines to test *in vitro* significant perturbations in toxicity pathways and to elucidate the extent of metabolism and whether potential reactive metabolites can be formed. Membrane penetration studies should also be conducted.[206] *Ex vivo* methods are gaining increasing attention for predicting *in vivo* behaviour while minimizing the use of animal tests; for instance, the Dermal Corrosion Test (Corrositex), the Bovine Corneal Opacity and Permeability (BCOP) Test and the Hen's Egg Test on the Chorioallantoic Membrane (HET-CAM) can provide useful information about the bio-compatibility of new copolymers.[207] If these preliminary tests did not indicate that the polymer can cause relevant toxicity problems, then the polymer can be tested *in vivo* in repeat-dose toxicity studies in rodent and non-rodent species.[206]

5.4 Pharmaceutical Applications

5.4.1 Drug Solubilization

Drugs solubility is usually higher in aqueous media containing polymeric micelles than in pure water. Hydrophobic molecules tend to move from the aqueous phase to the micellar cores, where they become thermodynamically stabilized, whereas the hydrophilic corona is responsible for maintaining the drug-loaded micelles dispersed in the medium. As a consequence, the apparent solubility of the drugs may increase considerably. For a given polymer, the amount of drug that can be incorporated into the micelles and their depth inside the micellar structure depend on drug polarity. Hydrophobic drugs are prone to incorporate into the apolar cores of polymeric micelles, whereas medium-polarity drugs remain in the core–corona inter-phase and the polar drugs interact weakly with the corona (Figure 5.11). Several methods have been reported for elucidating the localization of a compound inside micelles.[208]

Methods to load drugs into polymeric micelles are quite varied (Figure 5.12). The simplest approach is to disperse the polymer and the drug in the aqueous medium under mild agitation and/or stirring to facilitate the incorporation of the drug molecules into the micelles being formed (Figure 5.12, A). This approach is valid for moderately hydrophobic polymers and drugs and the amount encapsulated depends on the partition coefficient of the drug between the micelles and the aqueous medium. If the polymer or

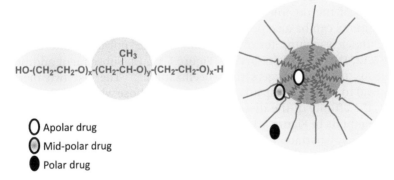

Figure 5.11 Scheme of the localization of different drugs in a PEO–PPO–PEO micelle as a function of their polarity.

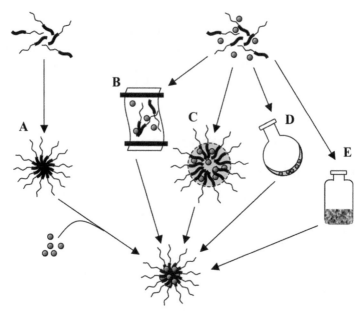

Figure 5.12 Common drug-loading procedures: (A) simple equilibrium, (B) dialysis, (C) oil-in-water emulsion, (D) solution casting and (E) freeze-
drying.
Reproduced from Ref. 209 with permission of Elsevier.

the drug cannot be easily dispersed in aqueous medium, they can be previously dissolved in a volatile or water-miscible organic solvent. In a subsequent step, the organic solvent is removed. Solutions of polymer and drug
in a water-miscible organic solvent can be dialysed against water for slow
replacement of the organic solvent inside the dialysis bag and the formation
of the drug-loaded micelles (Figure 5.12, B). If the solvent is not miscible
with water, incorporation of water into the organic solution may result in an
oil-in-water emulsion, with the drug molecules and the hydrophobic regions
of the copolymer inside the oily droplets (Figure 5.12, C). Slow diffusion and
evaporation of the organic solvent result in drug-loaded micelles. The organic solutions can be also processed in order first to evaporate the organic
solvent under vacuum, forming a film of the polymer and the drug
(Figure 5.12, D), or by freeze-drying (Figure 5.12, E). Subsequent addition of
an aqueous medium triggers the micellization.[209] The last two methods
usually lead to greater encapsulation yields than those which rely only on the
non-specific affinity of the dissolved drug for the preformed micellar cores
(*e.g.* the methods in Figure 5.12, A and B). In a comparative study carried out
with PEG–polyaspartate block copolymers, generically designated PEG–
P(Asp), the incorporation of the antitumour agent camptothecin was evaluated by applying three different methods.[210] The dialysis method involved
the preparation of a solution of the block copolymer and camptothecin
in dimethyl sulfoxide, which was dialysed against water and then filtered.

This approach rendered a clear micellar solution with low yields of camptothecin incorporation (1–2%). The emulsion method consisted in mixing the block copolymer and camptothecin solution in methylene chloride with distilled water. Then the methylene chloride was evaporated and the remaining aqueous solution was filtered, which resulted in cloudy solutions due to the presence of aggregates larger than polymeric micelles, with loading yields of 20–30%. The evaporation method involved the evaporation of chloroform from the block copolymer and camptothecin solution, followed by the addition of distilled water to the residue and sonication. The resultant micellar solution was clear and the drug incorporation yield ranged from 50 to 100%.[210]

A more elaborate approach consists in the covalent conjugation of the drug to the polymer by means of hydrolysable bonds before forming the micelles. Compared with conventional drug–polymer conjugates, the labile bonds are less exposed to the aqueous medium inside the micelles and, consequently, the hydrolysis rate is much slower.[211] When the drug is just loaded through non-covalent interactions (as in the methods shown in Figure 5.12), body distribution and cellular accumulation can be notably altered, but the pharmacological activity remains the same.[212] By contrast, covalent conjugates may exhibit totally different pharmacodynamic and toxicological profiles compared with those of the free drug. Therefore, additional preclinical and clinical studies are required to elucidate the extent and the repercussions of those changes.[213]

The ability of the polymeric micelles to solubilize drugs can be characterized using the following descriptors:[214]

(a) Molar solubilization capacity, χ, *i.e.* the number of moles of drug that can be solubilized by 1 mol of copolymer in the micellar state:

$$\chi = \frac{S_{\text{tot}} - S_{\text{w}}}{C_{\text{copolymer}} - \text{CMC}} \tag{5.16}$$

(b) Micelle–water partition coefficient, P, which is the ratio of drug concentration in the micelle to the drug concentration in water, for a fix copolymer concentration:

$$P = \frac{S_{\text{tot}} - S_{\text{w}}}{S_{\text{w}}} \tag{5.17}$$

(c) Standard free energy of solubilization, calculated from the molar micelle–water partition coefficient, P_{M} (*i.e.* P for $C_{\text{copolymer}} = 1$ M):

$$\Delta G_{\text{s}}^{\circ} = -RT \ln \frac{\chi(1 - \text{CMC})}{S_{\text{w}}} \tag{5.18}$$

(d) Solubilization capacity (mg) per gram of hydrophobic block calculated as the amount of drug dissolved in the copolymer solution in excess of that dissolved in an equivalent volume of solvent medium and referred to the mass fraction of hydrophobic blocks in the copolymer.

(e) Number of drug molecules solubilized per micelle calculated using the equation

$$n_{\mathrm{s}} = \frac{(S_{\mathrm{tot}} - S_{\mathrm{w}})N_{\mathrm{A}}}{n_{\mathrm{m}}} \qquad (5.19)$$

In the above equations, S_{tot} is the total drug solubility, S_{w} is the drug solubility in water (molar concentration) and n_{m} is the number of micelles per litre of solution, estimated as

$$n_{\mathrm{m}} = \frac{(C_{\mathrm{copolymer}} - \mathrm{CMC})N_{\mathrm{A}}}{N} \qquad (5.20)$$

where $C_{\mathrm{copolymer}}$ is the molar concentration of copolymer, N_{A} Avogadro's number and N the aggregation number. Since above the CMC the unimer concentration remains constant and equals to the CMC, the copolymer concentration in the micellar form can be estimated as $C_{\mathrm{copolymer}} - \mathrm{CMC}$.

In general, the more hydrophobic the core, the greater is the ability of the micelles to incorporate non-polar drugs.[215] For a given value of the hydrophobic–lipophilic balance (HLB), solubilization is more effective if the molecular weight is high. A rise in temperature usually favours the self-assembly of the copolymer and augments the number and/or the size of the micelles and, as a result, their capability to host drug molecules and to increase the apparent solubility of the solute also increases.[215] The affinity of the micelles for a given drug and the thermodynamic and kinetic stability of the drug-loaded micelles can be tuned by combining copolymers of different molecular weight and HLB. For example, polymeric micelles of poloxamers having different PEO/PPO block lengths can lead to the formation of mixed micelles that exhibit more efficient solubilization and are more stable against dilution.[216,217] When the micellar dispersions are diluted, separation of unimers proceeds slowly even when the final concentration is lower than the CMC. Such delayed disintegration is one of the main advantages of polymeric micelles as drug solubilizers and hence as carriers in the organism. In contrast to micelles of conventional surfactants or inclusion complexes with cyclodextrins, copolymer micelles with hydrophobic cores and PEO-based hydrophilic shells exhibit marked kinetic stability in aqueous media and therefore they do not release the encapsulated drug instantaneously, but in a sustained way that avoids the risk of drug precipitation in the physiological fluids. Moreover, the PEO shell also provides steric stabilization and makes the opsonization and internalization by macrophages difficult,[218,219] which in turn favours the circulation in the bloodstream of the drug-loaded micelles for up to several hours. The residence time in the bloodstream, the control of drug release and the targeting to the intended tissues can be modulated by adjusting the polymer architecture, molecular weight and HLB, as discussed in the following sections.[209,212,220] Some examples of the ability of the polymeric micelles to accommodate hydrophobic drugs, increasing their apparent solubility and protecting them from chemical or biological degradation, are presented below.

Poloxamine micelles have been tested as host agents for a variety of active substances such as antifungals, antivirals and hypolipidaemic drugs.[221–223] As explained in earlier sections, poloxamines (commercially available as Tetronics) are X-shaped copolymers formed by four PPO–PEO arms bonded to an ethylenediamine central group, which provides pH-responsive micellization properties in addition to the temperature sensitivity.[224,225] The physical–chemical conditions of the medium, particularly the pH and the ionic strength, alter the extent of protonation of the ethylenediamine central group, perturbing the hydrophobic interactions that govern the self-assembly phenomenon. As the pH of the medium decreases, the Coulombic repulsions among the positively charged amine groups make the aggregation more difficult, raising the CMC and decreasing the size of the micelles. The influence of pH on the process of micellization was evidenced by the changes undergone in the capability to solubilize griseofulvin, a pH-independent poorly soluble antifungal drug (4 mg per 100 mL). To carry out the experiments, dispersions of the poloxamine T904 (10%, 4 mL) were placed in glass ampoules containing an excess of drug (40 mg). The ampoules were flame-sealed and rocked at 25 °C and 50 rpm for 5 days and then filtered through 0.45 μm cellulose acetate membrane filters in order to remove the non-solubilized drug particles. Control experiments were carried out using as solvent the medium in which the micellar dispersions were prepared. As can be observed in Figure 5.13, the polymeric micelles enhanced the drug solubility up to threefold in 0.1 M HCl medium and up to

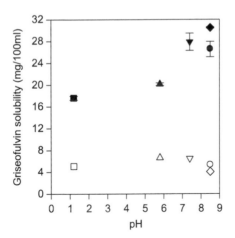

Figure 5.13 Solubility of griseofulvin in aqueous media of different pH and ionic strength in the absence (open symbols) and presence (full symbols) of poloxamine micelles (10% T904). Squares, 0.1 M HCl; up-triangles, pH 5.8 phosphate buffer; down-triangles, pH 7.4 phosphate buffer; circles, water; diamond, 0.9% NaCl solution.
Reproduced from Ref. 224 with permission of Elsevier.

Table 5.1 Fractions of drug-free (f_f) and incorporated in the micelles (f_m) and solubilization capacities of poloxamine micellar dispersions prepared with 10% T904 in different media.

Medium	Drug distribution		Solubilization capacity		
	f_f	f_m	mg g^{-1} T904	mol mol^{-1} T904	mg g^{-1} hydrophobic block
0.9% NaCl	0.134	0.866	2.64 (0.01)	0.050 (0.001)	4.48 (0.02)
Water	0.201	0.799	2.12 (0.14)	0.040 (0.002)	3.60 (0.24)
pH 7.4 buffer	0.250	0.750	2.09 (0.16)	0.040 (0.003)	3.54 (0.27)
pH 5.8 buffer	0.379	0.621	1.25 (0.02)	0.024 (0.001)	2.12 (0.04)
0.1 M HCl	0.288	0.712	1.25 (0.03)	0.024 (0.001)	2.13 (0.06)

Reproduced from Ref. 224 with permission of Elsevier.

sixfold in phosphate buffer (pH 7.4) compared with the medium without micelles.

Using the descriptors mentioned above [eqns (5.16)–(5.20)], a direct dependence on pH of the fraction of drug hosted by the micelles and also the amount of drug that is incorporated per gram of PPO blocks was observed (Table 5.1). The influence of pH on the protonation and hydrophobicity of the unimers and, consequently, on the concentration and properties of the micelles was reflected in drug release from the micellar dispersion. Using Franz–Chien diffusion cells, the diffusion coefficients obtained for copolymer systems prepared in 0.1 M HCl $(1.46 \times 10^{-4}$ cm^2 min^{-1}; s.d. 8.5×10^{-6} cm^2 min$^{-1})$ were almost twice those recorded for the same systems prepared in pH 7.4 phosphate buffer $(0.89 \times 10^{-4}$; s.d. 8.7×10^{-6} cm^2 min$^{-1})$.[224]

In most cases, the physical–chemical conditions for preparing and evaluating the polymeric micelles are the result of a compromise between the hosting capability of the micelles, the stability of the drug and the final application. For example, polymeric micelles designed to act as drug carriers should be evaluated in media that resemble the physiological environments with which the formulation will come into contact. The case of micelles designed as solubilizers of simvastatin can illustrate the interplay of all these factors. Simvastatin, as other statins, is widely used for the management of hypercholesterolaemia, but its effective adsorption after oral administration depends on the stability of the lactone ring in the gastric environment. The hydroxy acid form is more soluble but exhibits poor intestinal permeability. Hence the capability of the polymeric micelles of poloxamers and poloxamines to protect the lactone ring and to solubilize the drug was tested at acidic pH, which in turn represented the most unfavourable conditions for micellization. Under these adverse conditions, micellar dispersions of the block copolymers bearing larger PPO hydrophobic blocks were the most efficient in both the loading and protection.[222] The kinetic stability of the drug-loaded micelles in such an unfavourable environment should also be determined in order to gain an insight into the *in vivo* behaviour.[226] A simple experiment consists in adding a small volume of the drug-loaded micellar dispersion (*e.g.* 40 μL) in a thermostated quartz cell (suitable for UV–visible spectrophotometry) containing 1–2 mL of a buffered medium in order to monitor the absorbance of the drug as a function of time. In the case of simvastatin, slowly disintegrating micelles were those made by copolymers that combine long PPO and PEO blocks.[222]

Drug loading and micellar stability can be tuned by replacing the PPO block with a more hydrophobic species, such as poly(butylene oxide) (PBO), poly(styrene oxide) (PSO) or phenylglycidyl ether (PG). For example, PSO-based block copolymers self-assemble at very low concentrations into micelles of varied morphology and improved solubilization ability and stability.[90,111,227] Micelles of $EO_{137}SO_{18}EO_{137}$ have been shown to be able to solubilize five times more griseofulvin, quercetin and rutin than those of $EO_{62}PO_{39}EO_{62}$; mixed micelles led to intermediate solubilization capability.[228] PEO–PSO–PEO polymeric micelles with optimized features as drug solubilizers can be prepared from copolymers with an EO:SO ratio of ~1.5 and block lengths of 33–38 EO and 10–14 SO units to achieve a compromise between chain solubility and micellar core size. Shorter EO and longer SO lengths lead to reduced polymeric chain solubility, whereas copolymers with longer PEO blocks and/or very short PSO blocks self-assemble at high concentrations and form micelles with lower drug solubilization capabilities. $EO_{33}SO_{14}EO_{33}$ and $EO_{38}SO_{10}EO_{38}$ micelles showed improved griseofulvin encapsulation and release compared with poloxamers.[229] The influence of the amount of feeding drug on the encapsulation yield was evidenced in studies involving addition of excess griseofulvin to a 0.2 wt% copolymer solution and maintaining the systems under stirring for 3–5 days. The solubilization capacity increased with increase in the amount of feeding drug until a drug : copolymer weight ratio of 15% was attained. Subsequently, the solubilization capability decreased. This behaviour is a consequence of saturation of the inner core of the micelles, beyond which further addition of drug leads to precipitation.

5.4.2 Passive and Active Targeting

The size of polymeric micelles, similar to those of viruses, lipoproteins and other biological transport systems, make them suitable as drug carriers that can be administered intravenously and that circulate in the bloodstream without losing the cargo and eventually extravase to specific tissues. This tissue-specific distribution is being exploited for the delivery of drugs that require high concentrations in the affected tissues, but that could cause toxicity in healthy tissues, as is the case with antitumour drugs.[230]

Free drug molecules can easily pass through the vascular endothelium and indiscriminately distribute into all tissues, but also relatively rapidly the drug can return to the bloodstream and be eliminated from the body. The short residence time at the active site and the exposure of non-target tissues to the drug lead to limited therapeutic efficacy and notable side effects.[231] Ideally, the drug should be delivered only to the diseased tissues. Polymeric micelles can benefit from the leaky vasculature of tumours, inflamed tissues and infarcted areas. Vascular abnormalities due to newly formed vessels with poorly aligned endothelial cells, which are common to all tumours with the exception of hypovascular types such as prostate and pancreatic cancer, result in endothelial pores that can be as large as 100–1000 nm.[232] Additionally, the lymphatic drainage is defective and, as a consequence, the tumours present erratic fluid and molecular transport dynamics.[233] Overall, these pathological characteristics result in an enhanced permeability and retention (EPR) of colloidal carriers in tumour tissues (Figure 5.14). Therefore, polymeric micelles can be considered tumour-tropic by themselves.[234]

To benefit from the EPR effect for passive targeting of drugs to tumours, the polymeric micelles should not prematurely disassemble and release the drug into the systemic circulation and should circulate in the bloodstream for a prolonged time (more than 6 h) to be able to reach the tumour.[207]

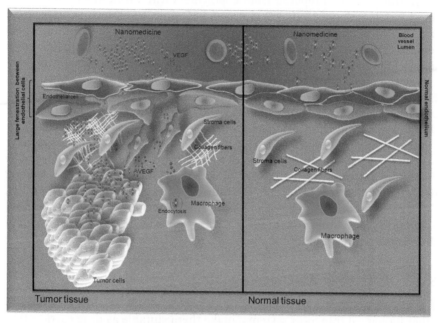

Figure 5.14   Differences between normal and tumour tissues in relation to targeting of nanocarriers (*e.g.* micelles) by the enhanced permeability and retention (EPR) effect. Normal tissue contains tightly connected endothelial cells, which prevents the diffusion of the nanocarrier outside the blood vessel. Tumour tissue exhibits large fenestration between the endothelial cells, allowing the nanocarrier to reach the extracellular matrix and the tumour cells. VEGF secreted by tumour cells, stroma cells and macrophages increases permeability and stimulates angiogenesis and the migration of endothelial cells towards the tumour. Nanocarriers tend to concentrate at the periphery of the tumour and only a small proportion can diffuse to the centre of the tumour.

Reproduced from Ref. 230 with permission of Elsevier.

In general, the more prolonged the circulation time, the greater is the likelihood of reaching the target tissue. The hydrophilic shell of the polymeric micelles should prevent opsonin adsorption and subsequent clearance by the mononuclear phagocyte system in the liver and the spleen.[219] Ideally, micelles should exhibit long and dense PEO blocks at the shell to be silent and should possess a sufficiently large size (10–100 nm) to avoid renal excretion (>50 kDa), but small enough (<200 nm) to bypass filtration by interendothelial cell slits in the spleen. Small silent micelles have been shown to accumulate in tumour tissue at concentrations 10–30 times (even up to 2000 times) higher than the concentration in plasma 24 h after intravenous injection and frequently at levels more than 10 times higher than in normal tissue.[235] Retention in the tumour can be prolonged for weeks or months. The usefulness of the EPR effect is particularly relevant for small-sized tumours. The targeting efficiency (% dose g^{-1}) declines as the tumour size increases owing to the intratumoral pressure and the necrotic areas in profound regions of the tumour.[236] The better the knowledge about the physiology of a given tumour, the greater are the possibilities of an adequate passive targeting.[232,237] Moreover, analytical tools that permit the visualization and the modelling of the movements of colloidal carriers into tumours may contribute substantially to the development of colloidal drug carriers that can fully exploit the EPR effect.[238,239]

The targeting can be notably improved if the micelle is 'decorated' with ligands that bind to receptors that are specific or that are over-expressed in the target cells. These ligands can form part of the initial block copolymer or be conjugated to the preformed micelles.[240] Molecules of different nature can be used as ligands. Small molecules such as folic acid or biotin recognize with high specificity over-expressed receptors at cancer cells.[241] Carbohydrates bind specifically to asialoglycoprotein receptors commonly found in liver cells.[241] The tripeptide Arg–Gly–Asp (RGD) can bind to the avb3 integrin receptor, which is largely expressed at the surface of malignant cells and in tumour-proliferating neovascular endothelial cells.[242] A large list of antibodies and aptamers has also proved successful for micelle targeting.[243] Readers interested in active targeting are referred to a comprehensive review.[240]

5.4.3 Stimuli-Responsive Drug Release

Either the micelles are intended for targeted or non-targeted delivery, control of drug release can be notably improved if the micelles are endowed with the capability to recognize certain variables of the site where the release should occur. Such variables can be physicochemical peculiarities of the body site (internal stimuli), *e.g.* pH, temperature, concentration of a certain biomolecules or redox potential. Alternatively, the changes in a certain parameter can be externally induced (external stimuli) by applying, for example, a source of heating, ultrasound or magnetic field. Stimuli-responsive micelles can be prepared with block copolymers that can undergo changes in

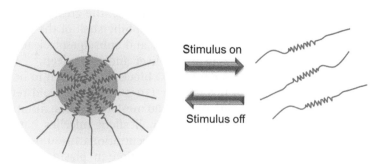

Stimulus on

Stimulus off

Figure 5.15 Stimuli-responsive micelles can disassemble by the action of a stimulus and re-form when the stimulus disappears.

the hydrophilicity or conformation of the unimers under a given stimulus,[244,245] or by incorporation of sensitive inorganic elements (*e.g.* gold or magnetite nanoparticles) that can recognize the stimulus leading to local perturbations that trigger the disassembly of the micelles.[246] Readers interested in materials useful for preparing responsive nanocarriers are referred to the comprehensive RSC book *Smart Materials for Drug Delivery*.[247] Ideally, stimuli-responsive micelles to be considered as smart carriers should disintegrate or destabilize in a number proportional to the intensity of the stimulus and, consequently, the drug release rate could be tuned. If the stimulus stops, the micelles could re-form and the release could be interrupted (Figure 5.15). However, for most applications, one-shot responsiveness at the site of interest is preferred for the rapid release of the whole encapsulated dose.

A variety of stimuli have already been evaluated as triggering agents of drug release from micelles. For example, pH gradients are common under both healthy and pathological conditions. Along the gastrointestinal tract, the pH varies from 1.00–3.00 in the stomach, to 4.80–8.20 in the upper gut and to 7.00–7.50 in the colon.[248] Inside cells, there are notably different pH environments at cytosol (7.4), Golgi apparatus (6.40), endosome (5.5–6.0) and lysosome (4.5–5.0).[249] The extracellular pH of inflamed tissues and tumour cells (6.5–7.0) is lower than that of the blood and the healthy tissues (7.4).[250,251] In contrast, wounds commonly show an increase in pH from 4–6 (normal skin) up to as high as 8.9.[252]

Block copolymers bearing weak acid or base groups can provide polymeric micelles with pH-sensitive drug release. In general, when the groups are not ionized, the block behaves as hydrophobic and is responsible for the self-assembly of the copolymer chains. Micelles are stable provided that the pH of the medium does not induce the ionization of the chemical groups. Ionization is accompanied by an increase in the polarity and also in the electrostatic repulsions among the ionic blocks, which results in the disintegration of the micelles. The triggering pH depends on the pK_a of the chemical groups; hence it can be tuned by means of the copolymer composition.[253] The copolymer should have a pK_a that allows sharp changes in the ionization state at the pH of interest.

Amphiphilic copolymers containing amino groups in one of the blocks, *e.g.* poly(2-vinylpyridine)-*b*-poly(ethylene oxide) (P2VP-*b*-PEO), poly-(*N*-vinylpyrrolidone)-*b*-poly(2-acrylamido-2-methyl-1-propanesulfonic acid) (PVP-*b*-PAMPS), poly(*N*-vinylpyrrolidone)-*b*-poly(*N*,*N*-dimethylaminoethyl methacrylate) (PVP-*b*-PDMAEMA), poly[2-(dimethylamino)ethyl methacrylate]-*b*-poly-(ethylene oxide) (PEO-*b*-DMAEMA), poly(ethylene glycol)-*b*-poly(L-histidine) (PEG-*b*-PLH) and methyl ether poly(ethylene glycol)-*b*-poly(β-amino ester) (MPEG-*b*-PAE), have been shown to be useful for drug release at acidic pH.[254–257] At pH 7.4 (well above the pK_a), the amino groups are not ionized and the block where they are localized can lead to hydrophobic micellar cores suitable for hosting drugs with a high affinity. After passive targeting *via* the EPR effect, the micelles can penetrate tumour cells, forming endocytic vesicles that subsequently fuse with lysosome (pH 5–6). A decrease in pH leads to the ionization of the amino groups and, as a consequence, to the disassembly of the micelles and to the release of the drug.[249,258] The micelles can also be designed to recognize specifically the slightly acidic extracellular pH typical of tissues affected by inflammation or tumoral processes. For example, MPEG–PAE micelles can undergo a brusque demicellization at pH 6.4–6.8 that makes the drug release be complete in a few hours.[257] Intravenous injection of MPEG–PAE micelles loaded with a fluorescent dye into MDA-MB231 human breast tumour-bearing mice has been shown to provide preferential accumulation of the dye in the tumour, with minimal distribution to the healthy tissues. MPEG–PAE micelles can also encapsulate Fe_3O_4 nanoparticles in the hydrophobic core for selective release in the cerebral ischaemic area, taking advantage of the acidic environment caused by this pathological condition. The Fe_3O_4 nanoparticles can then be used for magnetic resonance imaging of the ischaemic region.[259]

Moreover, pH-responsive polymeric micelles are particularly suitable as non-viral vectors of DNA and siRNA.[260,261] The nucleotides interact with the amino groups of the copolymer and form a complex that is included inside the micelle (micelleplex), providing protection against enzymatic attack.[262–265] Micelles of poly(ethylene glycol)-*b*-poly[(3-morpholinopropyl)-aspartamide]-*b*-poly(L-lysine) (PEG-*b*-PMPA-*b*-PLL) combine the buffering capacity of PMPA with the excellent aptitude of PLL to condense DNA, leading to high transfectional efficiency.[266] siRNA-loaded micelles of dimethylaminoethyl methacrylate (DMAEMA), acrylic acid and butyl methacrylate have shown enhanced transfection efficiency and lower toxicity compared with common polyplexes.[267]

Glutathione tripeptide (γ-glutamylcysteinylglycine; GSH)–glutathione disulfide (GSSG) is the major redox couple in animal cells.[268] Blood, the extracellular environment and cell surfaces possess a low concentration of GSH (2–20 μM). In contrast, the intracellular concentration of GSH is 0.5–10 mM, which is kept reduced by NADPH and glutathione reductase, maintaining a highly reducing environment. Endosomal and lysosomal compartments are also rich in reducing agents.[269] Moreover, tumour tissues contain fourfold higher concentrations of GSH than normal tissues.[270] These differences in GSH concentration prompted the development of micelles in which disulfide bonds provide stability while circulating in the bloodstream, but once inside healthy cells or in the surroundings of the tumour tissue the micelles disintegrate as the disulfide bonds break into thiol groups.[271] Three different approaches have been followed so far to obtain GSH-responsive micelles:

  (i) *Detachable shells.* Poly(ethylene glycol)-SS-poly(ε-caprolactone) (PEG-SS-PCL), dextran-SS-PCL or PEG-SS-poly(γ-benzyl L-glutamate) form stable micelles in aqueous medium, but inside cells the PEG shells detach owing to the reductive cleavage of the intermediate disulfide bonds, releasing the drug at a much faster rate than similarly prepared reduction-insensitive micelles (Figure 5.16).[272–274] Micelles based on hyaluronic acid-SS-deoxycholic acid (HA-SS-DOCA) conjugates have been shown to accumulate preferentially in human breast

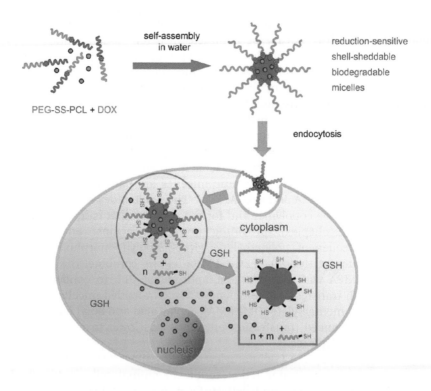

Figure 5.16   Schematic illustration of reduction-sensitive shell-sheddable biodegradable micelles based on PEG-SS–PCL block copolymer for efficient intracellular release of doxorubicin (DOX) triggered by GSH. Reproduced from Ref. 272 with permission of Elsevier.

adenocarcinoma cells (MDA-MB-231) *via* HA-receptor-mediated endocytosis and release the cargo in a relatively rapid manner.[275]

(ii) *Responsive cores.* Amphiphilic copolymers containing disulfide bonds in the hydrophobic segments, such as poly(ethylene oxide)-*b*-poly[*N*-methacryloyl-*N'*-(*tert*-butyloxycarbonyl)cystamine] (PEO-*b*-PMABC) and diselenide-containing block copolymers, can be broken when the GSH concentration rises, leading to micelle disintegration.[276,277]

(iii) *Cross-linkable copolymers.* Once the micelles are formed, the copolymers can be cross-linked with redox-responsive labile bonds. For example, core cross-linked micelles have been prepared from copolymers of poly(ethylene glycol methyl ether methacrylate), 5'-*O*-methacryloyluridine and bis(2-methacryloyloxyethyl) disulfide (DSDMA, bioreducible cross-linker). In a reductive environment, the core cross-linked micelles are readily hydrolysed in a short time, releasing the cargo.[278] Shell cross-linked micelles have been obtained through self-assembly of PEG-*b*-poly(L-lysine)-*b*-poly(L-phenylalanine) triblock copolymers followed by cross-linking of the poly(L-lysine) block with 3,3'-dithiobis(sulfosuccinimidylpropionate) (DTSSP).[279] Many other examples of reduction-responsive cross-linkable micelles can be found elsewhere.[280]

Several pathological conditions (*e.g.* inflammation, infarction or cancer) evolve with local increases in temperature.[281] Moreover, a site-specific increase in temperature can be achieved applying an external source to the skin or can be remotely induced through various types of radiation that can be absorbed by components of the formulation that transform the energy into heat. Commonly, temperature-sensitive polymers are hydrophilic below their critical temperature of dissolution (LCST). When the temperature is above LCST, the polymer becomes hydrophobic and its conformation changes from an expanded (soluble) to a globular (insoluble) state. A typical example is poly(*N*-isopropylacrylamide) (PNIPAM), which, when used as the hydrophilic block, leads to micelles at temperatures below LCST. PNIPAAm has been combined with hydrophobic blocks of polystyrene,[282] dimethylacrylamide,[283] poly(lactic acid),[284] butyl methacrylate,[285] poly(*N*-vinylimidazole),[286] methyl methacrylate[287] and octadecyl acrylate,[288] among others. Above LCST, the micelle destabilizes and the drug is released.[289] For example, block copolymers of poly(lactic acid) (PLA), PNIPAAm and poly-(dimethylacrylamide), PLA-*b*-(PNIPA-*co*-DMAAm), have LCST close to 40 °C. Once loaded with doxorubicin, the micelles slowly released the drug at 37 °C, but the release accelerated when the temperature rose to 42 °C.[290] External modulation of drug release can be achieved by combining temperature-sensitive polymers and gold nanoparticles. When a gold nanoparticle is irradiated with near-UV to near-IR light, the absorbed light *via* surface plasmon resonance (SPR) rapidly transforms into thermal energy, causing a local increase in temperature that may be high enough to cause irreversible damage to cancer cells, which are less resistant against temperature than the healthy cells. The photothermal therapy can be combined with pharmacological therapy if both the gold nanoparticles and the drug are encapsulated together in the micelles. The induced heat can trigger conformational changes in the structure of the block copolymers, which in turn causes the micelles to disintegrate.[291]

Light itself, particularly near-IR light that penetrates deeper in the body tissues, can be used directly to trigger drug release from micelles. Copolymers can respond to the light *via* several mechanisms (Figure 5.17):[292–294]

(i) *Photoisomerization. Trans* to *cis* isomerization upon irradiation, as occurs in polymers bearing azobenzene groups, or the generation of charged species, for example when spiropyran is converted to merocyanine, is usually accompanied by a change in the hydrophilic–hydrophobic balance of the photoexcitable molecules.

(ii) *Photodegradation.* The irradiation causes the rupture of labile bonds in the copolymer backbone.

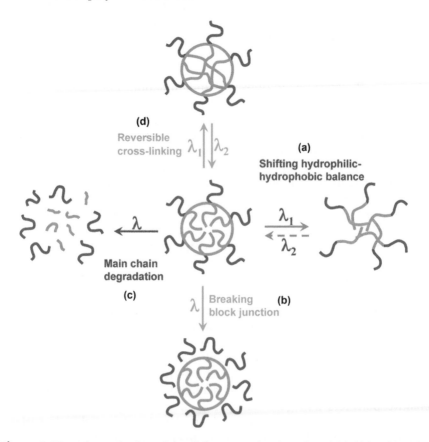

Figure 5.17 Schematic view of the different mechanisms by which light alters the structure of a sensitive copolymer micelle.
Reproduced with permission from Ref. 294. Copyright 2011 American Chemical Society.

(iii) *Photosensitization-induced oxidation.* Irradiation of a sensitizer molecule incorporated in the micelle may lead to strong oxidizing species that disrupt the nanocarrier.

(iv) *Photocross-linking and photodecross-linking.* Light modifies the cross-linking density of the polymers and, as a consequence, their conformation, which usually results in the formation of pores in the micelles.

Polymeric micelles can be reversibly destabilized by applying ultrasound.[295] The ultrasound should be applied when maximum micelle accumulation in the target tissue (*e.g.* tumour) is reached, the waiting period depending on the kinetics of the distribution of the micellar system. Rapoport's group found that the optimum time of ultrasound application after injection of doxorubicin-loaded Pluronic P-105 polymeric micelles (or PEO–diacylphospholipid mixed micelles) is between 4 and 8 h. The amount of drug released can be modulated through the control of the frequency, power density, pulse length and inter-pulse intervals.[296] Drug release from the micelles is reversible; during inter-pulse intervals longer than 0.5 s, the drug can be completely re-encapsulated into the restored micelles. The *in vivo* anti-tumour effectiveness of this approach is also promoted by the cell membrane perturbation caused by ultrasound (sonoporation), which enhances the intracellular uptake of micelles, drugs and genes.[297]

Materials that respond to several stimuli independently or in a synergistic way are particularly useful for the design of multi-responsive micelles.[298] The advantage of micelles that recognize changes in two or more variables is that they can regulate more precisely the site and the rate of the release process, since most pathological processes cause changes in several physicochemical parameters simultaneously. In that way, the risk of drug release in a non-target tissue is minimized. Moreover, multi-stimuli responsiveness can be exploited for controlling the release of several drugs in response to different internal or external signals.[299–301]

5.4.4 Biological Response Modulators

Drugs can penetrate as individualized molecules into the cells *via* different mechanisms. Passive diffusion, *i.e.* movement from a site where the drug concentration is high (the extracellular space) to a site where it is low (the intracellular space), is an unspecific and non-saturable process. This mechanism is available for all kinds of molecules and its efficiency depends on the lipophilic character and the size of the drug molecule. Transporter-mediated absorption is suitable for drugs that can utilize membrane proteins to bypass the cell membrane. The transporter takes up specific molecules from the extracellular space and releases them into the cytoplasm. However, the cells are also endowed with transporters that work in the opposite direction, namely they are responsible for the clearance of metabolites and other endogenous substances from the cytoplasm. The latter transporters are known as efflux pumps and contribute to the protection against adverse substances, accumulation of which inside cells should be avoided. A variety of drugs have also been shown to be substrates of efflux pumps, which in turn may result in subtherapeutic concentrations inside the cells. Efflux pumps work against the concentration gradient and therefore require energy, which is provided by means of ATP hydrolysis.[302]

The genes that codify for efflux pumps belong to the ATP Binding Cassette (ABC). In humans there are 49 ABC proteins that can be classified in the ABC1, MDR/TAP, MRP, ALD, OABP, GCN20 and White families.[303] These proteins are expressed up to different extents in most healthy tissues, being more abundant on the membrane of epithelial cells, particularly in the apical side of gut, liver and kidney cells, and also in the blood–tissue barriers to restrict the access of substances to the brain, the fetus or the testis.[304,305] Tumour cells over-express some efflux pumps, such as the permeability-glycoprotein (P-gp or ABCB1), the multidrug-resistant proteins (MRP) 1 and 2 (ABCC1 and ABCC2) or the breast cancer-resistant protein (BCRP or ABCG2).[306] The activity of efflux pumps results in low concentrations of anticancer agents and contribute to the multidrug resistance (MDR) and treatment failure observed in some tumour treatments (Figure 5.18). P-gp, MRP1 and BCRP are able to interact with anticancer agents, anti-HIV

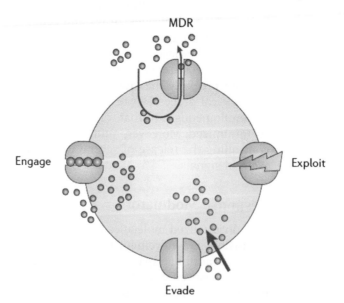

Figure 5.18   P-glycoprotein (P-gp) actively extrudes many types of drugs from cancer cells, keeping their intracellular levels below a cell-killing threshold. Strategies that circumvent P-gp-mediated multidrug resistance (MDR) include the co-administration of pump inhibitors and cytotoxic agents ('engage') and the use of cytotoxic agents that bypass P-gp-mediated efflux ('evade'). A third approach takes advantage of the collateral sensitivity of MDR cells ('exploit').
Reproduced from Ref. 308 with permission of Nature Publishing Group.

protease and reverse transcriptase inhibitors, immunosuppressants, iono-phores, peptides, fluorescent dyes, steroids, cardiac glycosides and many others.[306] The polyspecificity is due to the large and flexible binding site in the transporter, which contains multiple hydrophilic electron donor/acceptor groups, charged groups and aromatic amino acids where drugs can bind.[302] In fact, the interaction between the pump and different substrates may be through different epitopes.[307] To overcome the drug efflux mech-anism, it is necessary either to bypass transporter recognition, inhibit its expression or inhibit its function (Figure 5.18).[308,309]

Drugs formulated in micellar systems have two main routes to escape from the efflux pumps, depending on the way in which they enter into cells: (i) drugs that remain encapsulated in the micelles enter the cell mainly *via* endocytosis and evade efflux pumps; and (ii) drugs released in the extra-cellular space can benefit from the capability of certain block copolymers to inhibit the functioning of the efflux pumps.[16,310,311] In the first case, micellar endocytosis can bypass transporter recognition as the drug is released inside the cytoplasm, after the rupture of the endocytic vesicle, far from the cell membrane. This evasion mechanism can be exploited by any nanocarrier stable enough to pass through the membrane without losing the cargo.[309] The second case refers to drugs incorporated into stimuli-responsive mi-celles that disassemble in the extracellular space. The block copolymer unimers can be recognized by the efflux pump as preferential substrates, can alter the structure of the membrane and in turn the conformation of the pump, or can cut off the energy source (ATP hydrolysis).[312] Any of these phenomena result in the escape of the drug from the efflux pump and, as a consequence, in a greater accumulation of the drug inside the cell (Figure 5.18).

In addition to a few drugs (such as verapamil), several components of medicines (excipients) previously considered inert have been shown to be able to inhibit the efflux pump's function.[313] Among them, the most efficient seem to be copolymers containing PEO blocks and polymers with thiol moieties.[314,315] Nevertheless, it should be noted that not all drugs and cells respond equally to these polymeric inhibitors, since a given polymer could bind selectively to one of the epitopes of the efflux pump and effectively block the interaction of the substrates that interact similarly, but not of the drugs that interact through other epitopes.[311] Readers interested in com-parative studies of the performance of various polymeric inhibitors are re-ferred to comprehensive reviews.[311,312,316–318] PEO has been shown to be able to inhibit P-gp and MDR1 in a concentration-dependent manner.[319,320] Several studies have indicated that maximum drug accumulation inside cells occurs when the concentration of the PEO copolymers is close to the CMC, which indicates that unimers are responsible for the inhibition. Above the CMC, the drug is mainly kept in the micelle and there are less free drug molecules to permeate into the cell (*i.e.* most drug is 'sequestered' in the micelle). As mentioned above, drug-loaded micelles can evade efflux pumps as they enter *via* other mechanisms.[321]

Pluronics (PEO–PPO–PEO) have been widely evaluated as P-gp inhibitors.[322,323] Some Pluronic copolymers rapidly bind the cell membrane and cause its fluidization. They can also penetrate into the cells and co-localize with the mitochondria inhibiting the respiratory chain, which in turn decreases oxygen consumption and leads to ATP depletion.[324] These two mechanisms hinder the activity of the efflux pumps.[325,326] Experiments carried out with doxorubicin and rhodamine 123 in multidrug-resistant cancer cells have shown that the relative length of the PPO block of Pluronics strongly determines the efficiency of the unimers as efflux inhibitors. Pluronics of relatively low HLB that possess PPO blocks of intermediate length (30–60 units) and short PEO blocks are the most effective (Figure 5.19).[327] If the PPO block is larger, the CMC is lower and thus less free unimers will be available to interact with the cells. In contrast, if the PPO blocks are shorter, higher unimer concentration can be attained before micellization, but these more hydrophilic copolymers interact poorly with cell membranes and their inhibition activity is low. Hence efflux pump inhibition depends on a fine equilibrium between membrane affinity and unimer concentration.[328] Kabanov and co-workers classified Pluronic copolymers into four groups depending on their effects on P-gp (Figure 5.19).[329] Group I corresponds to hydrophilic copolymers with HLB ranging from 20 to 29 that have little effect on P-gp functional activity, because they are unable to penetrate into the cells and do not cause ATP depletion. Group II is formed by the most efficient inhibitors, which have intermediate-length PO blocks (30–60 units). These copolymers penetrate

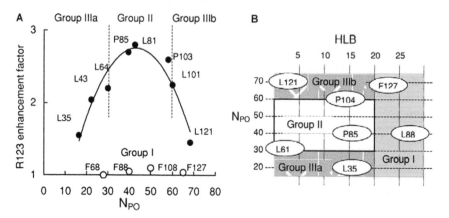

Figure 5.19   Efficacy of Pluronic block copolymer compositions displayed in inhibiting drug efflux function in bovine brain microvessel endothelial cells (BBMEC) monolayers. (a) Rhodamine 123 (R123) enhancement factors are defined here as the ratios of rhodamine 123 accumulation in the cells in the presence of the block copolymer to rhodamine 123 accumulation in the assay buffer. (b) A grid of Pluronic indicating four groups identified according to the copolymer activity on BBMEC monolayers as shown in (a).
Reproduced from Ref. 327 with permission of Elsevier.

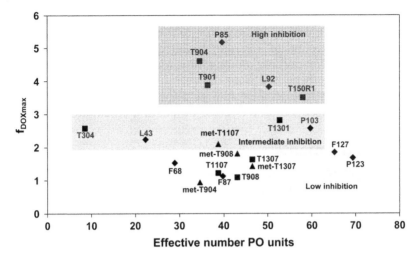

Figure 5.20 Enhanced accumulation of doxorubicin (DOX) in cancer cells due to the inhibitory efficacy of different poloxamers, pristine and *N*-methylated poloxamines *versus* the 'effective' number of PO units ($N_{PO}/2$ in the case of poloxamines).
Reproduced from Ref. 330 with permission of Future Medicine Ltd.

into the cells, cause high-energy depletion and fluidize membranes very efficaciously, inhibiting P-gp ATPase activity. Hydrophobic copolymers with shorter (<30 PO units; Group IIIa) or longer (>60 PO units; Group IIIb) PPO blocks were less efficient in membrane fluidization and in movement into the cytoplasm, respectively (Figure 5.19).

Enhanced accumulation of doxorubicin in tumour cells has been also demonstrated for related copolymers of the Tetronic family and their *N*-methylated derivatives[330] and also for block copolymers having hydrophobic blocks distinct from PPO.[331] For comparison with the performance of Pluronics, poloxamine molecules can be considered to be formed by two PEO–PPO–PEO triblocks covalently linked by the central ethylenediamine group. Therefore, the number of 'effective' PO units in the central PPO block of each of the 'triblocks' in poloxamine would be the half of the total PO units of each poloxamine molecule. A joint diagram of the inhibitory activity of different poloxamers and poloxamines according to the 'effective' number of PO units in the PPO blocks is shown in Figure 5.20. It is clear that the 'effective' PO number itself does not totally explain the differences in inhibition activity and that the HLB also has to be taken into account. The copolymers with effective numbers of PO units ranging from 30 to 50 and low-to-medium HLB were the most efficient inhibitors of P-gp activity.

5.5 Conclusion

Polymeric micelles strongly benefit from the versatility of the synthesis approaches available to produce amphiphilic polymers with very diverse architectures, molecular weights and functionalities. Knowledge of the self-assembly behaviour and of the structure of the resultant micelles is critical for the correct selection of a certain copolymer intended to fulfil the requirements of a given purpose. Among the diverse fields of application, drug therapy has greatly benefited from the performance of micellar carriers at various levels that involve their role as drug solubilizers and stabilizers, passive targeting agents to EPR-affected tissues and efflux pump inhibitors.

Acknowledgements

We thank the Ministerio de Economía y Competitividad (MINECO) for funding support through projects SAF2008-22771 and MAT 2010-17336, the Xunta de Galicia for research grants CN2012/045, CN2012/072 and EM2013/046 and the Fundación Ramón Areces for additional financial support. S.B. also thanks MINECO for a Ramon y Cajal Fellowship.

References

1. M. Szwarc, *Nature*, 1956, **178**, 1168.
2. M. Szwarc, M. Levy and R. Milkovich, *J. Am. Chem. Soc.*, 1956, **78**, 2656.
3. H. L. Hsieh and R. P. Quirk, *Anionic Polymerization: Principles and Practical Applications*, Marcel Dekker, New York, 1996.
4. M. Szwarc, *Carbanions, Living Polymers and Electron Transfer Processes*, Interscience, New York, 1968, p. 27.
5. G. Riess, G. Hurtrez and P. Bahadur, *Block Copolymers. Encyclopedia of Polymer Science and Engineering*, Vol. 2, Wiley, New York, 2nd edn, 1985, p. 324.
6. A. Vazaios, D. J. Lohse and N. Hadjichristidis, *Macromolecules*, 2005, **38**, 5468.
7. H. Iatrou, A. Avgeropoulos and N. Hadjichristidis, *Macromolecules*, 1994, **27**, 6232.
8. B. L. Laurent and S. M. Grayson, *Chem. Soc. Rev.*, 2009, **38**, 2202.
9. D. M. Knauss and T. Huang, *Macromolecules*, 2002, **35**, 2055.
10. C. Ramireddy, Z. Tuzar, K. Prochazka, S. E. Webber and P. Munk, *Macromolecules*, 1992, **25**, 2541.
11. L. Zhang and A. Eisenberg, *Science*, 1995, **268**, 1728.
12. K. S. Murthy, Q. Ma, E. E. Remsen, T. Kowaleski and K. L. Wooley, *J. Mater. Chem.*, 2003, **13**, 2785.
13. A. Choucair and A. Eisenberg, *J. Am. Chem. Soc.*, 2003, **125**, 11993.
14. A. V. Kabanov, P. Lemieux, S. Vinogradov and V. Alakhov, *Adv. Drug Deliv. Rev.*, 2002, **54**, 223.
15. E. V. Batrakova and A. V. Kabanov, *J. Control. Release*, 2008, **130**, 98.
16. C. Alvarez-Lorenzo, A. Sosnik and A. Concheiro, *Curr. Drug Targets*, 2011, **12**, 1112.
17. C. Alvarez-Lorenzo, A. Rey-Rico, A. Sosnik, P. Taboada and A. Concheiro, *Front. Biosci.*, 2010, **E2**, 424.

18. O. W. Webster, W. R. Hertler, D. Y. Sogah, W. B. Farnham and T. V. Rajanbabu, *J. Am. Chem. Soc.*, 1983, **105**, 5706.
19. R. P. Quirk and J. Ren, *Macromolecules*, 1992, **25**, 6612.
20. K. Matyjaszwski and M. Sawamoto, in *Cationic Polymerizations, Mechanisms, Synthesis and Applications*, ed. K. Matyjaszwski, Marcel Dekker, New York, 1996, p. 381.
21. G. Kaszas, J. E. Puskas, C. C. Chen, J. P. Kennedy and W. G. Hager, *J. Polym. Sci., Polym. Chem. Ed.*, 1991, **29**, 427.
22. T. Higashimura, Y. Kishimoto and S. Aoshima, *Polym. Bull.*, 1987, **18**, 111.
23. P. D. Hustad, *Science*, 2009, **325**, 707.
24. H. Markio, N. Kashiwa and T. Fujita, *Adv. Synth. Catal.*, 2002, **344**, 477.
25. D. J. Arriola, E. M. Carnahan, R. L. Kuhlman and T. T. Wenzel, *Science*, 2006, **312**, 714.
26. M. Kamigato, M. Sawamoto and T. Higashimura, *Macromolecules*, 1992, **25**, 2587.
27. R. T. Liggins and H. M. Burt, *Adv. Drug Deliv. Rev.*, 2002, **54**, 191.
28. J. M. Lü, X. Wang, C. Marin-Muller, H. Wang, P. H. Lin, Q. Yao and C. Chen, *Expert Rev. Mol. Diagn.*, 2009, **9**, 325.
29. F. Danhier, E. Ansorena, J. M. Silva, R. Coco, A. Le Breton and V. Préat, *J. Control. Release*, 2012, **161**, 505.
30. M. S. Kim, K. S. Seo, G. Khang, S. H. Cho and H. B. Lee, *J. Biomed. Mater. Res. A*, 2004, **70A**, 154.
31. O. S. Rabotyagova, P. Cebe and D. L. Kaplan, *Biomacromolecules*, 2011, **12**, 269.
32. A. K. Andrianov (ed.), *Polyphosphazenes for Biomedical Applications*, Wiley, Hoboken, NJ, 2009.
33. L. R. Gilliom and R. H. Grubbs, *J. Am. Chem. Soc.*, 1986, **108**, 733.
34. R. H. Grubbs (ed.), *Handbook of Metathesis*, Vol. 3, Wiley-VCH, Weinheim, 2003.
35. B. Chen and H. F. Sleiman, *Macromolecules*, 2004, **37**, 5866.
36. S. Hillf and A. F. M. Kilbinger, *Nat. Chem.*, 2009, **1**, 537.
37. G. Berger, M. Levy and D. Vofsi, *J. Polym. Sci., Polym. Lett.*, 1996, **4**, 183.
38. E. Esselborn, J. Fock and A. Knebelkamp, *Makromol. Chem. Macromol. Symp.*, 1996, **102**, 91.
39. H. W. Melville, *J. Chem. Soc.*, 1941, 414.
40. A. Ladousse, C. Filliatre, B. Maillard, C. Manigand and J. J. Villenave, *Eur. Polym. J.*, 1979, **15**, 987.
41. J. S. N. Su and J. J. Pirma, *Appl. Polym. Sci.*, 1987, **33**, 727.
42. T. Otsu and M. Yoshita, *Makromol. Chem. Rapid Commun.*, 1982, **3**, 127.
43. K. Matyjaszewski and A. H. E. Muller, *Polym. Prepr.*, 1997, **38**, 6.
44. M. Koto, M. Kamigaito, M. Sawamoto and T. Higashimura, *Macromolecules*, 1995, **28**, 1721.
45. J. S. Wang and K. Matyjaszewski, *J. Am. Chem. Soc.*, 1995, **117**, 5614.
46. H. Fisher, *Chem. Rev.*, 2001, **101**, 3581.
47. D. A. Shipp, J. L. Wang and K. Matyjaszewski, *Macromolecules*, 1998, **31**, 8005.

48. D. J. Siegwart, J. K. Oh and K. Matyjaszewski, *Prog. Polym. Sci.*, 2012, **37**, 18.

49. Y. Sheng, H. Tang and S. Ding, *Prog. Polym. Sci.*, 2004, **29**, 1053.

50. B. B. Wayland, L. Basickes, S. L. Mukerjee, M. Wei and M. L. Fryd, *Macromolecules*, 1997, **30**, 8109.

51. J. Chiefari, Y. K. Chong, F. Ercole, J. Krstina, J. Jeffery, T. P. T. Le, R. T. A. Mayadunne, G. F. Meijs, C. L. Moad, G. Moad, E. Rizzardo and S. H. Thang, *Macromolecules*, 1998, **31**, 5559.

52. T. P. T. Le, G. Moad, E. Rizzardo and S. H. Thang, *Int. Pat. Appl.*, WO 9801478, 1998.

53. G. Moad, E. Rizzardo and S. H. Thang, *Aust. J. Chem.*, 2005, **58**, 379.

54. G. Moad, E. Rizzardo and S. H. Thang, *Polymer*, 2008, **49**, 1079.

55. A. W. York, S. E. Kirkland and C. L. McCormick, *Adv. Drug Deliv. Rev.*, 2008, **60**, 1018.

56. D. Charmot, P. Corpart, H. Adam, S. Z. Zard, T. Biadatti and G. Bouhadir, *Macromol. Symp.*, 2000, **150**, 23.

57. S. Perrier and P. Takolpuckdee, *J. Polym. Sci., Polym. Chem.*, 2005, **43**, 5347.

58. C. L. McCormick and A. B. Lowe, *Acc. Chem. Res.*, 2004, **37**, 312.

59. Y. K. Chong, T. P. Le, G. Moad, E. Rizzardo and S. H. Thang, *Macromolecules*, 1999, **31**, 2071.

60. B. M. Mandal, *Fundamentals of Polymerization*, World Scientific, Singapore, 2013, p. 97.

61. P. Cohen, M. J. Abadie, F. Schue and D. H. Richards, *Polymer*, 1982, **23**, 1105.

62. K. Jankova, X. Chen, J. Kops and W. Batsberg, *Macromolecules*, 1998, **31**, 538.

63. Y. Yagci and M. K. Mishra, in *Polymeric Materials Encyclopedia*, Vol. 1, ed. J. C. Salamone, CRC Press, Boca Raton, FL, 1996, p. 789.

64. G. Riess, P. H. Dumas and G. Hurtrez, *Block Copolymer Micelles and Assemblies*, Citus Books, London, 2002, p. 69.

65. K. Matyjaszewski and T. Davis, *Handbook of Radical Polymerization*, Wiley, Chichester, 2002.

66. M. W. Weimer, O. A. Scherman and D. Y. Sogah, *Macromolecules*, 1998, **31**, 8425.

67. N. Hadjichristidis, S. Pispas and G. A. Floudas, *Block Copolymers: Synthetic Strategies, Physical Properties and Applications*, Wiley, Hoboken, NJ, 2003.

68. N. Hadjichristidis, M. Pitsikalis, S. Pispas and H. Iaotru, *Chem. Rev.*, 2001, **101**, 3747.

69. K. Khanna, S. Varshney and A. Kakkar, *Polym. Chem.*, 2010, **1**, 1171.

70. E. Fernandez-Megia, J. Correa and R. Riguera, *Biomacromolecules*, 2006, **7**, 3104.

71. F. Wurm and H. Frey, *Prog. Polym. Sci.*, 2011, **36**, 1.

72. A. Carlsen and S. Lecommandoux, *Curr. Opin. Colloid Interface Sci.*, 2009, **15**, 329.

73. H. Robson-Marsden and A. Kros, *Macromol. Biosci.*, 2009, **9**, 939.

74. F. C. Krebs and M. Jørgensen, *Polym. Bull.*, 2003, **50**, 359.
75. S. N. Patel, A. E. Javier and N. P. Balsara, *ACS Nano*, 2013, 7, 6056.
76. H. A. Klok and S. Lecommandoux, *Adv. Mater.*, 2001, **13**, 1217.
77. N. H. Aloorkar, A. S. Kulkarni, R. A. Patil and D. J. Ingale, *Int. J. Pharm. Sci. Nanotechnol.*, 2012, **5**, 1675.
78. I. Katime, *Química-Física Macromolecular, Servicio Editorial*, Universidad del País Vasco, Bilbao, 1994.
79. R. J. Young, *Introduction to Polymers*, CRC Press, Boca Raton, FL, 2011.
80. B. J. Hunt and M. I. James (eds), *Polymer Characterization*, Chapman & Hall, London, 1993.
81. R. A. Pethrick and J. V. Dawkins, *Modern Techniques for Polymer Characterization*, Wiley, Chichester, 1999.
82. L. H. Tung, in *Encyclopedia of Polymer Science and Engineering*, Vol. 7, ed. H. F. Mark, N. Bikales, C. G. Overberger, G. Menges and J. I. Kroschwitz, Wiley-Interscience, New York, 1987, p. 298.
83. I. W. Hamley, *The Physics of Block Copolymers*, Oxford University Press, New York, 1998, Chapters 3 and 4.
84. P. Alexandridis and B. Lindman (eds.), *Amphiphilic Block Copolymers: Self-Assembly and Applications*, Elsevier, Amsterdam, 2000.
85. J. F. Gohy, *Adv. Polym. Sci.*, 2005, **190**, 65–136.
86. Y. He, Z. Li, P. Simone and T. P. Lodge, *J. Am. Chem. Soc.*, 2006, **128**, 2745.
87. Z. Tyrrell, W. Winoto, Y. Shen and M. Radosz, *Ind. Eng. Chem. Res.*, 2009, **48**, 1928.
88. H. G. Elias, in *Light Scattering from Polymer Solutions*, ed. M. B. Huglin, Academic Press, London, 1982, p. 397.
89. Z. Tuzar and P. Kratochvil, *Adv. Colloid Interface Sci.*, 1976, **6**, 201.
90. C. Booth and D. Attwood, *Macromol. Rapid Commun.*, 2000, **21**, 501.
91. P. Taboada, V. Mosquera, D. Attwood, Z. Yang and C. Booth, *Phys. Chem. Chem. Phys.*, 2003, **5**, 2625.
92. F. Candau, F. Heatley, C. Price and R. B. Stubbersfield, *Eur. Polym. J.*, 1984, **20**, 685.
93. J. R. Quintana, M. Villacampa, M. Muñoz, A. Andrio and I. A. Katime, *Macromolecules*, 1992, **25**, 3125.
94. P. Alexandridis, J. E. Holzwarth and T. A. Hatton, *Macromolecules*, 1994, **27**, 2414.
95. M. Fernandez-Tarrio, C. Alvarez-Lorenzo and A. Concheiro, *J. Thermal Anal. Calor.*, 2007, **87**, 171.
96. I. W. Hamley, *Block Copolymers in Solution: Fundamentals and Applications*, Wiley, Hoboken, NJ, 2005.
97. K. Kataoka, A. Harada and Y. Nagasaki, *Adv. Drug Deliv. Rev.*, 2012, **64**, 37.
98. A. Rösler, G. W. M. Vandermeulen and H. A. Klok, *Adv. Drug Deliv. Rev.*, 2012, **64**, 270.
99. Z. Tuzar and P. Kratochvil, in *Surface and Colloid Science*, Vol. 15, ed. E. Matijevic, Plenum Press, New York, 1993, p. 1.

100. M. Moffit, K. Khougaz and A. Eisenberg, *Acc. Chem. Res.*, 1996, **29**, 95.

101. Z. Tuzar, P. Kratochvil, K. Prochazka and P. Munk, *Collect. Czech. Chem. Commun.*, 1993, **58**, 2362.

102. G. H. Hsiue, C. H. Wang, C. L. Lo, J. P. Li and J. L. Yang, *Int. J. Pharm.*, 2006, **317**, 69.

103. N. Adams and U. S. Schubert, *Adv. Drug Deliv. Rev.*, 2007, **59**, 1504.

104. R. Hoogenboom, *Angew. Chem. Int. Ed.*, 2009, **48**, 7978.

105. C. S. Cho, J. B. Cheon and Y. I. Jeong, *Macromol. Rapid Commun.*, 1997, **18**, 361.

106. A. Nykänen, M. Nuopponen, A. Laukkanen, S. P. Hirvonen, M. Rytelä, O. Turunen, H. Tenhu, R. Mezzenga, O. Ikkala and J. Ruokolainen, *Macromolecules*, 2007, **40**, 5827.

107. X. M. Zheng, T. Jiang and F. He, *Acta Polym. Sin.*, 2011, **8**, 895.

108. K. Skrabania, J. Kristen, A. Laschewsky, Ö. Akdemir, A. Hoth and J. F. Lutz, *Langmuir*, 2007, **23**, 84.

109. C. He, S. W. Kim and D. S. Lee, *J. Control. Release*, 2008, **127**, 189.

110. M. Crothers, D. Attwood, J. H. Collett, Z. Yang, C. Booth, P. Taboada, V. Mosquera, N. M. P. S. Ricardo and L. G. A. Martini, *Langmuir*, 2002, **18**, 8685.

111. P. Taboada, G. Velasquez, S. Barbosa, V. Castelletto, S. K. Nixon, Z. Yang, F. Heatley, I. W. Hamley, M. Ashford, V. Mosquera, D. Attwood and C. Booth, *Langmuir*, 2005, **21**, 5263.

112. A. Kelarakis, V. Havredaki, C. J. Rekatas and C. Booth, *Phys. Chem. Chem. Phys.*, 2001, **3**, 5550.

113. B. Chu and Z. Zhou, in *Nonionic Surfactants. Polyoxyalkylene Block Copolymers*, ed. V. Nace, Marcel Dekker, New York, 1996, p. 67.

114. C. Booth, D. Attwood and C. Price, *Phys. Chem. Chem. Phys.*, 2006, **8**, 3612.

115. D. Attwood, J. H. Collett and C. J. Tait, *Int. J. Pharm.*, 1985, **26**, 25.

116. A. Kelarakis, V. Havredaki, G. E. Yu, L. Derici and C. Booth, *Macromolecules*, 1998, **31**, 944.

117. C. Booth, G. Y. Yu and V. M. Nace, in *Amphiphilic Block Copolymers: Self-Assembly and Applications*, ed. P. Alexandridis and B. Lindman, Elsevier, Amsterdam, 2000, p. 57.

118. P. Linse, *Macromolecules*, 1993, **26**, 4437.

119. Z. K. Zhou and B. Chu, *Macromolecules*, 1994, **27**, 2025.

120. Y. W. Yang, N. J. Deng, G. E. Yu, Z. K. Zhou, D. Attwood and C. Booth, *Langmuir*, 1995, **11**, 4703.

121. Y. W. Yang, Z. Yang, Z. K. Zhou, D. Attwood and C. Booth, *Macromolecules*, 1996, **29**, 670.

122. G. Riess and D. Rogez, *ACS Polym. Prepr.*, 1982, **23**, 19.

123. A. Kelarakis, X. F. Yuan, S. M. Mai, Y. W. Yang and C. Booth, *Phys. Chem. Chem. Phys.*, 2003, **5**, 2628.

124. G. E. Yu, D. Mistry, S. Ludhera, F. Heatley, D. Attwood and C. Booth, *J. Chem. Soc., Faraday Trans.*, 1997, **93**, 3383.

125. K. Sahakaro, C. Chaibundit, Z. Kaligradaki, S. M. Mai, F. Heatley, C. Booth, J. C. Padget and I. M. Shirley, *Eur. Polym. J.*, 2000, **36**, 1835.
126. A. Kelarakis, S. M. Mai, V. Havredaki, A. Brett and C. Booth, *J. Colloid Interface Sci.*, 2004, **275**, 439.
127. T. L. Bluhm and M. D. Whitmore, *Can. J. Chem.*, 1985, **63**, 249.
128. A. Halperin, *Macromolecules*, 1987, **20**, 2943.
129. R. Nagarajan and K. Ganesh, *J. Chem. Phys.*, 1989, **90**, 5843.
130. R. Xu, M. A. Winnik, G. Riess, B. Chu and M. D. Croucher, *Macromolecules*, 1992, **25**, 644.
131. P. Taboada, G. Velasquez, S. Barbosa, Z. Yang, S. K. Nixon, Z. Zhou, F. Heatley, I. W. Hamley, M. Ashford, V. Mosquera, D. Attwood and C. Booth, *Langmuir*, 2006, **22**, 7465.
132. D. Attwood, C. Booth, S. G. Yeates, C. Chaibundit and N. M. P. S. Ricardo, *Int. J. Pharm.*, 2007, **345**, 35.
133. R. Ivanova, B. Lindman and P. Alexandridis, *Adv. Colloid Interface Sci.*, 2001, **89–90**, 351.
134. J. B. C. Amstrong, J. Mitchell, A. Beezer and S. Leharne, *J. Phys. Chem. B*, 1996, **100**, 1738.
135. P. Alexandridis and L. Yang, *Macromolecules*, 2000, **33**, 5574.
136. B. Bharatiya, C. Guo, H. P. Ma, P. A. Hassan and P. Bahadur, *Eur. Polym. J.*, 2007, **43**, 1883.
137. E. Castro, P. Taboada and V. Mosquera, *J. Phys. Chem. B*, 2006, **110**, 13113.
138. E. Castro, S. Barbosa, J. Juárez, P. Taboada, I. A. Katime and V. Mosquera, *J. Phys. Chem. B*, 2008, **112**, 5296.
139. J. P. Mata, P. R. Majhi, C. Guo, H. Z. Liu and P. Bahadur, *J. Colloid Interface Sci.*, 2005, **292**, 548.
140. K. Patel, P. Bahadur, C. Guo, J. H. Ma, H. Z. Liu, Y. Yamashita, A. Khnal and K. Nakashima, *Eur. Polym. J.*, 2007, **43**, 1699.
141. P. Alexandridis and J. F. Holzwarth, *Langmuir*, 1997, **13**, 6074.
142. A. G. Denkova, E. Mendes and M. O. Coopens, *J. Phys. Chem. B*, 2008, **112**, 793.
143. E. B Jørgensen, S. Hvidt, W. Brown and K. Schillén, *Macromolecules*, 1997, **30**, 2355.
144. N. J. Deng, Y. Z. Luo, S. Tanodekaew, N. Bingham, D. Attwood and C. Booth, *J. Polym. Sci. B: Polym. Phys.*, 1995, **33**, 1085.
145. P. Alexandridis, V. Athanassiou and T. A. Hatton, *Langmuir*, 1995, **11**, 2442.
146. A. Eisenberg and M. Rinaudo, *Polym. Bull.*, 1990, **24**, 671.
147. L. Zhang, K. Khougaz, M. Moffitt and A. Eisenberg, in *Amphiphilic Block Copolymers: Self-Assembly and Applications*, ed. P. Alexandridis and B. Lindman, Elsevier, Amsterdam, 2000, p. 87.
148. I. Astafieva, A. Khougaz and A. Eisenberg, *Macromolecules*, 1995, **28**, 7127.
149. M. Burkhardt, N. Martinez-Castro, S. Tea, M. Dreschler, I. Babin, I. Grishagin, R. Schweins, D. V. Pergushov, M. Gradzielski, A. B. Zezin and A. H. E. Müller, *Langmuir*, 2007, **23**, 12864.

150. A. Qin, C. Tian, C. Ramireddy, S. E. Webber, P. Munk and Z. Tuzar, *Macromolecules*, 1994, **27**, 120.
151. O. Colombani, M. Ruppel, M. Burkhardt, M. Dreschler, M. Schimacher, M. Gradzielski, R. Schweins and A. H. E. Müller, *Macromolecules*, 2007, **40**, 4351.
152. H. Cölfen, *Macromol. Rapid Commun.*, 2001, **22**, 219.
153. X. H. Dai, C. Y. Hong and C. Y. Pan, *Macromol. Chem. Phys.*, 2012, **213**, 2192.
154. V. Dordovic, M. Uchman, K. Prochazka, A. Zhigunov, J. Plestil, A. Nykanen, J. Ruokolainen and P. Matejicek, *Macromolecules*, 2013, **46**, 6881.
155. Z. Ge and S. Liu, *Chem. Soc. Rev.*, 2013, **42**, 7289.
156. G. Riess, *Prog. Polym. Sci.*, 2003, **28**, 1107.
157. R. Barreiro-Iglesias, C. Alvarez-Lorenzo and A. Concheiro, *Prog. Colloid Polym. Sci.*, 2003, **122**, 95.
158. D. F. Evans and H. Wennerström, *The Colloidal Domain. Where Physics, Chemistry, Biology and Technology Meet*, Wiley, New York, 1999.
159. T. Nivaggioli, P. Alexandridis, T. A. Hatton, A. Yetka and M. A. Winnik, *Langmuir*, 1995, **11**, 730.
160. G. Marinov, B. Michels and R. Zana, *Langmuir*, 1998, **14**, 2639.
161. R. Zana, in *Amphiphilic Block Copolymers: Self-Assembly and Applications*, ed. P. Alexandridis and B. Lindman, Elsevier, Amsterdam, 2000, p. 221.
162. N. F. Carnahan and K. E. Starling, *J. Chem. Phys.*, 1969, **51**, 635.
163. A. J. Vrij, *Chem. Phys.*, 1978, **69**, 1742.
164. S. W. Provencher, *Makromol. Chem.*, 1979, **180**, 201.
165. K. Mortensen, in *Amphiphilic Block Copolymers: Self-Assembly and Applications*, ed. P. Alexandridis and B. Lindman, Elsevier, Amsterdam, 2000, p. 191.
166. V. Castelletto and I. W. Hamley, *Curr. Opin. Colloid Sci.*, 2002, 7, 167.
167. M. Nakano, H. Matsuoka, H. Yamaoka, A. Poppe and D. Richter, *Macromolecules*, 1999, **32**, 697.
168. J. Godward, F. Heatley and C. Price, *J. Chem. Soc., Faraday Trans.*, 1993, **89**, 3471.
169. H. Vink, *J. Chem. Soc., Faraday Trans. 1*, 1985, **81**, 1725.
170. G. Fleischer, D. Geschke, J. Karger and W. Heink, *J. Magn. Reson.*, 1985, **65**, 429.
171. C. Chachaty, *Prog. Nucl. Magn. Reson. Spectrosc.*, 1987, **19**, 183.
172. Q. G. Wang, G. E. Yu, Y. L. Deng, C. Price and C. Booth, *Eur. Polym. J.*, 1993, **29**, 665.
173. A. D. Bedells, R. M. Arafeh, Z. Yang, D. Attwood, F. Heatley, J. C. Padget, C. Price and C. Booth, *J. Chem. Soc., Faraday Trans.*, 1993, **89**, 1235.
174. R. Brydson and C. Hammond, in *Nanoscale Science and Technology*, ed. R. W. Kelsall, M. Geoghegan and I. W. Hamley, Wiley, Hoboken, NJ, 2005, p. 56.

175. M. Godraich and Y. Talmon, in *Amphiphilic Block Copolymers: Self-Assembly and Applications*, ed. P. Alexandridis and B. Lindman, Elsevier, Amsterdam, 2000, p. 253.

176. I. W. Hamley, S. D. Connell and S. Collins, *Macromolecules*, 2004, **37**, 5337.

177. O. Glatter, G. Scherf, K. Schillen and W. Brown, *Macromolecules*, 1994, **27**, 6046.

178. J. Rassing, S. Bandyopahyay and E. M. Eyring, *J. Mol. Liq.*, 1983, **26**, 97.

179. P. Munk, in *Solvents and Self-Organization of Polymers*, ed. S. E. Webber, Z. Tuzar and P. Munk, Kluwer, Dordrecht, 1996, p. 367.

180. R. K. O'Reylly, C. J. Hawker and K. L. Wooley, *Chem. Soc. Rev.*, 2006, **35**, 1068.

181. J. Juárez, P. Taboada, M. A. Valdez and V. Mosquera, *Langmuir*, 2008, **24**, 7107.

182. M. Elsabahy, M. E. Perron, N. Bertrand, G. Y. Yu and J. C. Leroux, *Biomacromolecules*, 2007, **8**, 2250.

183. A. Blanazs, S. P. Armes and A. J. Ryan, *Macromol. Rapid Commun.*, 2009, **30**, 267.

184. R. Ganguly, V. K. Aswal and P. A. Hassan, *J. Colloid Interface Sci.*, 2007, **315**, 693.

185. V. Patel, J. Dey, R. Ganguly, S. Kumar, S. Nath, V. K. Aswal and P. Bahadur, *Soft Matter*, 2012, **9**, 7583.

186. G. Landazauri, V. V. A. Fernandez, J. F. A. Soltero and Y. Rharbi, *J. Phys. Chem. B*, 2012, **116**, 11720.

187. R. B. Grubbs and Z. Sun, *Chem. Soc. Rev.*, 2013, **42**, 7436.

188. A. Blanazs, J. Madsen, G. Battaglia, A. J. Ryan and S. P. Armes, *J. Am. Chem. Soc.*, 2011, **133**, 16581.

189. X. C. Pang, L. Zhao, M. Akinc, J. K. Kim and Z. Q. Lin, *Macromolecules*, 2011, **44**, 3746.

190. X. Liu, Z. C. Tian, C. Chen and H. R. Allcock, *Polym. Chem.*, 2013, **4**, 1115.

191. D. Mistry, T. Annable, X. F. Yuan and C. Booth, *Langmuir*, 2006, **22**, 2992.

192. T. Zheltonozhskaya, S. Partsevskaya, S. Fedorchuk, D. Klymchuk, Y. Gomza, N. Permyakova and L. Kunitskaya, *Eur. Polym. J.*, 2013, **49**, 405.

193. L. Lei, J. F. Gohy, N. Willet, J. X. Zhang, S. Varshney and R. Jérome, *Macromolecules*, 2004, **37**, 1089.

194. P. Petrov, C. B. Tsvetanov and R. Jerome, *Polym. Int.*, 2008, 57, 1258.

195. Z. Zhou, Z. Li, Y. Ren, M. A. Hillmyer and T. P. Lodge, *J. Am. Chem. Soc.*, 2003, **125**, 10182.

196. X. Z. Jiang, G. Y. Zhang, R. Narain and S. Y. Liu, *Soft Matter*, 2009, **5**, 1530–1538.

197. T. P. Lodge, M. A. Hillmyer, Z. Zhou and Y. Talmon, *Macromolecules*, 2004, **37**, 6680.

198. I. K. Voets, A. de Keizer, P. de Waard, P. M. Frederik, P. H. H. Bomans, H. Schmalz, A. Walther, S. M. King, F. A. M. Leermakers and M. A. Cohen-Stuart, *Angew. Chem. Int. Ed.*, 2006, **45**, 6673.
199. Z. Li, E. Kesselman, Y. Talmon, M. A. Hillmyer and T. P. Lodge, *Science*, 2004, **306**, 98.
200. R. Erhardt, A. Böker, H. Zettl, H. Kaya, W. Pychout-Hintzen, G. Krausch, V. Abetz and A. H. E. Müller, *Macromolecules*, 2001, **34**, 1069.
201. I. K. Voets, F. A. Leermakers, A. de Keizer, M. Charlaganov and M. A. C. Stuart, in *Self Organized Nanostructures of Amphiphilic Block Copolymers. Advances in Polymer Science*, Vol. 241, ed. A. H. E. Muller and O. V. Borisov, Springer, Berlin, 2011, p. 163.
202. P. Linse, in *Amphiphilic Block Copolymers: Self-Assembly and Applications*, ed. P. Alexandridis and B. Lindman, Elsevier, Amsterdam, 2000, p. 13.
203. O. V. Borisov, E. B. Zhulina, F. A. M. Leermarkers and A. H. E. Muller, in *Self Organized Nanostructures of Amphiphilic Block Copolymers. Advances in Polymer Science*, Vol. 241, ed. A. H. E. Muller and O. V. Borisov, Springer, Berlin, 2011, p. 57.
204. Q. Wang, *Soft Matter*, 2011, 7, 3711.
205. FDA, *Guidance for Industry: Nonclinical Studies for the Safety Evaluation of Pharmaceutical Excipients*, US Food and Drug Administration, Rockville, MD, 2005.
206. Committee on Toxicity Testing and Assessment of Environmental Agents, *Toxicity Testing in the 21st Century. A Vision and A Strategy*, National Academies Press, Washington, DC, 2007.
207. E. C. L. Cazedey, F. C. Carvalho, F. A. M. Fiorentino, M. P. D. Gremiao and H. R. N. Salgado, *Braz. J. Pharm. Sci.*, 2009, **45**, 759.
208. I. E. Borissevitch, C. P. Borges, V. E. Yushmanov and M. Tabak, *Biochim. Biophys. Acta*, 1995, **1238**, 57.
209. G. Gaucher, M. H. Dufresne, V. P. Sant, N. Kang, D. Maysinger and J. C. Leroux, *J. Control. Release*, 2005, **109**, 169.
210. M. Yokoyama, P. Opanasopit, T. Okano, K. Kawano and Y. Maitani, *J. Drug Target.*, 2004, **12**, 373.
211. Y. Li and G. S. Kwon, *Pharm. Res.*, 2000, **17**, 607.
212. J. Yue, S. Liu, Z. Xie, Y. Xing and X. Jing, *J. Mater. Chem. B*, 2013, **1**, 4273.
213. M. Yokoyama, *J. Exp. Clin. Med.*, 2001, **3**, 151.
214. C. O. Rangel-Yagui, A. Pessoa Jr and L. C. T. Costa-Tavares, *J. Pharm. Pharm. Sci.*, 2005, **8**, 147.
215. P. N. Hurter and T. A. Hatton, *Langmuir*, 1992, **8**, 1291.
216. A. B. E. Attia, Z. Y. Ong, J. L. Hedrick, P. P. Lee, P. L. R. Ee, P. T. Hammond and Y. Y. Yang, *Curr. Opin. Colloid Interface Sci.*, 2011, **16**, 182.
217. A. Ribeiro, A. Sosnik, D. A. Chiappetta, F. Veiga, A. Concheiro and C. Alvarez-Lorenzo, *J. R. Soc., Interface*, 2012, **9**, 2059.
218. S. M. Moghimi, I. S. Muir, L. Illum, S. S. Davis and V. Kolb-Bachofen, *Biochim. Biophys. Acta*, 1993, **1179**, 157.

219. G. Barratt, *Cell Mol. Life Sci.*, 2003, **60**, 21.
220. A. V. Kabanov, E. V. Batrakova and V. Y. Alakhov, *J. Control. Release*, 2002, **82**, 189.
221. D. A. Chiappetta, J. Degrossi, S. Teves, M. D'Aquino, C. Bregni and A. Sosnik, *Eur. J. Pharm. Biopharm.*, 2008, **69**, 535.
222. J. Gonzalez-Lopez, C. Alvarez-Lorenzo, P. Taboada, A. Sosnik, I. Sandez-Macho and A. Concheiro, *Langmuir*, 2008, **24**, 10688.
223. A. Ribeiro, I. Sandez-Macho, M. Casas, S. Alvarez-Pérez, C. Alvarez-Lorenzo and A. Concheiro, *Colloids Surf. B*, 2013, **103**, 550.
224. C. Alvarez-Lorenzo, J. González-López, M. Fernández-Tarrío, M. I. Sández-Macho and A. Concheiro, *Eur. J. Pharm. Biopharm.*, 2007, **66**, 244.
225. R. Ganguly, Y. Kadam, N. Choudhury, V. K. Aswal and P. Bahadur, *J. Phys. Chem. B.*, 2011, **115**, 3425.
226. Y. H. Bae and H. Yin, *J. Control. Release*, 2008, **131**, 2.
227. S. Barbosa, M. A. Cheema, P. Taboada and V. Mosquera, *J. Phys. Chem. B*, 2007, **111**, 10920.
228. M. E. N. P. Ribeiro, I. G. P. Vieira, I. M. Cavalcante, N. M. P. S. Ricardo, D. Attwood, S. G. Yeates and C. Booth, *Int. J. Pharm.*, 2009, **378**, 211.
229. A. Cambón, S. Barbosa, A. Rey-Rico, E. Figueroa-Ochoa, J. F. A. Soltero, S. G. Yeates, C. Alvarez-Lorenzo, A. Concheiro, P. Taboada and V. Mosquera, *J. Colloid Interface Sci.*, 2012, **387**, 275.
230. S. Taurin, H. Nehoff and K. Greish, *J. Control. Release*, 2012, **164**, 265.
231. S. Modi, J. P. Jain, A. J. Domb and N. Kumar, *Curr. Pharm. Design*, 2006, **12**, 4785.
232. M. Yokoyama, *Expert Opin. Drug Deliv.*, 2010, 7, 145.
233. H. Maeda, J. Wu, T. Sawa, Y. Matsamura and K. Hori, *J. Control. Release*, 2000, **65**, 271.
234. G. S. Kwon, *Crit. Rev. Ther. Drug Carrier Syst.*, 2003, **20**, 357.
235. H. Maeda, G. Y. Bharate and J. Daruwalla, *Eur. J. Pharm. Biopharm.*, 2009, **71**, 409.
236. V. Torchilin, *Adv. Drug Deliv. Rev.*, 2011, **63**, 131.
237. A. K. Iyer, G. Khaled, J. Fang and H. Maeda, *Drug Discov. Today*, 2006, **11**, 812.
238. K. Hori, M. Nishihara and M. Yokoyama, *J. Pharm. Sci.*, 2010, **99**, 549.
239. T. Hara, S. Iriyama, K. Makino, H. Terada and M. Ohya, *Colloids Surf. B*, 2010, **75**, 42.
240. J. Nicolas, S. Mura, D. Brambilla, N. Mackiewicz and P. Couvreur, *Chem. Soc. Rev.*, 2013, **42**, 1147.
241. P. S. Low, W. A. Henne and D. D. Doorneweerd, *Acc. Chem. Res.*, 2008, **41**, 120.
242. Z. Hu, F. Luo, Y. Pan, C. Hou, L. Ren, J. Chen, J. Wang and Y. Zhang, *J. Biomed. Mater. Res. A*, 2008, **85**, 797.
243. M. M. Cardoso, I. N. Peça and A. C. Roque, *Curr. Med. Chem.*, 2012, **19**, 3103.

244. C. J. F. Rijcken, O. Soga, W. E. Hennink and C. F. van Nostrum, *J. Control. Release*, 2007, **120**, 131.
245. N. Rapoport, *Prog. Polym. Sci.*, 2007, **32**, 962.
246. S. Pearson, W. Scarano and M. H. Stenzel, *Chem. Commun.*, 2012, **48**, 4695.
247. C. Alvarez-Lorenzo and A. Concheiro (eds), *Smart Materials for Drug Delivery*, Royal Society of Chemistry, Cambridge, 2013.
248. N. Washington, C. Washington and C. G. Wilson, *Physiological Pharmaceutics: Barriers to Drug Absorption*, Taylor & Francis, London, 2nd edn, 2001.
249. N. Nishiyama, Y. Bae, K. Miyata, S. Fukushima and K. Kataoka, *Drug Discov. Today Technol.*, 2005, **2**, 21.
250. A. S. E. Ojugo, P. M. J. Mesheedy, D. J. O. McIntyre, C. McCoy, M. Stubbs, M. O. Leach, I. R. Judson and J. R. Griffiths, *NMR Biomed.*, 1999, **12**, 495.
251. S. Ganta, A. Iyer and M. Amiji, in *Targeted Delivery of Small and Macromolecular Drugs*, ed. A. S. Narang and R. I. Mahato, CRC Press/ Taylor & Francis, Boca Raton, FL, 2010, p. 555.
252. L. A. Schneider, A. Korber, S. Grabbe and J. Dissemond, *Arch. Dermatol. Res.*, 2007, **298**, 413.
253. E. R. Gillies and J. M. J. Fréchet, *Pure Appl. Chem.*, 2004, **76**, 1295.
254. T. J. Martin, K. Prochazka, P. Munk and S. E. Webber, *Macromolecules*, 1996, **29**, 6071.
255. Y. L. Luo, J. F. Yuan, X. J. Liu, H. Xie and Q. Y. Gao, *J. Bioactive Compat. Polym.*, 2010, **25**, 292.
256. S. Lui, J. V. M. Weaver, Y. Tang, N. C. Billingham, S. P. Armes and K. Tribe, *Macromolecules*, 2002, **35**, 6121.
257. K. H. Min, J. H. Kim, S. M. Bae, H. Shin, M. S. Kim, S. Park, H. Lee, R. W. Park, I. S. Kim, K. Kim, I. C. Kwon, S. Y. Jeong and D. S. Lee, *J. Control. Release*, 2011, **144**, 259.
258. Y. Bae, N. Nishiyama, S. Fukushima, H. Koyama, M. Yasuhiro and K. Kataoka, *Bioconjug. Chem.*, 2005, **16**, 122.
259. G. H. Gao, J. W. Lee, M. K. Nguyen, G. H. Im, J. Yang, H. Heo, P. Jeon, T. G. Park, J. H. Lee and D. S. Lee, *J. Control. Release*, 2011, **155**, 11.
260. K. Miyata, R. J. Christie and K. Kataoka, *React. Funct. Polym.*, 2011, **71**, 227.
261. S. Fukushima, K. Miyata, N. Nishiyama, N. Kanayama, Y. Yamasaki and K. Kataoka, *J. Am. Chem. Soc.*, 2010, **127**, 2810.
262. L. Bromberg, S. Deshmukh, M. Temchenko, L. Iourtchenko, V. Alakhov, C. Alvarez-Lorenzo, R. Barreiro-Iglesias, A. Concheiro and T. A. Hatton, *Bioconjug. Chem.*, 2005, **16**, 626.
263. L. Bromberg, S. Raduyk, T. A. Hatton, A. Concheiro, C. Rodriguez-Valencia, M. Silva and C. Alvarez-Lorenzo, *Bioconjug. Chem.*, 2009, **20**, 1044.

264. T. M. Sun, J. Z. Du, Y. D. Yao, C. Q. Mao, S. Dou, S. Y. Huang, P. Z. Zhang, K. W. Leong, E. W. Song and J. Wang, *ACS Nano*, 2011, **5**, 1483.
265. D. J. Gary, H. Lee, R. Sharma, J. S. Lee, Y. Kim, Z. Y. Cui, D. Jia, V. D. Bowman, P. R. Chipman, L. Wan, Y. Zou, G. Mao, K. Park, B. S. Herbert, S. F. Konieczny and Y. Y. Won, *ACS Nano*, 2011, **5**, 3493.
266. S. Fukushima, K. Miyata, K. Nishiyama, N. Kanayama, Y. Yamasaki and K. Kataoka, *J. Am. Chem. Soc.*, 2005, **127**, 2810.
267. A. J. Convertine, C. Diab, M. Prieve, A. Paschal, A. S. Hoffman, P. H. Johnson and P. S. Stayton, *Biomacromolecules*, 2010, **11**, 2904.
268. G. Wu, Y. Z. Fang, S. Yang, J. R. Lupton and N. D. Turner, *J. Nutr.*, 2004, **134**, 489.
269. T. Kurz, J. W. Eaton and U. T. Brunk, *Antioxid. Redox Signal.*, 2010, **13**, 511.
270. P. Kuppusamy, H. Li, G. Ilangovan, A. J. Cardounel, J. L. Zweier, K. Yamada, M. C. Krishna and J. B. Mitchell, *Cancer Res.*, 2002, **62**, 307.
271. W. Chen, P. Zhong, F. Meng, R. Cheng, C. Deng, J. Feijen and Z. Zhong, *J. Control. Release*, 2013, **169**, 171.
272. H. L. Sun, B. N. Guo, R. Cheng, F. H. Meng, H. Y. Liu and Z. Y. Zhong, *Biomaterials*, 2009, **30**, 6358.
273. H. L. Sun, B. N. Guo, X. Q. Li, R. Cheng, F. H. Meng, H. Y. Liu and Z. Y. Zhong, *Biomacromolecules*, 2010, **11**, 848.
274. T. Thambi, H. Y. Yoon, K. Kim, I. C. Kwon, C. K. Yoo and J. H. Park, *Bioconjug. Chem.*, 2011, **22**, 1924.
275. J. Li, M. Huo, J. Wang, J. Zhou, J. M. Mohammad, Y. Zhang, Q. Zhu, A. Y. Waddad and Q. Zhang, *Biomaterials*, 2012, **33**, 2310.
276. P. Sun, D. Zhou and Z. Gan, *J. Control. Release*, 2011, **155**, 96.
277. N. Ma, Y. Li, H. P. Xu, Z. Q. Wang and X. Zhang, *J. Am. Chem. Soc.*, 2010, **132**, 442.
278. L. Zhang, W. G. Liu, L. Lin, D. Y. Chen and M. H. Stenzel, *Biomacromolecules*, 2008, **9**, 3321.
279. A. N. Koo, H. J. Lee, S. E. Kim, J. H. Chang, C. Park, C. Kim, J. H. Park and S. C. Lee, *Chem. Commun.*, 2008, 6570.
280. H. Wang, L. Tang, C. Tu, Z. Song, Q. Yin, L. Yin, Z. Zhang and J. Cheng, *Biomacromolecules*, 2013, **14**, 3706.
281. D. Sutton, N. Nasongkla, E. Blanco and J. M. Gao, *Pharm. Res.*, 2007, **24**, 1029.
282. S. Cammas, K. Suzuki, C. Sone, Y. Sakurai, K. Kataoka and T. Okano, *J. Control. Release*, 1997, **48**, 157.
283. M. Liu and L. S. Wang, *Biomaterials*, 2004, **25**, 1929.
284. N. V. Economidis, D. A. Pena, P. G. Smirniotis, F. Kohori, K. Sakai, T. Aoyagi, M. Yokoyama, M. Yamato, Y. Sakurai and T. Okano, *Colloids Surf. B*, 1999, **16**, 195.

285. J. E. Chung, M. Yokoyama, M. Yamato, T. Aoyagi, Y. Sakurai and T. Okano, *J. Control. Release*, 1999, **62**, 115.
286. Z. S. Ge, D. Xie, D. Y. Chen, X. Z. Jiang, Y. F. Zhang, H. W. Liu and S. Liu, *Macromolecules*, 2007, **40**, 3538.
287. H. Wei, C. Cheng, C. Chang, W. Q. Chen, S. X. Cheng, X. Z. Zhang and R. X. Zhuo, *Langmuir*, 2008, **24**, 4564.
288. J. Taillefer, M. C. Jones, N. Brasseur, J. E. Vanlier and J. C. Leroux, *J. Pharm. Sci.*, 2000, **89**, 52.
289. J. E. Chung, M. Yokoyama and T. Okano, *J. Control. Release*, 2000, **65**, 93.
290. N. V. Economidis, D. A. Pena, P. G. Smirniotis, F. Kohori, K. Sakai, T. Aoyagi, M. Yokoyama, M. Yamato, Y. Sakurai and T. Okano, *Colloids Surf. B*, 1999, **16**, 195.
291. Y. N. Zhong, C. Wang, L. Cheng, F. H. Meng, Z. Y. Zhong and Z. Liu, *Biomacromolecules*, 2013, **14**, 2411.
292. C. Alvarez-Lorenzo, L. Bromberg and A. Concheiro, *Photochem. Photobiol.*, 2009, **85**, 848.
293. B. Yan, J. C. Boyer, N. R. Branda and Y. Zhao, *J. Am. Chem. Soc.*, 2011, **133**, 19714.
294. Y. Zhao, *Macromolecules*, 2012, **45**, 3647.
295. G. A. Husseini, G. D. Myrup, W. G. Pitt, D. A. Christensen and N. Y. Rapoport, *J. Control. Release*, 2006, **69**, 43.
296. A. H. Ghaleb, D. Stevenson-Abouelnasr, W. G. Pitt, K. T. Assaleh, L. O. Farahat and J. Fahadi, *Colloids Surf. A*, 2010, **359**, 18.
297. P. Kamev and N. Rapoport, *Am. J. Phys.*, 2006, **829**, 543.
298. G. Pasparakis and M. Vamvakaki, *Polym. Chem.*, 2011, **2**, 1234.
299. X. Wang, G. Jiang, X. Li, B. Tang, Z. Wei and C. Mai, *Polym. Chem.*, 2013, **4**, 4574.
300. J. Dong, Y. Wang, J. Zhang, X. Zhan, S. Zhu, H. Yang and G. Wang, *Soft Matter*, 2013, **9**, 370.
301. P. Schattling, F. D. Jochum and P. Theato, *Polym. Chem.*, 2014, **5**, 25–36.
302. T. W. Loo and D. M. Clarke, *J. Membr. Biol.*, 2005, **206**, 173.
303. A. H. Schinkel and J. W. Jonker, *Adv. Drug Deliv. Rev.*, 2003, **55**, 3.
304. F. Thiebaut, T. Tsuruo, H. Hamada, M. M. Gottesman, I. Pastan and M. C. Willingham, *Proc. Natl. Acad. Sci. U. S. A.*, 1987, **84**, 7735.
305. S. Gil, R. Saura, F. Forestier and R. Farinotti, *Placenta*, 2005, **26**, 268.
306. P. D. W. Eckford and F. J. Sharom, *Chem. Rev.*, 2009, **109**, 2989.
307. M. Garrigos, L. M. Mir and S. Orlowski, *Eur. J. Biochem.*, 1997, **244**, 664.
308. G. Szakács, J. K. Paterson, J. A. Ludwig, C. Booth-Genthe and M. M. Gottesman, *Nat. Rev.*, 2006, **5**, 219.
309. N. R. Patel, B. S. Pattni, A. H. Abouzeid and V. P. Torchilin, *Adv. Drug Deliv. Rev.*, 2013, **65**, 1748.
310. A. Cambón, A. Rey-Rico, D. Mistry, J. Brea, M. I. Loza, D. Attwood, C. Alvarez-Lorenzo, A. Concheiro, S. Barbosa, P. Taboada and V. Mosquera, *Int. J. Pharm.*, 2013, **445**, 47.
311. A. Sosnik, *Adv. Drug Deliv. Rev.*, 2013, **65**, 1828.

312. A. V. Kabanov, E. V. Batrakova and V. Y. Alakhov, *Adv. Drug Deliv. Rev.*, 2002, **54**, 759.
313. M. Werle, *Pharm. Res.*, 2008, **25**, 500.
314. V. Grabovac and A. Bernkop-Schnürch, *Sci. Pharm.*, 2006, **74**, 75.
315. A. K. Sharma, L. Zhang, S. Li, D. L. Kelly, V. Y. Alakhov, E. V. Batrakova and A. V. Kabanov, *J. Control. Release*, 2008, **131**, 220.
316. T. Demina, I. Grozdova, O. Krylova, A. Zhirnov, V. Istratov, H. Frey, H. Kautz and N. Melik-Nubarov, *Biochemistry*, 2005, **44**, 4042.
317. F. Föger, H. Hoyer, K. Kafedjiiski, M. Thaurer and A. Bernkop-Schnürch, *Biomaterials*, 2006, **27**, 5855.
318. A. Cambón, J. Brea, M. I. Loza, C. Alvarez-Lorenzo, A. Concheiro, S. Barbosa, P. Taboada and V. Mosquera, *Mol. Pharm.*, 2013, **10**, 3232.
319. Q. Shen, Y. Lin, T. Handa, M. Doi, M. Sugie, K. Wakayama, N. Okada, T. Fujita and A. Yamamoto, *Int. J. Pharm.*, 2006, **313**, 49.
320. J. A. Zastre, J. K. Jackson, W. Wong and H. M. Burt, *Mol. Pharm.*, 2008, **5**, 643.
321. D. W. Miller, E. V. Batrakova, D. O. Waltner, V. Alakhov and A. V. Kabanov, *Bioconjug. Chem.*, 1997, **8**, 649.
322. V. Alakhov, E. Klinksi, S. Li, G. Pietrzynski, A. Venne, E. Batrakova, T. Bronitch and A. V. Kabanov, *Colloids Surf. B*, 1999, **16**, 113.
323. A. V. Kabanov, E. V. Batrakova and V. Y. Alakhov, *J. Control. Release*, 2003, **91**, 75.
324. D. Y. Alakhova, N. Y. Rapoport, E. V. Batrakova, A. A. Timoshin, S. Li, D. Nicholls, V. Y. Alakhov and A. V. Kabanov, *J. Control. Release*, 2010, **142**, 89.
325. E. V. Batrakova, S. Li, S. V. Vinogradov, V. Alakhov, D. W. Miller and A. V. Kabanov, *J. Pharmacol. Exp. Ther.*, 2001, **299**, 483.
326. E. V. Batrakova, S. Li, W. F. Elmquist, D. W. Miller, V. Y. Alakhov and A. V. Kabanov, *Br. J. Cancer*, 2001, **85**, 1987.
327. A. V. Kabanov, E. V. Batrakova and D. W. Miller, *Adv. Drug Deliv. Rev.*, 2003, **55**, 151.
328. E. Batrakova, S. Lee, S. Li, A. Venne, V. Alakhov and A. Kabanov, *Pharm. Res.*, 1999, **16**, 1373.
329. E. V. Batrakova, S. Li, V. Y. Alakhov, D. W. Miller and A. V. Kabanov, *J. Pharmacol. Exp. Ther.*, 2003, **304**, 845.
330. C. Alvarez-Lorenzo, A. Rey-Rico, J. Brea, M. I. Loza, A. Concheiro and A. Sosnik, *Nanomedicine (London)*, 2010, **5**, 1371.
331. A. Cambón, A. Rey-Rico, S. Barbosa, J. F. A. Soltero, S. G. Yeates, J. Brea, M. I. Loza, C. Alvarez-Lorenzo, A. Concheiro, P. Taboada and V. Mosquera, *J. Control. Release*, 2013, **167**, 68.

CHAPTER 6
DNA Particles

M. CARMEN MORÁN[a,b]

[a] Departament de Fisiologia, Facultat de Farmàcia, Universitat de Barcelona, Avda Joan XXIII, 08028 Barcelona, Spain; [b] Interaction of Surfactants with Cell Membranes, Unit Associated with CSIC, Facultat de Farmàcia, Universitat de Barcelona, Avda. Joan XXIII, 08028 Barcelona, Spain
Email: mcmoranb@ub.edu

6.1 Introduction

Gene therapy represents a new paradigm of therapy for diseases, where the disease is treated at the molecular level by restoring defective biological functions or reconstituting homeostatic mechanisms within cells. The delivery of therapeutic nucleic acids, normally in the form of plasmids, but increasingly also as smaller oligomers, remains one of the major obstacles currently hampering the further exploitation of genetic therapies. Specific and efficient delivery of genetic material to diseased sites and to particular cell populations is the challenge that is being addressed using a variety of viral and non-viral delivery systems, all of which have distinct advantages and disadvantages.[1,2] Compared with viral vectors, the synthetic (non-viral) systems are in general reputed to lack efficiency while offering flexibility and safety. However, this simplistic view ignores the fact that the suitability of any gene delivery system will always have to be matched with the clinical situation, the specific disease and the chosen therapeutic strategy.[3]

RSC Nanoscience & Nanotechnology No. 34
Soft Nanoparticles for Biomedical Applications
Edited by José Callejas-Fernández, Joan Estelrich, Manuel Quesada-Pérez and Jacqueline Forcada
© The Royal Society of Chemistry 2014
Published by the Royal Society of Chemistry, www.rsc.org

Nucleic acid-based therapies take two conceptually different approaches. The first is the delivery of plasmid DNA or related constructs[4,5] to express the gene of interest under the control of a suitable promoter, which will result in increased activity of the target, *i.e.* by production of a therapeutic protein. In contrast, the second approach involves the expression of oligomeric genetic material such as antisense oligonucleotides, siRNA or DNAzyme, which in general will lead to a decrease in target activity. In deciding on the appropriate genetic therapy for a given clinical problem, key factors to be taken into consideration include the number of genes involved in the pathogenesis (monogenetic/polygenetic), the required duration of therapy (temporary *versus* permanent), potency of the therapeutic product and the need for targeting or regulation of the genetic warhead. Clearly, none of the current vector systems is able to satisfy these potentially disparate needs and it is therefore important to appreciate the strengths and weaknesses of synthetic vector systems in the appropriate therapeutic context.[6,7]

Owing to the size and charge of naked DNA and the enzymatic and membrane barriers imposed by the cell, the entry of DNA molecules into the cells and subsequent expression represent a very wasteful process.[8] The observation that free plasmid DNA is able to transfect the skeletal muscle,[9] the liver[10] or a tumour[11] when given in the appropriate way, but will normally be degraded in the systemic circulation,[12] provides the rationale for 'packaging' of the plasmid DNA. This packaging occurs with the help of a delivery system that tends to compact and protect the nucleic acid. Furthermore, the delivery system should help to target the therapeutic nucleic acid to the desired site of action and facilitate efficient intracellular trafficking, typically to the nucleus.[3] Effective gene therapy requires that the DNA successfully gains access to the target cell, is taken up for internalization into the cell, is trafficked through the cell after escaping the degradative pathway to the nucleus and is subsequently transcribed and translated to produce the desired gene product[13] (Figure 6.1).

The most common strategy employed for the 'packaging' of DNA is based on electrostatic interaction between the anionic nucleic acid and the positive charges of the synthetic vector, which will complex and condense the nucleic acid into nanoparticles. Commonly used classes of synthetic vectors are based on various cationic lipids and polymers and, depending on the synthetic vector material used, the resulting particles have also been termed lipoplex or polyplex.[14]

Suspensions of such particles tend to be colloidally stable only if the particles are charged, that is, the cationic carrier will be present in excess to create particles that repel one another. This positive charge is also important because it facilitates cell adsorption and mediates efficient endosomal uptake into cells.[15] However, its non-specific nature is thought to contribute to the discrepancies commonly observed between *in vivo* and *in vitro* experiments. Whereas promiscuous binding may be advantageous in the simplified *in vitro* environment, it translates into extensive non-specific binding to cells, biological surfaces and blood components when the charged particles

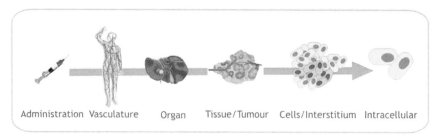

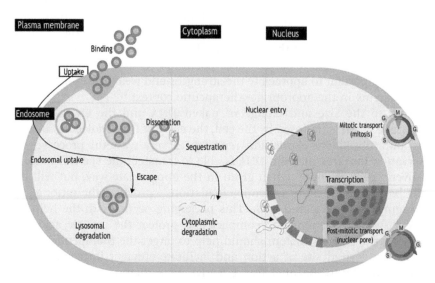

Figure 6.1 The systemic delivery of gene medicines to organs and disease sites that are not directly accessible is particularly challenging (top panel). In order to reach a remote target cell, the synthetic vector system needs to travel in the bloodstream to the organ. The nanoparticles need to extravasate to finally reach the target cells, where they need to be taken up efficiently. Current systems are invariably taken up into endosomes where they would eventually be degraded. Therefore, a mechanism that allows endosomal escape, *e.g.* by disruption after osmotic swelling (proton sponge), is required. After escaping into the cytoplasm, the nucleic acid (plasmid DNA) needs to gain entry into the nucleus to be able to utilize the nuclear transcription machinery and initiate gene expression. Access to the nuclear machinery can in principle occur during cell division when the nuclear envelope disappears through the nuclear pores, which allow shuffling of suitable molecules between nucleus and cytoplasm.
Adapted with permission from Ref. 30.

are administered *in vivo*. This non-specific binding is thought to modify the complexes and thus make them less stable and difficult to target to organs and remote sites.[16,17]

The rapidly rising demand for therapeutic-grade DNA molecules requires corresponding improvements in encapsulation and delivery technologies.

This includes the formulation of DNA molecules into synthetic delivery systems for enhanced cellular transformation efficiencies.

6.2 Colloidal Delivery Systems for DNA

Colloids are defined as materials with length scales below 1 μm and often in the sub-100 nm regime. The huge surface area of colloidal materials offers the flexibility to tailor their surface properties and particle behaviour to achieve new delivery modes for drugs and improve their therapeutic profiles.[18]

Research studies on colloidal delivery systems in genetic therapeutics are based on the molecular level, focusing on the interdisciplinary development of pharmaceutical DNA delivery approaches. Colloidal delivery systems modify many physicochemical properties, aiming to protect the DNA from degradation, minimize DNA loss, prevent harmful side effects, enhance DNA targeting, increase drug bioavailability and stimulate the immune systems.[13,19,20]

Various colloidal systems have been studied for decades and there have been promising approaches in improving the delivery of problematic DNA candidates. In this chapter, recent advances in the major colloidal delivery carriers are reviewed. The structure/synthesis, biological properties and cellular transfection capabilities of the different colloidal systems are discussed. With all the ongoing efforts, improved colloidal delivery techniques have become one of the promising delivery methods that may open up new markets and offer great potential benefits to DNA genetic therapeutics.

Currently and in the forthcoming years, simulations represent an efficient tool to explore complex nanostructures such as these DNA particles. Owing to their importance, Chapter 10 describes a study using computer simulation techniques.

6.2.1 Polymeric Nanoparticles

6.2.1.1 Structures and Synthesis

Polymeric nanoparticles are solid, colloidal particles consisting of macromolecular substances that vary in size from 10 to 1000 nm. The term *nanoparticle* is a collective name for both nanospheres and nanocapsules. Nanospheres have a matrix type of structure. Drugs may be adsorbed on the sphere surface or encapsulated within the particle.

Nanocapsules are vesicular systems in which the drug is confined to a cavity consisting of an inner oily core surrounded by a polymeric membrane. In this case, the active substances are usually dissolved in the inner core but may also be adsorbed on the capsule surface.[21]

Polymeric nanoparticles are used mostly in medical applications and have attracted considerable attention as targeted drug-delivery carriers owing to their biocompatibility, physical stability, protection of incorporated labile drugs from degradation and controlled release.[22,23]

Cationic polymers such as poly(L-lysine) (PLL),[24] polyethylenimine (PEI),[25–28] polyamidoamine (PAMAM) dendrimers,[29,30] and chitosan[31,32] can

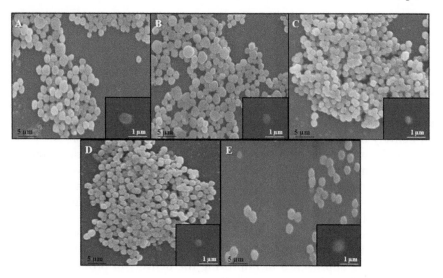

Figure 6.2 Polyplexes obtained by simple complexation at pH 4.5 from five different polycations and sRNA visualized by scanning electron microscopy
(SEM). (A) sRNA–PEI-LMW, (B) sRNA–PEI-HMW, (C) sRNA–CS-LMW,
(D) sRNA–CS-MMW and (E) sRNA–PAA, where PEI-LMW and PEI-HMW
are polyethylenimine with low molecular weight (LMW, $M_w = 1.3$ kDa)
and high molecular weight (HMW, $M_w = 10$ kDa), respectively, CS-LMW
and CS-MMW are chitosan with low molecular weight (LMW, $M_w = 50$–
190 kDa) and medium molecular weight (MMW, $M_w = 190$–310 kDa),
respectively, and PAA is polyallylamine with molecular weight 17 kDa.
Adapted with permission from Ref. 33.

be applied as carriers for converting gene vectors into complex polyplexes to
form defined sizes similar to the virus-like forms[33] (Figure 6.2).

6.2.1.2 Biological Properties

With the assistance of cationic polymers, it can be observed that transfection
activity parallels the membrane toxicity.[34,35] The general association between cytotoxicity and transfection suggests that a degree of membrane
damage must be caused for the cells to gain access to the cytoplasm.[36,37]
Successful transfection relies on achieving the correct balance between
gaining adequate accesses by the DNA molecules into the cytoplasm and
causing excessive and lethal damage to the cells. Cell culture experiments
showed that PEI-induced high transfection efficiency, however, entailed a
higher cell death rate. The high density of amino groups confers significant
buffering capacity to the PEI, especially in the endosome when the pH decreases from 7 to 5. The 'proton sponge effect' explains the high transfection
efficiencies obtained with PEI, as it appears to be an endosomolytic reagent.
However, the large number of positive charges leads to a high toxicity profile
and it is directly affected by many factors such as molecular weight, degree

of branching, ionic strength of the solution, zeta potential and particle size.[38] Coupling poly(ethylene glycol) (PEG) to mask the surface charge of the PEI–pDNA polyplexes is a popular approach for lowering the cellular toxicity by reducing the non-specific interactions of polyplexes in the bloodstream.[39] Additionally, poly(β-amino ester)s have been developed by incorporating moieties that can be hydrolysed, allowing this type of cationic polymer to degrade readily into non-toxic metabolites for long-term and repeated application in *in vivo* use. This biodegradable cationic polymer provides a mechanism for polymer–DNA dissociation following cellular uptake, which is critical for efficient transfection and gene expression.

6.2.1.3 *Polymeric Nanoparticles as Cellular Transfection Agents*

When applied to cells, the positively charged polyplexes mediate the transfection *via* a multistage process that includes cationic binding to the negatively charged cell membrane, to facilitate entry into the cytoplasm.[26,40] As a result, formulated or encapsulated therapeutics have been shown to exhibit greater therapeutic efficacy compared with unformulated biomolecules.[41]

Polymeric gene carriers can accommodate large-sized DNA, can be conjugated with appropriate functionalities and can be administered repeatedly. Functionalization of polymeric carriers with specific ligands is a very promising and highly selective strategy to improve specificity for the target sites of interest.[38] The ligands could be recognized by receptors on the surface of the cells of interest. On the other hand, block copolymers, especially poly(lactide-*co*-glycolide) (PLGA), attract DNA delivery applications. This is because their chemical composition, total molecular weight and block length ratios can easily be changed to allow control of the size and morphology of the polymeric carriers.[42] Polymeric particles with encapsulated DNA are vesicular systems in which the DNA is confined to a cavity surrounded by polymers, whereas particles adsorbed with DNA are matrix systems in which the DNA is physically and uniformly dispersed in the surface.[43] Encapsulation of pDNA protects it from nuclease degradation and the controlled gene delivery system can be readily designed to exhibit varying degradation times and release times of pDNA for prolonged gene expression over a required duration.

6.2.2 Polymeric Micelles

6.2.2.1 *Structures and Synthesis*

In aqueous media, amphiphilic block copolymers, which have a large solubility difference between the hydrophilic and hydrophobic segments, spontaneously form polymeric micelles. Their individual particle size is normally <100 nm in diameter with a distinct core–shell structure, in which hydrophobic segments are segregated from the aqueous exterior for a hydrophilic exterior (shell), formed by self-assembly of block or graft copolymers in

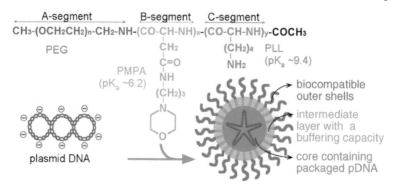

Figure 6.3 Chemical structure of PEG-*b*-PMPA-*b*-PLL triblock copolymers and schematic illustration of the three-layered polyplex micelles with spatially regulated structure.
Reproduced with permission from Ref. 46.

aqueous media. As indicated in Chapter 5, modifiable properties of micelles are of great interest for gene and drug-delivery applications.[44]

Micellar delivery systems including graft, diblock and multiblock copolymers have attracted attention as promising gene carriers[45,46] (Figure 6.3) Specifically, the graft copolymers have a comb-like structure with hydrophilic segments attached on the side of the cationic segments to improve the solution properties of the polyion complex.[47] The block copolymer is a linear copolymer with one segment covalently joined to the head of another segment and provides a core–shell structure, where pharmaceutical complexes are surrounded by a protective and stabilizing hydrophilic corona.[47]

6.2.2.2 Biological Properties

The thermodynamic and kinetic stability of the polymeric micelles is an important issue in their application to systemic administration, because their rapid dissociation in the blood compartment results in the burst release of loaded drugs, inducing systemic toxicity. In contrast to micelles from small surfactant molecules, polymeric micelles are generally more stable and can retain the loaded drug for a prolonged period even in much diluted conditions in the body owing to the appreciably lowered critical micelle concentration, particularly for polymeric micelles from highly regulated block copolymers with distinct core–shell structure.[48] Functional groups, such as amines and carboxylic acids in the core-forming segments, are useful for introducing drugs into the micelle core. PEG is often used as the hydrophilic segment by virtue of its biocompatibility. A polar PEG shell of the micelle facilitates solubility in water through steric stabilization and also provides biocompatibility and a stealth effect on the polymeric micelles, since PEG prevents the adsorption of proteins and is generally believed to be non-interactive with biological components.[49,50] In this sense, PEG–polypeptide block copolymers, prepared by the polymerization of the appropriate

N-carboxy anhydrides from a terminal amine on PEG, are advantageous because a series of block copolymers with different functional groups in the side chain can be prepared from the same platform. To facilitate micelle formation, a combination of intermolecular forces is available, such as hydrophobic interactions,[51-53] electrostatic interactions,[54-57] metal complexation[58,59] and hydrogen bonding.[60] The variation of the intermolecular force of core-forming segments enables one to regulate micelle stability for prolonged circulation and also for controlled drug release properties.

The sizes of these micelles are determined by thermodynamic parameters and are typically in the range of several tens of nanometres with a relatively narrow size distribution, similar to the size range of viruses and lipoproteins. Owing to the combination of the high molecular weight and biocompatibility of polymeric micelles, avoidance of renal glomerular filtration and reticuloendothelial system (RES) uptake is expected, thus providing longevity in blood circulation. This characteristic of polymeric micelles may facilitate tumour accumulation by the enhanced permeability and retention (EPR) effect[61] (see Chapter 5).

6.2.2.3 Polymeric Micelles as Cellular Transfection Agents

As seen also in Chapter 5, a polymeric micelle comprising nucleic acids is formed by polyion complexation between anionically charged DNA or RNA and a block copolymer having a hydrophilic segment and a cationic segment.[57,62-64] Complexation of plasmid DNA (pDNA) with a PEG–polycation such as PEG–poly(lysine) (PEG–PLys) occurs spontaneously, resulting in a polyion complex micelle (PIC micelle or polyplex micelle) with a size of approximately 100 nm.[62,65,66] The polyplex micelles exhibit a near-neutral zeta potential due to the PEG shell even in the presence of an excess amount of PEG–polycations.[67] Therefore, non-specific interaction with serum proteins or cells in the blood compartment is expected to be suppressed. Eventually, long-circulation properties and thus tumour accumulation by the EPR effect will be readily expected. Polyplex micelles containing pDNA indeed exhibited efficient gene introduction in cultured cells and also showed gene expression in the liver following intravenous injection into a mouse tail vein.[68] By packaging DNA into a polymeric micelle, prolonged blood circulation was achieved in which pDNA remained intact after 3 h, whereas naked pDNA was immediately degraded by nucleases in a few minutes. These results demonstrate that polymeric micelles also exhibit great feasibility as gene carriers. More detailed information can be found in Chapter 5.

6.2.3 Dendrimers

6.2.3.1 Structures and Synthesis

Dendrimers are macromolecular nanoparticles that comprise a series of branches around an inner core. As can be seen more widely in Chapter 7, dendrimers can be synthesized starting from the central core and working out towards the periphery (divergent synthesis) or in a top-down approach starting from the outermost residues (convergent synthesis).[69]

The use of dendrimers in gene delivery draws on a much narrower range of chemical architectures, predominantly those with a cationic net surface charge, of which the PAMAM and PPI dendrimers are commercially available. The first exploration of dendrimers as molecules for gene delivery focused on the PAMAM dendrimers.[70] The PAMAM dendrimers are normally based on an ethylenediamine or ammonia core with four and three branching points, respectively.[71-73] Using a divergent approach, the molecule is built up iteratively from the core through addition of methyl acrylate followed by amidation of the resulting ester with ethylenediamine. Each complete reaction sequence results in a new 'full' dendrimer generation (*e.g.* G3, G4, ...) with a terminal amine functionality, whereas the intermediate 'half' generations (*e.g.* G2.5, G3.5, ...) terminate in anionic carboxylate groups. The structure of these polymers involves repeated branching around the central core that results in a nearly perfect 3D geometric pattern. At higher generations, dendrimers resemble spheres, with a number of cavities within its branches to hold both drug and diagnostic agents. The size of dendrimers used in targeted drug delivery is usually 10–100 nm.[70,74]

Their unique molecular architecture means that dendrimers have a number of distinctive properties that differentiate them from other polymers; specifically, the gradual stepwise method of synthesis means that they in general have a well-defined size and structure with a comparatively low polydispersity index. Key properties in terms of the potential use of these materials in drug and gene delivery are defined by the high density of terminal groups. These contribute to the molecules' surface characteristics, offer multiple attachment sites, *e.g.* for conjugation of drugs or targeting moieties, and determine the molecular volume, which is important for the ability to sequester other molecules within the core of the dendrimer.

6.2.3.2 Biological Properties

The initial evaluation of the biological properties of PAMAM dendrimers *in vitro* found them to be relatively non-toxic.[75] In cytotoxicity assays, they compare favourably with some of the other transfection agents, in particular cationic polymers of higher molecular weight such as PEI (600–1000 kDa), PLL (36.6 kDa) and DEAE–dextran (500 kDa), which in these assays are around three orders of magnitude more toxic.[76] In contrast to the high molecular weight PEI and PLL polymers, the toxicity of PAMAM dendrimers did not seem to stem from membrane damage as assayed by LDH release or haemolysis.[76] Nevertheless, dendrimers interact effectively with cell membranes and the electrostatic interactions of cationic polymer and anionic cell surfaces are highly important for the cellular uptake of charged DNA complexes.[77]

Size is a key determinant of dendrimer cytotoxicity for both PAMAM[78] and PPI dendrimers.[79,80] The cytotoxicity of PAMAM dendrimers increases with generation, independent of surface charge, for both full-generation cationic dendrimers (G2–G4) and the 'half-generation' anionic intermediates (G2.5, G3.5).[80,81] The nature and density of charged groups are other factors that determine dendrimer toxicity.[82] Cationic (surface) charges are in general more toxic but details depend on the specific groups involved, that is, for amines it has been proposed that primary amines are relatively more toxic than secondary or tertiary amines. A concentration-dependent tendency to cause haemolysis and changes in erythrocyte morphology has been linked to the presence of $-NH_2$ groups.[82] In general, dendrimers were found to interact significantly less than PEI with erythrocytes.[83–86]

Clearly, these observations do not necessarily hold true for complexes made from DNA and dendrimers. In general, the toxicity of cationic polymers bound to DNA decreases in *in vitro* assays but the particulate nature of complexes is likely to have a major influence on their biodistribution, *e.g.* their involvement with enhanced permeation and retention effect to target tumours.[87] Macromolecules are also expected to be able to utilize this effect,[88] which has been exploited for the targeting of drug-loaded dendrimers.[89,90]

6.2.3.3 Dendrimers as Cellular Transfection Agents

The complexation process between dendrimers and nucleic acids (dendriplexes in Chapter 7) does not seem to differ fundamentally from that of other cationic polymers with high charge density: dendrimers interact with various forms of nucleic acids, such as plasmid DNA or antisense oligonucleotides, to form complexes that protect the nucleic acid from degradation.[89,91,92] The interaction between dendrimer and nucleic acids is based on electrostatic interactions[93] and lacks any sequence specificity.[92]

The morphology of the aggregates formed between DNA and PAMAM dendrimers depends on the dendrimer generation.[94] The morphological transition from rods and toroids to globular aggregates with increasing dendrimer generation has been observed (Figure 6.4). The aggregates produced by the higher generation dendrimers were polydisperse and more disordered. Toroidal structures were found to form when the attractive electrostatic forces were well balanced by the repulsive terms opposing DNA condensation. The range of morphological structures detected could have an effect on the transcription feasibility, *i.e.* the ease with which the genetic activity could be switched on as the DNA is decondensed and also how easily the aggregate could be taken up in the cell during gene transfection.

The transfer from the cytoplasm to the nucleus is a critical step in the transfection process. Studies using fluorescence microscopy suggest that the dendrimer itself has the ability to accumulate to some extent in the nucleus[95] The first report of the use of Starburst PAMAM dendrimers as transfection agents demonstrated that these agents could efficiently induce

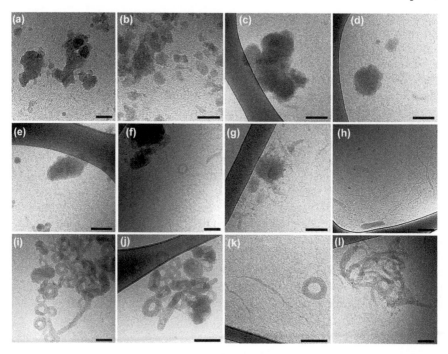

Figure 6.4 Cryo-TEM images of dendrimer–DNA aggregates where the morphology is seen to vary depending on dendrimer generation. Displayed is G8/DNA (a, b), G6/DNA (c, d), G4/DNA (e–h), G2/DNA (i, j) and G1/DNA (k, l). Scale bars are 100 nm. Free DNA molecules are observed in all samples (*e.g.* a, k) and semi-condensed DNA can also be seen in the lower generation samples (*e.g.* h, k).
Reproduced with permission from Ref. 94.

expression of reporter genes in adherent and suspension cell cultures with the G6 (NH$_3$) dendrimer having optimum efficiency.[96] Relatively small dendrimer–DNA complexes with a significant excess of positive to negative charge (6:1) were most efficient but strongly affected by the presence of serum.[96] Interestingly, it was also demonstrated that these materials, in contrast to poly(L-lysine), were not dependent on the presence of lysosomotropic agents, suggesting that they had an intrinsic ability to escape from the endosome. This ability may be related to the ability of the dendrimer amine groups to buffer pH changes in the endosome.[96–98] This has been proposed as a general mechanism that facilitates escape from the endosome because of the accumulation of Cl$^-$ and subsequent osmotic swelling of the endosome.[99]

PAMAM dendrimers were able to complex and deliver not only plasmid DNA but also antisense oligonucleotides.[100] Although Starburst PAMAM dendrimers of generation G3–G10 were found to form stable complexes with DNA, their ability to transfect different cell lines varies. Overall, the higher generation dendrimers (G5–G10) were found to be of superior efficiency, showing a near-exponential increase in efficiency with generation in Rat2 cells.[100] The nature of the core, ammonia (NH_3) or ethylenediamine (EDA), was found to be less significant, highlighting the greater importance of the surface to the nature of the complex.[101] This may be less clear, however, for smaller dendrimers where access to the core groups is sterically less restricted. More recently, a comparison of PAMAMs derived from pentaerythritol (DP), inositol (GI) and trimesyl (DT) core architectures demonstrated an effect of core structure on both the optimum dendrimer generation for condensation and *in vitro* transfection, with DT having an optimum of G6 rather G5.[102]

One of the key advantages of synthetic transfection agents is their sequence independence and the ease with which even large DNA constructs can be accommodated. An extreme example is the successful transfection of a 60 Mb artificial mammalian chromosome into cells using a PAMAM dendrimer (SuperFect).[101] In a comparative evaluation of various polyplexes based on linear, branched and dendritic polymer structures, Gebhart and Kabanov demonstrated that the transfection activity between these polymers varied by three orders of magnitude.[103]

6.2.4 Liposomes

6.2.4.1 Structures and Synthesis

Liposomes consist of an aqueous core entrapped by one or more bilayers composed of natural or synthetic lipids. Liposomes composed of natural phospholipids are biologically inert and weakly immunogenic and they possess low intrinsic toxicity. Further, drugs with different lipophilicities can be encapsulated into liposomes: strongly lipophilic drugs are entrapped almost completely in the lipid bilayer, strongly hydrophilic drugs are located exclusively in the aqueous compartment and drugs with intermediate log$P$ easily partition between the lipid and aqueous phases, both in the bilayer and in the aqueous core.[104,105]

Liposomes can be classified according to their lamellarity (uni-, oligo- and multilamellar vesicles), size (small, intermediate or large) and preparation method [such as reverse-phase evaporation vesicles (VETs)]. Unilamellar vesicles comprise one lipid bilayer and generally have diameters of 50–250 nm. They contain a large aqueous core and are preferentially used to encapsulate water-soluble drugs. Multilamellar vesicles comprise several concentric lipid bilayers in an onion-skin arrangement and have diameters of 1–5 μm. The high lipid content allows these multilamellar vesicles to passively entrap lipid-soluble drugs.

Since liposomes were first developed (around 1980), the related technology has made considerable progress and several important formulations for the treatment of different diseases are now available commercially or in advanced clinical trials. The potency of liposomes depends on composition, electric charge and methods of formulation and can be categorized as conventional, pH-sensitive, cationic, immune and long-circulating lipo-somes.[18,105] There is no limit to the length of DNA that can be incorporated during liposomal composition, in contrast to viral vectors. However, con-ventional liposomes prepared with a neutral zwitterionic lipid or an anionic lipid showed low entrapment efficiency for pDNA with high molecular weight.[106,107] This property was improved by the inclusion of cationic lipids to promote the pDNA condensation by forming lipoplexes and to encourage cellular uptake.[108-111]

6.2.4.2   Biological Properties

Two of the key problems in drug therapy (biodistribution throughout the body and targeting to specific receptors) can be overcome by using liposomal formulations: liposomes protect encapsulated molecules from degradation and can passively target tissues or organs that have a discontinuous endo-thelium, such as the liver, spleen and bone marrow. On intravenous ad-ministration, liposomes are rapidly captured by the mononuclear phagocyte system (MPS) and removed from the blood circulation.[112] However, when the target site is beyond the MPS, efficient liposome uptake by the macrophages and their consequent removal from circulation are one of the main dis-advantages for possible use of liposomes as drug-delivery systems. Binding of selected serum proteins (opsonins) is the first signal for removal of liposomes: the MPS does not recognize the liposomes themselves but, ra-ther, recognizes opsonins, which are bound to the surface of the liposomes. A limited number of possible opsonizing proteins that affect the fate of liposomes have been identified, *e.g.* immunoglobulins,[113] fibronectin,[113,114] β_2-glycoprotein,[115] C-reactive protein (CRP)[116] and β_2-macroglobulin.[117]

Complement components[114,118,119] comprise another important system able to recognize liposomes, which evolved as an immediate host defence against invading pathogens. This system acts through initiating membrane lysis and enhancing uptake by the MPS cells (neutrophils, monocytes, macrophages). In particular, the assembly of C5b-9 complexes [membrane attack complex (MAC)] of the complement system is able to produce lytic pores, which induce cell lysis or, in the case of liposomes, the release of their contents. The complement-dependent release of liposomal contents appears to be one of the dominant factors in determining the biological fate of liposomes. However, serum components that inhibit the phagocytosis of pathogens or particles, referred to as dysopsonins, have also been identified. Human serum albumin and IgA possess dysopsonic properties and their presence on particle surfaces has been shown to reduce recognition and phagocytosis. A balance between blood opsonic proteins and suppressive proteins has been found to regulate the rate of liposome clearance.[120]

The instability of liposomes in plasma due to their interaction with high-(HDL) and low-density (LDL) lipoproteins is another limitation, since this interaction results in the rapid release of the encapsulated drug into the plasma. The physicochemical properties of liposomes, such as net surface charge, hydrophobicity, size, fluidity, and packing of the lipid bilayers, influence their stability and the type of proteins that bind to them.[121,122] One of the first attempts to overcome these problems was focused on manipulation of lipid membrane components in order to modify bilayer fluidity. Damen[123] demonstrated that incorporation of cholesterol (CHOL), by causing increased packing of phospholipids in the lipid bilayer, reduces transfer of phospholipids to HDL. Senior[124] reported that liposomes obtained from phosphatidylcholine (PC) with saturated fatty acyl chains (with a high liquid crystalline transition temperature) or from sphingomyelin (SM) are more stable in the blood than liposomes prepared from PC with unsaturated fatty acyl chains.

Several approaches have also involved modulating liposome size and charge, so as to reduce MPS uptake. In general, larger liposomes are eliminated from the blood circulation more rapidly than smaller liposomes.[124] Small unilamellar vesicles (SUVs) have a half-life longer than that of multi-lamellar liposomes (MLVs) (500–5000 nm), which suggests that phagocytes can distinguish between the sizes of foreign particles. Based on these observations, it is evident that the binding of opsonins to liposomes depends on the size of the liposomes and that in consequence the enhanced MPS uptake of liposomes by the liver is likewise size dependent.[117] Negatively charged liposomes have a shorter half-life in the blood than do neutral liposomes, although the opposite has also been found;[125,126] positively charged liposomes are toxic and thus quickly removed from circulation.[127]

6.2.4.3   Liposomes as Cellular Transfection Agents

Among various synthetic carriers currently in use in gene therapy, cationic liposomes are the most suitable transfecting vectors. Gene encapsulation in liposomal vesicles allows condensation of DNA plasmid into a highly organized structure and protects DNA against degradation during storage and in the systemic circulation of the gene encoding a therapeutic protein. Numerous cationic lipids have been tested in the formulation of liposomes for gene delivery. Transfection efficiency is strongly affected by the presence of three components in the structure of these lipids: a positively charged headgroup that interacts with negatively charged DNA, a linker group (which determines the lipid's chemical stability and biodegradability) and a hydrophobic region to anchor the cationic lipid into the bilayer. Among these, the most often used are N-[1-dioleyloxy)propyl]-N,N,N-trimethylammonium (DOTMA) and dioleoylphophatidylethanolamine (DOPE) in a 1:1 phospholipid mixture (Lipofectin; Invitrogen, Carlsbad, CA, USA). Other commercially-available lipids include 2,3-dioleyloxy-N-[2(sperminecarboxamido)ethyl]-N,N-dimethyl-1-propanammonium trifluoroacetate (DOSPA; Lipofectamine; Invitrogen, Carlsbad, CA, USA), 1,2-bis(oleoyloxy)-3-(trimethylammonio)-propane (DOTAP), 1,2-dimyristyloxypropyl-3-dimethylhydroxyethylammonium bromide (DMRE), 3β-[N-(N',N'-dimethylaminoethane)carbamoyl]cholesterol (DC-CHOL) and dioctadecylaminoglycylspermine (DOGS; Transfectam; Promega Corporation).[128]

Nevertheless, the clinical use of cationic liposomes is limited by their instability, rapid clearance, large particle size, toxicity on repeated administration and induction of immunostimulation and complement activation. Water-soluble lipopolymers, obtained by conjugating different fatty acid chains to branched PEI of 25 kDa or above, have been shown to be effective for gene delivery; they can be delivered into the cytoplasm after endosomal disruption. Similarly, phosphatidylethylene glycol (PhosEG) has been linked to the amino group of branched PEI.[128]

On the other hand, PEGylation of cationic liposomal vesicles is a promising alternative approach to overcome these problems, prolonging circulation time *in vivo* and increasing accumulation at the disease site, even if the transfecting efficiency might be significantly reduced. In liposomes composed of a cationic lipid [DOTAP, DOGS, dimethyldioctadecylammonium bromide (DDAB)], a neutral lipid (DOPE) and a phospholipid derivative of PEG (PEG-PE), complexing 18-mer phosphothioate as a model for active oligodeoxyribonucleotide (ODN), surface modification with a relatively large amount of PEG (5.7 mol%) has been shown to improve ODN loading without losing structural activity or stability of the resulting complexes, retaining size without vesicle aggregation see (Figure 6.5). Moreover, the hydrophilic shell of PEG enhances the *in vitro* stability by avoiding mononuclear phagocyte clearance and retains a high level of the originally loaded ODN in the complex after plasma incubation. Only after modification of PEG cationic liposomes with targeting agents can cytoplasmatic delivery of DNA material be observed. The PEG-modified complex conjugated anti-HER2 F(ab′) dramatically enhanced cell uptake, increasing diffuse cytoplasmatic and nuclear localization of ODN in SK-BR-3 cells.[129]

In liposomes composed of DODAC–DOPE, the inclusion of 5 mol% of PEG–lipid conjugate did not inhibit uptake by the cell membrane of lipid–DNA complex, but substantially modified the ability of the cationic liposomal carrier to disrupt the endosomal membrane. Endosomal escape into the cytoplasm depended on the acyl chain of the lipid complex and on the molecular weight of the PEG. Optimizing the desorption rate of PEG–lipids may be one approach to overcoming the inhibitory effect on intracellular delivery of plasmid.[130]

To contrast the low transfection efficiency of PEG-modified cationic liposomes due to the absence of a net positive charge on the vesicle surface, a series of cationic PEG–lipids with one or more positive charges were synthesized and designed for post-insertion in preformed stabilized plasmid–lipid particles. Incorporation of cationic poly(ethylene glycol)–lipid conjugates (CPL4) in DOPE–DODAC–PEG-CerC20 (PEG-CerC20 is PEG moieties attached to a ceramide anchor containing an arachidoylacyl group) liposomes resulted in both improved uptake into BHK cells and dramatically enhanced transfection potency in the presence of Ca^{2+}, which assists in destabilizing the endosomal membrane following uptake. However, in this type of liposomal preparation, aggregation of vesicles was observed, probably due to the formation of H-bonding between the amino and carbonyl

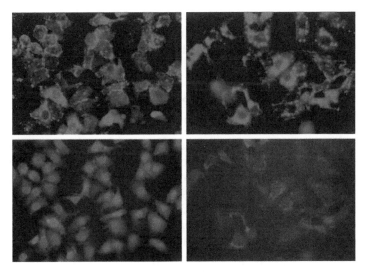

Figure 6.5 Intracellular distribution of PEG-modified cationic immunoliposome–oligodeoxyribonucleotide (ODN) complexes studied by fluorescence microscopy. Liposomes of DOTAP, DOPE, PEG-PE and maleimido-PEG-PE were conjugated to F(ab′) fragments of rhuMAbHER2 antibody and complexed with FITC-ODN ([ODN/LIP]$_{bound}$ = 0.007, [+/−] = 4:1). Top panels, localization of the lipid marker; bottom panels, localization of the ODN marker. Incubation conditions were: without free anti-HER2 F(ab′) (left) or in the presence of a 10-fold excess of free F(ab′) (right). Reproduced with permission from Ref. 129. Copyright 1998 American Society for Biochemistry and Molecular Biology.

groups present in the distal headgroup at the end of the PEG chain.[131] In order to optimize CPL–liposomes for systemic delivery, the length of the PEG linker in the CPL can be modulated. When the PEG3400 linker extending beyond the PEG-CerC20 'cloud' was employed for liposomal insertion, charged liposomal systems were produced that rapidly cleared from the circulation; it was suggested that a shorter PEG linker might be used, such as PEG1000, allowing the PEG-CerC20 to shield the positive charge of CPL. Moreover, PEG-CerC20 can be designed to dissociate slowly at the disease site, achieving exposure of the CPL at the target area with retention of long-circulation properties and interaction between liposomes and targeting cell.[132]

Overall, the most suitable use of PEG is as a tether for a specific ligand on the surface of these systems, in order to obtain a target-specific gene delivery facilitating internalization in cells and endosomal escape. Cell-penetrating peptides (CPPs), such as trans-activating transcriptional activator (TAT), homeodomain of Antennapedia (Antp), herpex simplex virus type I protein VP22 and transportan, have been reported to guarantee direct cytosolic delivery when coupled with several carriers, including liposomes. Multiple TATp molecules can be attached on the surface of liposomes *via* the spacer group of *p*-nitrophenylcarbonyl–PEG–phosphatidylethanolamine.

TATp–liposome–DNA complexes were found to be capable of transfection of both normal and cancer cells *in vitro* and *in vivo* with lower cytotoxicity that the commonly used lipid-based gene-delivery systems.[133]

6.2.5 Solid Lipid Nanoparticles

6.2.5.1 Structures and Synthesis

Nanoparticles made from solid lipids are attracting increasing attention as colloidal drug carriers. The nanoparticles are in the submicron size range (50–1000 nm) and are composed of physiological lipids. Solid lipid nanoparticles represent aqueous colloidal dispersions of biodegradable substances that are solid at room temperature.[134] Solid lipid nanoparticles have been obtained by many researchers using different preparation approaches, which include high-pressure and ultrasonication/high-speed homogenization.[135]

With production by high-pressure homogenization they can be obtained on a large industrial scale. In addition, this production method avoids the use of organic solvents. The greatest difficulty for their application is the complexity of the physical state of lipids, associated with the nanometric size.[136] Most of the lipids present different crystalline states. Polymorphism transitions are able to occur with time or with storage conditions. Drug-loading capacities decrease with increasing crystallinity, which expels drug molecules. Therefore, appropriate characterization is necessary and requires several analytical methods, particularly on the molecular level.

6.2.5.2 Biological Properties

Solid lipid nanoparticles of below 200 nm in size have an increased blood circulation and thus an increased time during which the drug remains in contact with the blood–brain barrier and for the drug to be taken up by the brain. These carriers can gain access to the blood compartment easily (because of their small size and lipophilic nature). Opsonization of solid lipid nanoparticles can be prevented by coating the particles with a hydrophilic or a flexible polymer and/or a surfactant. Targeting ligands that specifically bind to surface epitopes or receptors on the target sites can be coupled to the surface of solid lipid nanoparticles for targeting the brain trough the blood–brain barrier.[137,138]

6.2.5.3 Solid Lipid Nanoparticles as Cellular Transfection Agents

Only a few reports about the use of solid lipid nanoparticles for delivery of genes[139–142] have been published since Olbrich *et al.*[143] introduced these particles as a non-viral transfection system. From the point of view of application, solid lipid nanoparticles have good stability,[144] which facilitates their industrial elaboration and their manipulation for different processes such as lyophilization.

In most cases, the elaboration of solid lipid nanoparticles in addition to the matrix lipid and the cationic lipid requires additional surfactants. Tween 80 is one of the most commonly employed surfactants in the pharmaceutical industry and it has some interesting characteristics that can be used in formulations for gene therapy, because of the presence of poly(ethylene glycol) (PEG) chains in its structure. Several research groups have observed that the presence of PEG in cationic lipid emulsions[145,146] and in liposomes[147] improves their transfection capacity. Liu *et al.*[145] showed that Tween 80 was the most effective nonionic surfactant to prevent the formation of aggregates. When complexes are formed, each molecule of DNA may bind more than one emulsion particle such that large aggregates are formed. However, Tween 80 may prevent sterically each DNA molecule from binding to more than one particle and the formation of such large aggregates does not occur. In addition, Tween 80 has another important characteristic for the transfection of these systems *in vivo*: it creates a steric barrier,[148] which neutralizes the excess of positive charges of the systems and reduces the interaction with blood components, such as serum proteins, which could limit the arrival of the gene therapy system at the cell surface.

In order to form lipoplexes, a positive superficial charge of the systems is necessary to bind electrostatically the DNA, which has a negative charge. When DNA binds with these systems, it is condensed and that condensation increases as the charge ratio increases.[149] Condensation is necessary to facilitate the mobility of DNA molecules, which is limited by their large size, and to protect the DNA from agents present inter- and intracellularly. Condensation reduces the exposure of the DNA to those agents and improves its protection. However, DNA condensation may limit the transfection efficiency of non-viral systems because the larger the condensation, the more difficult is the release of the DNA from the complexes.[149]

Cationic solid lipid nanoparticles for gene transfer were formulated using the same cationic lipid as for liposomal transfection agents. The differences and similarities in the structure and performance between solid lipid nanoparticles and liposomes were investigated.[150] In the main experiment, the solid lipid nanoparticles formulation S1 and its counterpart formulation without matrix lipid L1 were further compared with the commercial liposomal reagents DLTR, consisting of DOTAP only, and Escort, which further contains DOPE, in different transfection media (Figure 6.6). The overview indicates comparable transfection activities for the formulations containing only DOTAP (S1, L1 and DLTR) in the different media. No statistically significant difference in transfection efficiency was found between SLN (S1), liposomes (DLTR) and the formulation L1. Neither the presence of the matrix lipid (S1 *versus* L1) nor the different core structure (solid lipid *versus* liposomes in S1 and DLTR) significantly affected the transfection activity. Escort, the liposomal preparation containing the helper lipid DOPE, showed statistically significant higher transfection efficiencies than S1, L1 and DLTR

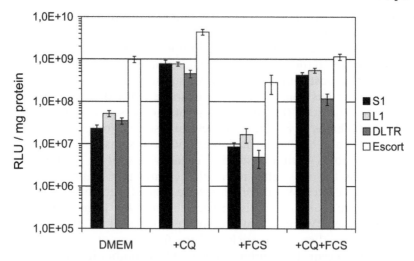

Figure 6.6 Comparison of transfection efficiencies (RLU per mg protein) of S1, L1, DLTR and Escort in various incubation media [DMEM, DEMEM + 100 μM chloroquine (CQ), DEMEM and 10% FBS, DEMEM + FCS + CQ]. Adapted with permission from Ref. 150.

in all incubation media tested. As solid lipid nanoparticles can be produced on a large scale and under favourable technological parameters, they may become a valuable addition to the well-established repertoire of non-viral transfection agents led by cationic liposomes.

6.2.6 Hydrogels

6.2.6.1 Structures and Synthesis

Hydrogels are typically composed of a hydrophilic organic polymer component that is cross-linked into a network by either covalent or non-covalent interactions.[151–153] It is the cross-linking that provides dimensional stability, whereas the high solvent content gives rise to the fluid-like transport properties (see Chapter 4).

The synthetic and natural physically cross-linked hydrogels have led to the concept of reversible or degradable hydrogels that undergo a transition from the three-dimensionally stable structure to a polymer solution. Most often these hydrogels have been used to encapsulate proteins,[154] cells[155] or drugs[156] and then release them through the dissolution of the hydrogel structure. The cross-links in this class of hydrogels arise from non-covalent attractive forces between the polymer chains. These forces are often hydrophobic interactions, hydrogen bonding or ionic interactions.

Alginate can be cross-linked by divalent calcium ions.[157,158] Another example of cross-linking by ionic interactions is that of dextran, which lacks charged regions, but forms hydrogels in the presence of potassium ions.[159]

Non-covalent cross-links can also be formed in blends and interpenetrating networks of two dissimilar polymers. For example, poly(acrylic acid) and poly(methacrylic acid) form hydrogen bonds with poly(ethylene glycol), which results in the formation of hydrogels.[160,161] Oligonucleotides have also been used in the formation of hydrogels. Nagahara and Matsuda coupled water-soluble poly(*N,N*-dimethylacrylamide-*co*-*N*-acryloylsuccinimide) to a single-stranded DNA.[162]

We have recently prepared novel DNA gel particles based on associative phase separation and interfacial diffusion. By mixing solutions of DNA [either single-stranded (ssDNA) or double-stranded (dsDNA)] with solutions of different cationic agents, such as surfactants, proteins and polysaccharides, the possibility of the formation of DNA gel particles without adding any kind of cross-linker or organic solvent was confirmed.[163–166]

Chemically cross-linked hydrogels are usually more stable than the physically cross-linked hydrogels because the cross-links are formed by covalent bonds.[157] The hydrogels formed by such cross-links have a permanent structure unless labile chemical bonds have been intentionally added to the network. Chemically cross-linked gels are usually formed by polymerizing monomers in the presence of cross-linking agents.

Hydrogels can also be classified based on their size as either macrogels or microgels. Macrogels[167–169] are bulk gels where the size can be millimetres and larger, whereas microgels[170,171] are colloidally stable hydrogels and their size can vary from tens of nanometres to micrometres. Hydrogels can be further classified as stimuli-responsive or non-responsive gels. Non-responsive gels, as the name suggests, are merely materials that swell upon water absorption. On the other hand, stimuli-responsive gels have been called 'smart' materials because they respond by a change in swelling to subtle changes in the environment.[172,173] These hydrogels can be made responsive to temperature,[174] pH,[171,175] ionic strength,[176–178] light[179–183] and electric field.[184]

6.2.6.2 Biological Properties

An ideal drug carrier should not induce an immune response in the host. This property is commonly achieved by making the surface of the particle hydrophilic, which can prevent opsonization (*i.e.* adhesion-enhanced phagocytosis) by macrophages.[185] For example, Gaur *et al.* synthesized cross-linked polyvinylpyrrolidine hydrogel nanoparticles (~ 100 nm diameter).[186] The surface of these particles was then made hydrophilic by attaching poloxamers and poloxamines, which are PEG–poly(propylene glycol) block copolymers. *In vivo* studies in mice indicated that less than 1% of the dose was retained by the macrophages in the liver and even after 8 h of injection around 5–10% of these particles were still circulating in the vasculature. This enhanced circulation time and the lack of liver accumulation could allow the use of such particles in drug delivery. They also reported that an increase in size and hydrophobicity of the particles increased the uptake by the reticuloendothelial system, suggesting that both factors may play a role in the ability of the body's defence mechanisms to recognize the particles as foreign invaders.

Mixtures of two cationic proteins have been used to prepare protein–DNA gel particles, employing associative phase separation and interfacial diffusion.[187] From a haemolytic point of view, these protein–DNA gel particles were demonstrated to be promising long-term blood-contacting medical devices. Safety evaluation with the established cell lines revealed that, in comparison with proteins in solution, the cytotoxicity was reduced when administered in the protein–DNA systems. The cytotoxic responses of small-sized protein–DNA gel particles were shown to be strongly dependent on both the protein composition and the cell line, the tumour cell line HeLa being more sensitive to the deleterious effects of the mixed protein-based particles. Fluorescence microscopy studies indicated cellular uptake and internalization of these protein–DNA particles, a prerequisite for subsequent DNA delivery (Figure 6.7).

6.2.6.3   Hydrogels as Cellular Transfection Agents

Encapsulation of pDNA into the flexible polymer networks of microgels or nanogels (effective diameter <100 nm) can protect the pDNA from enzymatic degradation by extracellular and intracellular nucleases. Loading of biological agents is usually achieved spontaneously through electrostatic, van der Waals or hydrophobic interactions between the agent and the polymer matrix.[188]

The macroporous networks of swollen nanogels with a small number of cross-links can accommodate larger biomolecules and form polymeric envelopes.[189] The nanocarrier surface is often modified with biospecific targeting groups to enhance the delivery of genes into specific cells; and also with inert hydrophilic polymers, such as PEG, to extend the circulation time in the bloodstream. A human transferrin-modified PEG-cl-PEI nanogel loaded with pDNA was shown to increase the transfection efficiency of human breast carcinoma cells in the presence of serum.[190] Indeed, the delivery of pDNA from a wide range of hydrogel materials has been investigated, from natural biopolymers such as gelatin,[191–193] hyaluronan,[194–196] pullulan[197] and silk elastin-like protein (SELP)[198] to PEG-based synthetic polymers.[190,199,200] PEGylated nanogels that bear a lactose group at the PEG end showed significant endosomolytic abilities, achieving the pronounced transfection efficiency of the PEG-b-PLL–pDNA polyplex without any significant toxicity.[200] Precise control of pDNA release from natural polymer hydrogels can be difficult to achieve, because it depends on enzymatic degradation and ion exchange, which may vary within subjects and sites. Alternatively, synthetic polymer hydrogels offer a higher degree of control over pDNA dosing through tailoring the properties of the gel such as mesh size and degradation kinetics.[199]

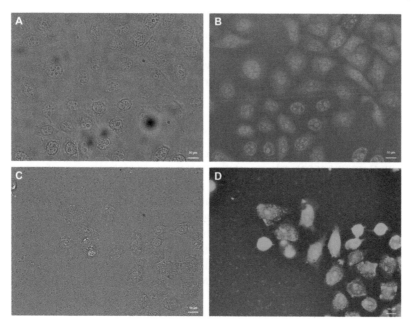

Figure 6.7 Protamine sulfate (PS)–DNA gel particle uptake in HeLa cells. Representative contrast microscopy images (A) and fluorescence images (B) of HeLa control. Contrast microscopy (C) and fluorescence images (D) of HeLa cells incubated for 2 h in the presence of PS–DNA particles. Original magnification ×40.
Adapted with permission from Ref. 187.

6.3 Prospects

Over the last two decades, gene therapy has brought human medical prospects into a new phase, whereby genetic defects on cells can be regulated and also a range of diseases can be prevented. DNA-based molecules are being employed to prevent, treat and cure diseases by changing the expression of genes that are responsible for the pathology. Since its inception, plasmid DNA-mediated gene therapy has seen significant growth and undergone fruitful clinical trials. However, the major underlying challenge is the development of a carrier system that can facilitate the safe and efficient delivery of plasmid DNA to the target site, followed by cellular uptake, internalization and processing and the production of a therapeutic level of gene products for a desired period. Colloidal delivery systems are being recognized as a viable carrier to provide solutions in gene therapy, from the perspective of formulation and clinical problems, by improving the therapeutic outcomes. Colloidal carriers offer the opportunity to design surface properties to enable them to traverse biological barriers such as skin, mucous barriers and leaky vasculature. The smart design of the colloidal carrier can protect DNA-based molecules from deleterious degradation and may provide sustained release of the payload in a therapeutically advantageous fashion. Although these nanomedicines may offer various advantages over conventional drug-delivery systems, their safety should not be ignored. The toxicity of these nanomedicines may be due to their large surface area. Continuous efforts focused on improving the safety, feasibility and efficacy of colloidal carriers for DNA gene therapy are required.

Acknowledgements

The author acknowledges the support of the MICINN (Ramon y Cajal contract RyC 2009-04683). The work on DNA gel particles was performed in collaboration with several colleagues and co-workers in Coimbra and Barcelona; in this context, the author would like to mention especially Dr Björn Lindman and Dr Maria da Graça Miguel of the University of Coimbra for their support in recent years. Furthermore, discussions with Dr Maria Pilar Vinardell and Dr Montserrat Mitjans of the University of Barcelona on *in vitro* biocompatibility characterization are gratefully acknowledged. This research was supported by the Project MAT2012-38047-C02-01 of the Spanish Ministry of Science and Innovation.

References

1. M. Nishikawa and L. Huang, *Hum. Gene Ther.*, 2001, **12**, 861.
2. W. F. Anderson, *Nature*, 1998, **392**, 25.
3. M. D. Brown, A. G. Schätzlein and I. F. Uchegbu, *Int. J. Pharm.*, 2001, **229**, 1.
4. A. M. Darquet, B. Cameron, P. Wils, D. Scherman and J. Crouzet, *Gene Ther.*, 1997, **4**, 1341.
5. F. Schakowski, M. Gorschluter, C. Junghans, M. Schroff, P. Buttgereit, C. Ziske, B. Schottker, S. A. Konig-Merediz, T. Sauerbruch, B. Wittig and I. G. Schmidt-Wolf, *Mol. Ther.*, 2001, **3**, 793.
6. A. G. Schätzlein, *Anticancer Drugs*, 2001, **12**, 275.
7. E. Wagner, R. Kircheis and G. F. Walker, *Biomed. Pharmacother.*, 2004, **58**, 152.
8. M. A. Liu, *J. Intern. Med.*, 2003, **253**, 402.
9. J. A. Wolff, R. W. Malone, P. Williams, W. Chong, G. Acsadi, A. Jani and P. L. Felgner, *Science*, 1990, **247**, 1465.
10. M. A. Hickman, R. W. Malone, K. Lehmann-Bruinsma, T. R. Sih, D. Knoell, F. C. Szoka, R. Walzem, D. M. Carlson and J. S. Powell, *Hum. Gene Ther.*, 1994, **5**, 1477.
11. P. Yang and L. Huang, *Gene Ther.*, 1996, **3**, 542.
12. R. Niven, R. Pearlman, T. Wedeking, J. Mackeigan, P. Noker, L. Simpson-Herren and J. G. Smith, *J. Pharm. Sci.*, 1998, **87**, 1292.
13. F. D. Ledley, *Pharm. Res.*, 1996, **13**, 1595.
14. A. L. Parker, C. Newman, S. Briggs, L. Seymour and P. J. Sheridan, *Expert Rev. Mol. Med.*, 2003, **5**, 1.

15. K. A. Mislick and J. D. Baldeschwieler, *Proc. Natl. Acad. Sci. U. S. A.*, 1996, **93**, 12349.
16. J. Cassidy and A. G. Schätzlein, *Expert Rev. Mol. Med.*, 2004, **6**, 1.
17. A. G. Schätzlein, *J. Biomed. Biotechnol.*, 2003, **2003**, 149.
18. W. E. Bawarski, E. Chidlowsky, D. J. Bharali and S. A. Mousa, *Nanomed. Nanotechnol. Biol. Med.*, 2008, **4**, 273.
19. D. T. O'Hagan, M. Singh and J. B. Ulmer, *Immunol. Rev.*, 2004, **199**, 191.
20. S. D. Patil, D. G. Rhodes and D. J. Burgess, *AAPS J.*, 2005, 7, E61.
21. E. Allemann, R. Gurny and E. Doelker, *Eur. J. Pharm. Biopharm.*, 1993, **39**, 173.
22. R. Gref, Y. Minamitake, M. T. Peracchia, V. S. Trubetskoy, V. P. Torchilin and R. Langer, *Science*, 1994, **263**, 1600.
23. V. Labhasetwar, C. Song and R. J. Levy, *Adv. Drug Deliv. Rev.*, 1997, **24**, 63.
24. W. Zauner, M. Ogris and E. Wagner, *Adv. Drug Deliv. Rev.*, 1998, **30**, 97.
25. O. Boussif, F. Lezoualch, M. A. Zanta, M. D. Mergny, D. Scherman, B. Demeneix and J. P. Behr, *Proc. Natl. Acad. Sci. U. S. A.*, 1995, **96**, 7297.
26. U. Lungwitz, M. Breuning, T. Blunk and A. Gopferich, *Eur. J. Pharm. Biopharm.*, 2005, **60**, 247.
27. A. F. Jorge, R. S. Dias, J. C. Pereira and A. A. C. C. Pais, *Biomacromolecules*, 2010, **11**, 2399.
28. A. F. Jorge, R. S. Dias and A. A. C. C. Pais, *Biomacromolecules*, 2012, **13**, 3151.
29. J. F. KukowskaLatallo, A. U. Bielinska, J. Johnson, R. Spindler, D. A. Tomalia and J. R. Baker, *Proc. Natl. Acad. Sci. U. S. A.*, 1996, **93**, 4897.
30. C. Dufes, I. F. Uchegbu and A. G. Schatzlein, *Adv. Drug Deliv. Rev.*, 2005, 57, 2177.
31. F. C. MacLaughlin, R. J. Mumper, J. J. Wang, J. M. Tagliaferri, I. Gill, M. Hinchcliffe and A. P. Rolland, *J. Control. Release*, 1998, **56**, 259.
32. G. Borchard, *Adv. Drug Deliv. Rev.*, 2001, **52**, 145.
33. P. Pereira, A. F. Jorge, R. Martins, A. A. C. C. Pais, F. Sousa and A. Figueiras, *J. Colloid Interface Sci.*, 2012, **387**, 84.
34. E. Wagner, M. Cotton, R. Foisner and M. L. Birnstiel, *Proc. Natl. Acad. Sci. U. S. A.*, 1991, **88**, 4255.
35. M. A. Wolfert and L. W. Seymour, *Gene Ther.*, 1996, **3**, 269.
36. W. T. Godbey, K. K. Wu and A. G. Mikos, *Biomaterials*, 2001, **22**, 471.
37. J. Panyam and V. Labhasetwar, *Adv. Drug Deliv. Rev.*, 2003, 55, 329.
38. M. Elfinger, S. Uezguen and C. Rudolph, *Curr. Nanosci.*, 2008, **4**, 322.
39. G. P. Tang, J. M. Zeng, S. J. Gao, Y. X. Ma, L. Shi, Y. Li, H. P. Too and S. Wang, *Biomaterials*, 2003, **24**, 2351.
40. H. Petersen, K. Kunath, A. L. Martin, S. Stolnik, C. J. Roberts, M. C. Davies and T. Kissel, *Biomacromolecules*, 2002, **3**, 926.
41. C. W. Pouton and L. W. Seymour, *Adv. Drug Deliv. Rev.*, 2001, **46**, 187.
42. D. N. Nguyen, J. J. Green, J. M. Chan, R. Langer and D. G. Anderson, *Adv. Mater.*, 2009, **21**, 847.

43. D. T. O'Hagan, M. Singh and J. B. Ulmer, *Methods*, 2006, **40**, 10.
44. V. P. Torchilin, *Pharm. Res.*, 2007, **24**, 1.
45. J. S. Choi, D. K. Joo, C. H. Kim, K. Kim and J. S. Park, *J. Am. Chem. Soc.*, 2000, **122**, 474.
46. N. Ninhiyama and K. Kataoka, *Pharmacol. Ther.*, 2006, **112**, 630.
47. Y. Kakizawa and K. Kataoka, *Adv. Drug Deliv. Rev.*, 2002, **54**, 203.
48. G. Kwon, M. Naito, M. Yokoyama, T. Okano, Y. Sakurai and K. Kataoka, *Langmuir*, 1993, **9**, 945.
49. S. I. Jeon, J. H. Lee, J. D. Andrade and P. G. Degennes, *J. Colloid Interface Sci.*, 1991, **142**, 149.
50. H. Otsuka, Y. Nagasaki and K. Kataoka, *Curr. Opin. Colloid Interface Sci.*, 2001, **6**, 3.
51. H. R. Bader, H. Ringsdorf and B. Schmidt, *Angew. Makromol. Chem.*, 1984, **123/124**, 457.
52. M. Yokoyama, S. Inoue, K. Kataoka, N. Yui, T. Okano and Y. Sakurai, *Makromol. Chem. Macromol. Chem. Phys.*, 1989, **190**, 2041.
53. R. Gref, Y. Minamitake, M. T. Peracchia, V. Trubetskoy, V. Torchilin and R. Langer, *Science*, 1994, **263**, 1600.
54. A. Harada and K. Kataoka, *Macromolecules*, 1995, **28**, 5294.
55. A. Harada and K. Kataoka, *Science*, 1999, **283**, 65.
56. A. V. Kabanov, T. K. Bronich, V. A. Kabanov, K. Yu and A. Eisenberg, *Macromolecules*, 1996, **29**, 6797.
57. K. Kataoka, H. Togawa, A. Harada, K. Yasugi, T. Matsumoto and S. Katayose, *Macromolecules*, 1996, **29**, 8556.
58. M. Yokoyama, T. Okano, Y. Sakurai, S. Suwa and K. Kataoka, *J. Control. Release*, 1996, **39**, 351.
59. N. Nishiyama, M. Yokoyama, T. Aoyagi, T. Okano, Y. Sakurai and K. Kataoka, *Langmuir*, 1999, **15**, 377.
60. K. Kataoka, A. Harada, D. Wakebayashi and Y. Nagasaki, *Macromolecules*, 1999, **32**, 6892.
61. Y. Matsumura, *Adv. Drug Deliv. Rev.*, 2008, **60**, 899.
62. S. Katayose and K. Kataoka, *Bioconjug. Chem.*, 1997, **8**, 702.
63. K. Itaka, N. Kanayama, N. Nishiyama, W. D. Jang, Y. Yamasaki, K. Nakamura, H. Kawaguchi and K. Kataoka, *J. Am. Chem. Soc.*, 2004, **126**, 13612.
64. S. Matsumoto, R. J. Christie, N. Nishiyama, K. Miyata, A. Ishii, M. Oba, H. Koyama, Y. Yamasaki and K. Kataoka, *Biomacromolecules*, 2009, **10**, 119.
65. S. Katayose and K. Kataoka, *J. Pharm. Sci.*, 1998, **87**, 160.
66. D. Oupicky, C. Konak, K. Ulbrich, M. A. Wolfert and L. W. Seymour, *J. Control. Release*, 2000, **65**, 149.
67. K. Itaka, K. Yamauchi, A. Harada, K. Nakamura, H. Kawaguchi and K. Kataoka, *Biomaterials*, 2003, **24**, 4495.
68. M. Harada-Shiba, K. Yamauchi, A. Harada, I. Takamisawa, K. Shimokado and K. Kataoka, *Gene Ther.*, 2002, **9**, 407.

69. T. M. Fahmy, P. M. Fong, J. Park, T. Constable and W. M. Saltzman, *AAPS J.*, 2007, **9**, E171.
70. D. A. Tomalia, H. Baker, J. Dewald, M. Hall, G. Kallos, S. Martin, J. Roeck, J. Ryder and P. Smith, *Polym. J.*, 1985, **17**, 117.
71. C. Worner and R. Mulhaupt, *Angew. Chem. Int. Ed. Engl.*, 1993, **32**, 1306.
72. E. M. M. de Brabander-van den Berg and E. W. Meijer, *Angew. Chem. Int. Ed. Engl.*, 1993, **32**, 1308.
73. E. M. M. de Brabander-van den Berg, A. Nijenhuis, M. Mure, J. Keulen, R. Reintjens, F. Vandenbooren, B. Bosman, R. de Raat, T. Frijns, S. van der Wal, M. Castelijns, J. Put and E. W. Meijer, *Macromol. Symp.*, 1994, 77, 51.
74. Y. Li, Y. Cheng and T. Xu, *Curr. Drug Discov. Technol.*, 2007, **4**, 246.
75. D. Fischer, Y. Li, B. Ahlemeyer, J. Krieglstein and T. Kissel, *Biomaterials*, 2003, **24**, 1121.
76. M. Ruponen, S. Yla-Herttuala and A. Urtti, *Biochim. Biophys. Acta*, 1999, **1415**, 331.
77. M. F. Ottaviani, P. Favuzza, B. Sacchi, N. J. Turro, S. Jockusch and D. A. Tomalia, *Langmuir*, 2002, **18**, 2347.
78. B. H. Zinselmeyer, S. P. Mackay, A. G. Schatzlein and I. F. Uchegbu, *Pharm. Res.*, 2002, **19**, 960.
79. J. C. Roberts, M. K. Bhalgat and R. T. Zera, *J. Biomed. Mater. Res.*, 1996, **30**, 53.
80. R. Jevprasesphant, J. Penny, R. Jalal, D. Attwood, N. B. McKeown and A. D'Emanuele, *Int. J. Pharm.*, 2003, **252**, 263.
81. R. Jevprasesphant, J. Penny, D. Attwood, N. B. McKeown and A. D'Emanuele, *Pharm. Res.*, 2003, **20**, 1543.
82. N. Malik, R. Wiwattanapatapee, R. Klopsch, K. Lorenz, H. Frey, J. W. Weener, E. W. Meijer, W. Paulus and R. Duncan, *J. Control. Release*, 2000, **65**, 133.
83. T. Yoshimura, J. Fukai, H. Mizutani and K. Esumi, *J. Colloid Interface Sci.*, 2002, **255**, 428.
84. J. H. Lee, Y. B. Lim, J. S. Choi, Y. Lee, T. I. Kim, H. J. Kim, J. K. Yoon, K. Kim and J. S. Park, *Bioconjug. Chem.*, 2003, **14**, 1214.
85. I. F. Uchegbu, L. Sadiq, A. Pardakhty, M. El-Hammadi, A. I. Gray, L. Tetley, W. Wang, B. H. Zinselmeyer and A. G. Schatzlein, *J. Drug Target.*, 2004, **12**, 527.
86. A. Brownlie, I. F. Uchegbu and A. G. Schatzlein, *Int. J. Pharm.*, 2004, **274**, 41.
87. R. Duncan, *Nat. Rev. Drug Discov.*, 2003, **2**, 347.
88. N. Malik, E. G. Evagorou and R. Duncan, *Anticancer Drugs*, 1999, **10**, 767.
89. A. U. Bielinska, J. F. Kukowska-Latallo and J. R. Baker Jr, *Biochim. Biophys. Acta*, 1997, **1353**, 180.
90. H. G. Abdelhady, S. Allen, M. C. Davies, C. J. Roberts, S. J. Tendler and P. M. Williams, *Nucleic Acids Res.*, 2003, **31**, 4001.
91. A. Bielinska, J. F. Kukowska-Latallo, J. Johnson, D. A. Tomalia and J. R. Baker Jr, *Nucleic Acids Res.*, 1996, **24**, 2176.

92. M. X. Tang and F. C. Szoka, *Gene Ther.*, 1997, **4**, 823.

93. W. Chen, N. J. Turro and D. A. Tomalia, *Langmuir*, 2000, **16**, 15.

94. M. L. Ainalem, A. M. Carnerup, J. Janiak, V. Alfredsson, T. Nylander and K. Schillén, *Soft Matter*, 2009, **5**, 2310.

95. W. T. Godbey, K. K. Wu and A. G. Mikos, *Proc. Natl. Acad. Sci. U. S. A.*, 1999, **96**, 5177.

96. J. Haensler and F. C. Szoka Jr, *Bioconjug. Chem.*, 1993, **4**, 372.

97. J. P. Behr, *Chimia*, 1997, **51**, 34.

98. J. P. Behr, *Bioconjug. Chem.*, 1994, **5**, 382.

99. N. D. Sonawane, F. C. Szoka Jr and A. S. Verkman, *J. Biol. Chem.*, 2003, **278**, 44826.

100. J. F. Kukowska-Latallo, A. U. Bielinska, J. Johnson, R. Spindler, D. A. Tomalia and J. R. Baker Jr, *Proc. Natl. Acad. Sci. U. S. A.*, 1996, **93**, 4897.

101. X. Q. Zhang, X. L. Wang, S. W. Huang, R. X. Zhuo, Z. L. Liu, H. Q. Mao and K. W. Leong, *Biomacromolecules*, 2005, **6**, 341.

102. T. Kim, H. J. Seo, J. S. Choi, H. S. Jang, J. Baek, K. Kim and J. S. Park, *Biomacromolecules*, 2004, **5**, 2487.

103. C. L. Gebhart and A. V. Kabanov, *J. Control. Release*, 2001, **73**, 401.

104. M. Gulati, M. Grover and S. Singh, *Int. J. Pharm.*, 1998, **165**, 129.

105. J. Smith, Y. L. Zhang and R. Niven, *Adv. Drug Deliv. Rev.*, 1997, **26**, 135.

106. R. J. Mannino, E. S. Allebach and W. A. Strohl, *FEBS Lett.*, 1979, **101**, 229.

107. L. H. Lindner, R. Brock, D. Arndt-Jovin and H. Eibl, *J. Control. Release*, 2006, **110**, 444.

108. P. L. Felgner, *Hum. Gene Ther.*, 1996, **7**, 1791.

109. S. May, D. Harries and A. Ben-Shaul, *Biophys. J.*, 2000, **78**, 1681.

110. M. C. P. de Lima, S. Simoes, P. Pires, H. Faneca and N. Duzgunes, *Adv. Drug Deliv. Rev.*, 2001, **47**, 277.

111. D. Lundberg, H. Faneca, M. C. Morán, M. C. P. de Lima, M. G. Miguel and B. Lindman, *Mol. Membr. Biol.*, 2011, **28**, 42.

112. G. L. Scherphof, *Ann. N. Y. Acad. Sci.*, 1985, **446**, 368.

113. H. M. Patel, *Crit. Rev. Ther. Drug Carrier Syst.*, 1992, **9**, 39.

114. D. J. Falcone, *J. Leukoc. Biol.*, 1986, **39**, 1.

115. A. Chonn, S. C. Semple and P. R. Cullis, *J. Biol. Chem.*, 1995, **270**, 25845.

116. J. E. Volanakis and A. J. Narkates, *J. Immunol.*, 1981, **126**, 1820.

117. M. Murai, Y. Aramki and S. Tsuchiya, *Immunology*, 1995, **86**, 64.

118. D. V. Devine, K. Wong, K. Serrano, A. Chonn and P. R. Cullis, *Biochim. Biophys. Acta*, 1994, **1191**, 43.

119. H. Harashima, K. Sakata, K. Funato and H. Kiwada, *Pharm. Res.*, 1994, **11**, 402.

120. T. Ishida, H. Harashima and H. Kiwada, *Biosci. Rep.*, 2002, **22**, 197.

121. A. Chonn, S. C. Semple and P. R. Cullis, *J. Biol. Chem.*, 1992, **267**, 18759.

122. C. D. M. Oja, S. C. Semple, A. Chonn and P. R. Cullis, *Biochim. Biophys. Acta*, 1996, **1281**, 31.

123. J. Damen, *Biochim. Biophys. Acta*, 2005, **665**, 538.
124. J. Senior, *FEBS Lett.*, 1982, **145**, 109.
125. K. Nishikawa, *J. Biol. Chem.*, 1990, **265**, 5226.
126. K. Funato, *Biochim. Biophys. Acta*, 1992, **1103**, 204.
127. J. H. Senior, *Crit. Rev. Ther. Drug Carrier Syst.*, 1987, **3**, 123.
128. R. I. Mahato, *Adv. Drug Deliv. Rev.*, 2005, **57**, 699.
129. O. Meyer, D. Kirpotin, K. Hong, B. Sternberg, J. W. Park and M. C. Woodle, *J. Biol. Chem.*, 1998, **273**, 15621.
130. L. Y. Song, Q. F. Ahkong, Q. Rong, Z. Wang, S. Ansell, M. J. Hope and B. Mui, *Biochim. Biophys. Acta*, 2002, **1558**, 1.
131. L. R. Palmer, T. Chen, A. M. Lam, D. B. Fenske, K. F. Wong, I. MacLachlan and P. Cullis, *Biochim. Biophys. Acta*, 2003, **1611**, 204.
132. T. Chen, L. R. Palmer, D. B. Fenske, A. M. I. Lam, K. F. Wong and P. R. Cullis, *J. Liposome Res*, 2004, **14**, 155.
133. B. Gupta, T. S. Levchenko and V. P. Torchilin, *Adv. Drug Deliv. Rev.*, 2005, **57**, 637.
134. R. H. Müller, W. Mehnert, J. S. Lucks, C. Schwarz, A. zur Mühlen, H. Weyhers, C. Freitas and D. Rühl, *Eur. J. Pharm. Biopharm.*, 1995, **41**, 62.
135. W. Mehnert and K. Mäder, *Adv. Drug Deliv. Rev.*, 2001, **47**, 165.
136. H. Bunjes, K. Westesen and M. H. J. Koch, *Int. J. Pharm.*, 1996, **129**, 159.
137. S. C. Yang, L. F. Lu, Y. Cai, J. B. Zhu, B. W. Liang and C. Z. Yang, *J. Control. Release*, 1999, **59**, 299.
138. S. B. Tiwari and M. M. Amiji, *Curr. Drug Deliv.*, 2006, **3**, 219.
139. K. Tabatt, C. Kneuer, M. Sameti, C. Olbrich, R. H. Muller, C. M. Lehr and U. Bakowsky, *J. Control. Release*, 2004, **97**, 321.
140. K. Tabatt, M. Sameti, C. Olbrich, R. H. Muller and C. M. Lehr, *Eur. J. Pharm. Biopharm.*, 2004, **57**, 155.
141. C. Rudolph, U. Schillinger, A. Ortiz, K. Tabatt, C. Plank, R. Müller and J. Rosenecker, *Pharm. Res.*, 2004, **21**, 1662.
142. N. Pedersen, S. Hansen, A. V. Heydenreich, H. G. Kristensen and H. S. Poulsen, *Eur. J. Pharm. Biopharm.*, 2006, **62**, 155.
143. C. Olbrich, U. Bakowsky, C. M. Lehr, R. H. Muller and C. Kneuer, *J. Control. Rel.*, 2001, 77, 345.
144. C. Freitas and R. H. Müller, *Eur. J. Pharm. Biopharm.*, 1999, **47**, 125.
145. F. Liu, J. Yang, L. Huang and D. Liu, *Pharm. Res.*, 1996, **13**, 1642.
146. F. Liu, J. Yang, L. Huang and D. Liu, *Pharm. Res.*, 1996, **13**, 1856.
147. O. Meyer, D. Kirpotin, K. L. Hong, B. Sternberg, J. W. Park, M. C. Woodle and D. Papahadjopoulos, *J. Biol. Chem.*, 1998, **273**, 15621.
148. P. Harvie, F. M. Wong and M. B. Bally, *J. Pharm. Sci.*, 2006, **89**, 652.
149. H. Faneca, S. Simoes and M. C. P. de Lima, *Biochim. Biophys. Acta*, 2002, **1567**, 23.
150. K. Tabatt, C. Kneuer, M. Sameti, C. Olbrich, R. H. Müller, C. Lehr and U. Bakowsky, *J. Control.Release*, 2004, **97**, 321.
151. S. H. Gehrke, *Adv. Polym. Sci.*, 1993, **110**, 81.
152. A. S. Hoffman, *Adv. Drug Deliv. Rev.*, 2002, **54**, 3.

153. K. Yeomans, *Chem. Rev.*, 2000, **100**, 2.

154. W. R. Gombotz and S. Wee, *Adv. Drug Deliv. Rev.*, 1998, **31**, 267.

155. M. F. A. Goosen, G. M. O'Shea, H. M. Gharapetian, S. Chou and A. M. Sun, *Biotechnol. Bioeng.*, 1985, **27**, 146.

156. M. P. Lutolf, G. P. Raeber, A. H. Zisch, N. Tirelli and J. A. Hubbell, *Adv. Mater.*, 2003, **15**, 888.

157. W. E. Hennink and C. F. van Nostrum, *Adv. Drug Deliv. Rev.*, 2002, **54**, 13.

158. P. Gacesa, *Carbohydr. Polym.*, 1988, **8**, 161.

159. T. Watanabe, A. Ohtsuka, N. Murase, P. Barth and K. Gersonde, *Magn. Reson. Med.*, 1996, **35**, 697.

160. D. Eagland, N. J. Crowther and C. J. Butler, *Eur. Polym. J.*, 1994, **30**, 767.

161. A. M. Mathur, K. F. Hammonds, J. Klier and A. B. Scranton, *J. Control. Release*, 1998, **54**, 17.

162. S. Nagahara and T. Matsuda, *Polym. Gels Netw.*, 1996, **4**, 111.

163. M. C. Morán, M. G. Miguel and B. Lindman, *Soft Matter*, 2010, **6**, 3143.

164. M. C. Morán, F. R. Baptista, A. Ramalho, M. G. Miguel and B. Lindman, *Soft Matter*, 2009, **5**, 2538.

165. M. C. Morán, M. Mitjans, V. Martínez, D. R. Nogueira and M. P. Vinardell, *Current directions in DNA gel particles, in Recent Advances in Pharmaceutical Sciences III*, ed. D. Muñoz-Torrero, A. Cortés and E. L. Mariño, Transworld Research Network, Kerala, 2013, 145–162.

166. M. C. Morán, M. P. Vinardell, M. R. Infante, M. G. Miguel and B. Lindman, *Adv. Colloid Interface Sci.*, 2014, **205**, 240–256.

167. Y. Li and T. Tanaka, *J. Chem. Phys.*, 1989, **90**, 5161.

168. Y. Li and T. Tanaka, *J. Chem. Phys.*, 1990, **92**, 1365.

169. T. Tanaka, *Physica A*, 1986, **140**, 261.

170. D. Gan and L. A. Lyon, *J. Am. Chem. Soc.*, 2001, **123**, 7511.

171. C. D. Jones and L. A. Lyon, *Macromolecules*, 2000, **33**, 8301.

172. T. Tanaka, *Phys. Rev. Lett.*, 1978, **40**, 820.

173. K. Dusek and K. Patterson, *J. Polym. Sci., Polym. Phys. Ed.*, 1968, **6**, 1209.

174. Y. Li and T. Tanaka, *J. Chem. Phys.*, 1990, **92**, 1365.

175. J. Moselhy, X. Y. Wu, R. Nicholov and K. Kodaria, *J. Biomater. Sci. Polym. Ed.*, 2000, **11**, 123.

176. D. Duracher, F. Sauzedde, A. Elaissari, A. Perrin and C. Pichot, *Colloid Polym. Sci.*, 1998, **276**, 219.

177. D. Duracher, F. Sauzedde, A. Elaissari, C. Pichot and L. Nabzar, *Colloid Polym. Sci.*, 1998, **276**, 920.

178. M. J. Snowden, B. Z. Chowdhry, B. Vincent and G. E. Morris, *J. Chem. Soc., Faraday Trans.*, 1996, **92**, 5013.

179. S. R. Sershen, S. L. Westcott, N. J. Halas and J. L. West, *J. Biomed. Mater. Res.*, 2000, **51**, 293.

180. S. R. Sershen, S. L. Westcott, N. J. Halas and J. L. West, *Appl. Phys. Lett.*, 2002, **80**, 4609.

181. S. R. Sershen, S. L. Westcott, J. L West and N. J. Halas, *Appl. Phys. B*, 2001, **73**, 379.

182. A. Suzuki, T. Ishii and Y. Maruyama, *J. Appl. Phys.*, 1996, **80**, 131.

183. A. Suzuki and T. Tanaka, *Nature*, 1990, **346**, 345.

184. T. Tanaka, I. Nishio, S. T. Sun and S. Ueno-Nishio, *Science*, 1982, **218**, 467.

185. P. K. Ghosh, *Indian J. Biochem. Biophys.*, 2000, **37**, 273.

186. U. Gaur, S. K. Sahoo, T. K. De, P. C. Ghosh, A. Maitra and P. K. Ghosh, *Int. J. Pharm.*, 2000, **202**, 1.

187. M. C. Morán, D. R. Nogueira, M. P. Vinardell, M. G. Miguel and B. Lindman, *Int. J. Pharm.*, 2013, **454**, 192.

188. P. Lemieux, S. V. Vinogradov, C. L. Gebhart, N. Guerin, G. Paradis, H. K. Nguyen, B. Ochietti, Y. G. Suzdaltseva, E. V. Bartakova, T. K. Bronich, Y. St-Pierre, V. Y. Alakhov and A. V. Kabanov, *J. Drug Target.*, 2000, **8**, 91.

189. A. V. Kabanov and S. V. Vinogradov, *Angew. Chem. Int. Ed.*, 2009, **48**, 5418.

190. S. V. Vinogradov, *Curr. Pharm. Des.*, 2006, **12**, 4703.

191. Y. Fukunaka, K. Iwanaga, M. Morimoto, M. Kakemi and Y. Tabata, *J. Control. Release*, 2002, **80**, 333.

192. T. Kushibiki, R. Tomoshige, Y. Fukunaka, M. Kakemi and Y. Tabata, *J. Control. Release*, 2003, **90**, 207.

193. T. Kushibiki, R. Tomoshige, K. Iwanaga, M. Kakemi and Y. Tabata, *J. Control. Release*, 2006, **112**, 249.

194. A. Kim, D. M. Checkla, P. Dehazya and W. L. Chen, *J. Control. Release*, 2003, **90**, 81.

195. T. Segura, P. H. Chung and L. D. Shea, *Biomaterials*, 2005, **26**, 1575.

196. J. A. Wieland, T. L. Houchin-Ray and L. D. Shea, *J. Control. Release*, 2007, **120**, 233.

197. M. Gupta and A. K. Gupta, *J. Control. Release*, 2004, **99**, 157.

198. Z. Megeed, M. Haider, D. Q. Li, B. W. O'Malley, J. Cappello and H. Ghandehari, *J. Control. Release*, 2004, **94**, 433.

199. F. K. Kasper, S. K. Seidlitis, A. Tang, R. S. Crowther, D. H. Carney, M. A. Barry and A. G. Mikos, *J. Control. Release*, 2005, **104**, 521.

200. M. Oishi and Y. Nagasaki, *React. Funct. Polym.*, 2007, **67**, 1311.

CHAPTER 7

Dendrimers

A. J. PERISÉ-BARRIOS,[a,†] D. SEPÚLVEDA-CRESPO,[a,†]
D. SHCHARBIN,[b] B. RASINES,[c] R. GÓMEZ,[c,d]
B. KLAJNERT-MACULEWICZ,[e] M. BRYSZEWSKA,[e]
F. J. DE LA MATA[c,d] AND M. A. MUÑOZ-FERNÁNDEZ*[a,d]

[a] Laboratorio InmunoBiología Molecular, Hospital General Universitario Gregorio Marañón, Instituto de Investigación Sanitaria Gregorio Marañón, 28007 Madrid, Spain; [b] Institute of Biophysics and Cell Engineering, National Academy of Sciences of Belarus, 220072 Minsk, Belarus; [c] Departamento de Química Inorgánica, Universidad de Alcalá, 28801 Alcalá de Henares, Spain; [d] Networking Research Center on Bioengineering, Biomaterials and Nanomedicine (CIBER-BBN), 28029 Madrid, Spain; [e] Department of General Biophysics, Faculty of Biology and Environmental Protection, University of Łódź, 90–131, Łódź, Poland
*Email: mmunoz.hgugm@gmail.com

7.1 Introduction

Dendritic structures are considered a new class of polymers that possess interesting properties, they are well-defined structures with a high degree of molecular uniformity and they have a surface with high functionality.[1-5] Particulate systems with well-defined sizes and shapes are synthesized maintaining a high level of control over the design (their size, shape, branching length/density and surface functionality). This distinguishes

---

[†] These two authors contributed equally to this work.

RSC Nanoscience & Nanotechnology No. 34
Soft Nanoparticles for Biomedical Applications
Edited by José Callejas-Fernández, Joan Estelrich, Manuel Quesada-Pérez and Jacqueline Forcada

these structures as optimum carriers in medical applications, such as drug delivery, gene transfection and imaging.

In this chapter, we discuss different aspects concerning dendritic structures. We present different types of dendritic systems and summarize the most representative types of dendrimers, including polyamidoamine dendrimers, polypropylenimine dendrimers, peptide dendrimers, carbosilane dendrimers, phosphorus dendrimers, polyglycerol dendrimers, triazine dendrimers and Fréchet-type dendrimers. Second, we describe the structural characterization of dendrimers and dendriplexes using different techniques, such as spectroscopic methods, scattering techniques, microscopy, gel electrophoresis or capillary electrophoresis and isothermal titration calorimetry (ITC), among others. Lastly, we focus on some applications of dendrimers in nanomedicine, because they can be used as vectors for gene therapy and for targeted therapy, and also as therapeutic agents, and they can be used in diagnostics.

7.2 Dendritic Structures

Dendritic structures are macromolecules that were first described in 1978,[1] although it was not until years later[2] when the term 'dendrimer' was used to refer to this type of molecule. These hyperbranched materials are considered as a new class of polymers that possess many very interesting properties, and differ from the classical polymers that have well-defined structures with a high degree of molecular uniformity, in addition to having a surface with high functionality.[2-5] Their versatility lies in the ease with which the end-groups can be functionalized with different reactive groups and incorporated into cross-linked networks or modified to alter and tune the behaviour according to specific needs.[6]

Dendrimers consist of three components: a central core, an interior dendritic structure or branches and a surface with functional groups. Varying the combination of these components is possible to obtain molecules with different shapes and sizes that are good candidates for applications in both biological and materials sciences. The surface groups are implicated in the solubility and chelation ability, whereas the different cores affect certain properties such as the cavity size, absorption capacity and capture–release characteristics.

Monodisperse dendrimers are synthesized through iterative reactions to prepare increasing generations (G_0, G_1, G_2, ...) of molecules with low polydispersity, uniform size and shape and a multivalent surface. In contrast, hyperbranched polymers are polydisperse dendritic macromolecules that are synthesized using a low-cost polymerization method. Figure 7.1 shows one type of globular dendritic architecture. These macrostructures are built through a convergent or divergent synthesis route (see below), which provides a structure containing three distinct parts: the core, the interior and the surface.

Other structures consist of dendrons (Figure 7.2), which are monodisperse wedge-shaped dendrimer sections with multiple terminal groups and a

Interior with multiple branches (generation G), well-suited
for encapsulation of drugs and nanomaterials.

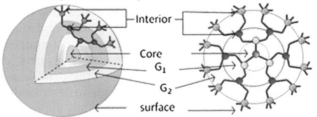

Surface modified with multiple surface groups
for host-guest interaction and functionalization.

Figure 7.1 Globular dendritic architecture.

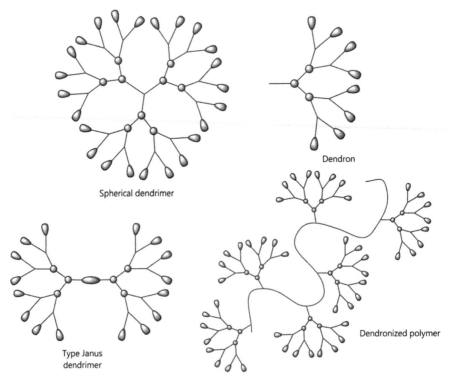

Figure 7.2 Different dendritic architectures based on dendrons.

single reactive function at the focal point. Multiple dendrons can be combined in a convergent method leading to a spherical dendrimer. If the
dendrons have a different functionalized surface and they are attached to a
core, the result is an asymmetric Janus-type dendrimer.[7–9] Polymers can
also be dendronized by reaction of the focal point of the dendron with
the polymeric backbone.[10,11]

7.2.1 Synthesis of Dendrimers

Dendrimers are synthesized mainly by two general methods, the divergent[12-15] and convergent[16] approaches. The first examples of dendrimers were prepared *via* the divergent method. In this approach, the dendrimer grows from the inside to the surface, generation to generation, from a core molecule containing one reactive and two dormant groups that react with monomer molecules, giving the first-generation dendrimer. For reacting with new monomers, the periphery has to be activated. The process is repeated for several generations (Scheme 7.1).

In contrast to the above, for the production of large quantities of dendrimers, the convergent approach is a successful method. In the first step, the growing branched polymeric arms, called dendrons, are extended. When they are large enough, they are attached to a multifunctional core by a coupling reaction (Scheme 7.2). Using this strategy, it is possible to minimize the structural defects of the dendrimer because it avoids the use of large amounts of reagents, which facilitates purification. However, if the dendrons are very bulky, the final assembly around a central core can be hindered.

Other techniques aimed at minimizing the synthetic steps for obtaining dendrimers of higher generations have been described. They combine the principles of the convergent and divergent methods,[17-20] so that one can speak of a mixed synthesis (Scheme 7.3).

The limitations of these strategies are overcome by eliminating the deactivation iterative steps by chemoselective reactions (*e.g.* click chemistry), orthogonal coupling strategy,[21] one-pot strategy multi-click cascade, self-assembly strategy,[22] *etc.*

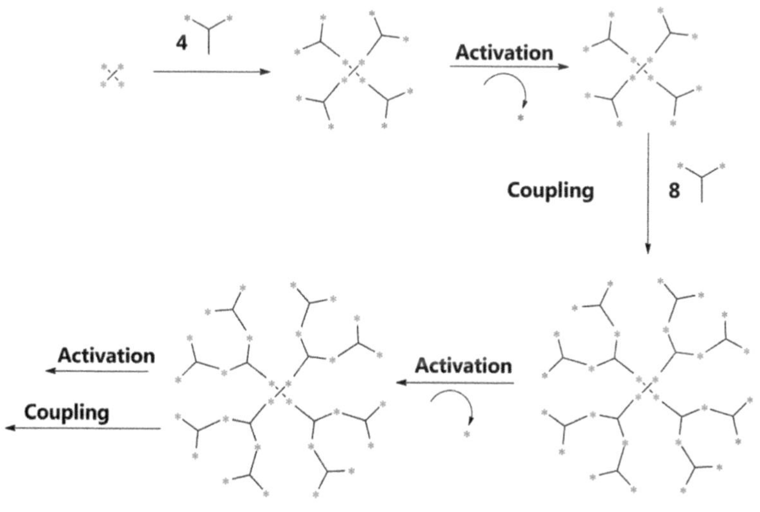

Scheme 7.1   Divergent approach.

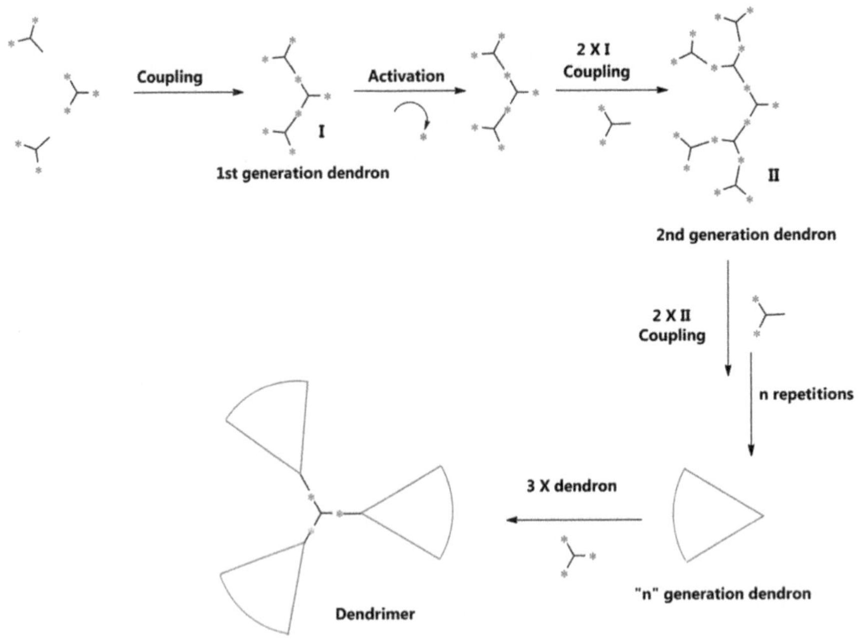

Scheme 7.2 Convergent approach.

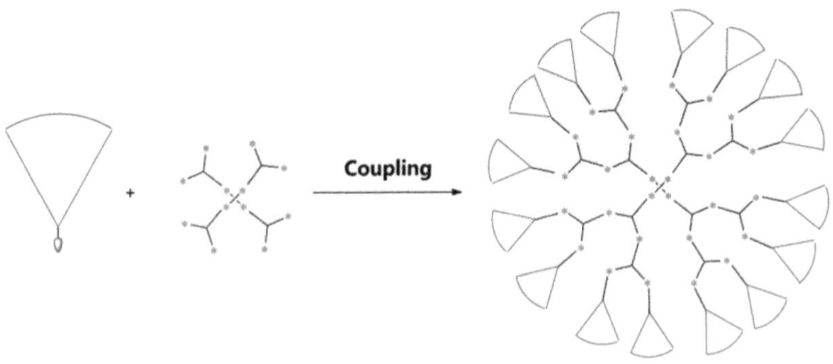

Scheme 7.3 Mixed synthesis of dendrimers.

7.2.2 Types of Dendritic Systems

7.2.2.1 *Polyamidoamine Dendrimers*

Tomalia *et al.*[23] reported the synthesis of polyamidoamine (PAMAM) dendrimers through a divergent method using an ethylenediamine initiator core. This method involves a two-step iterative reaction sequence that leads different generations of dendritic β-alanine units. The repetitive steps are a Michael-type addition of amines to the double bond of methylacrylate and the formation of amide bonds between the terminal esters and

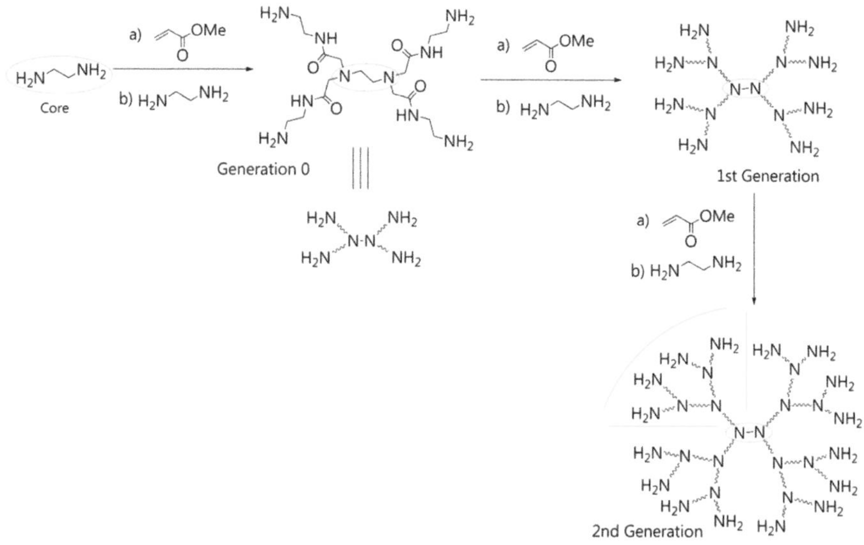

Scheme 7.4 Synthesis of polyamidoamine (PAMAM) dendrimers.

ethylenediamine (Scheme 7.4). The repetition of these two steps produces large dendrimers up to generation 10. PAMAM dendrimers have been extensively studied as non-viral vectors for gene therapy and drug delivery,[24-27] among others.

At physiological pH, PAMAM dendrimers can interact electrostatically with cationic biomolecules such as DNA molecules and plasmids due to the protonation of primary terminal and tertiary internal amines, this interaction being stronger on increasing the degree of quaternization.[28]

There is currently a large number of ethylenediamine-core PAMAM dendrimers that are commercially available as Starburst dendrimers in series of generations with several surface functional groups.

7.2.2.2 Polypropylenimine Dendrimers

In 1978, Vögtle and co-workers developed the synthesis of PPI dendrimers.[29] This family of dendrimers can be found in the literature under different names: PPI (for polypropylenimine), DAB (for diaminobutane) and POPAM (for polypropylenamine). The synthesis of these macromolecules is based on a divergent method (Scheme 7.5).[29] In the first step, acrylonitrile is added to a primary mono- or oligodiamine in a Michael reaction. The most commonly used reducing agents are Co(II)–borohydride complexes or diisobutylaluminium hydride.[30] Subsequent iterative reaction cycles permit repeated addition of acrylonitrile followed by reduction until the limiting generation is reached.

On the basis of this principle, the preparative synthesis of higher generation monodisperse PPI dendrimers was accomplished in 1993 simultaneously by the groups of Meijer and Mülhaupt.[31] In this synthesis, a highly

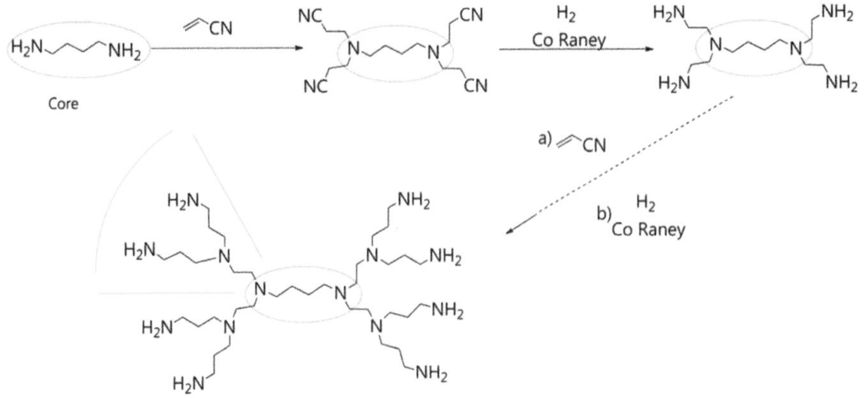

Scheme 7.5 Synthesis of PPI dendrimers.

symmetrical, spherically shaped macromolecule with a multitude of unique properties and functions as a result of the three-dimensional propagation is generated. The addition of each layer of propylenimine branches to the PPI dendrimer framework doubles the number of amino substituents on the surface, so the steric hindrance is substantially enhanced. Consequently, the reaction will stop at a certain generation for the dendrimer synthesis. So far, the highest generation of PPI dendrimers that has been synthesized is the fifth. Large amounts of pure PPI dendrimers are currently commercially available from Aldrich (USA) and DSM (The Netherlands).

7.2.2.3   Peptide Dendrimers

Peptide dendrimers are macromolecules consisting of a peptidyl branching core and/or covalently attached surface functional units. These are radial or wedge-like branched macromolecules. They have various biotechnological and biochemical applications thanks to the multimeric nature of these structures, their unambiguous composition and their ease of production. In the early 1980s, Denkewalter *et al.*[32] published a well-defined poly(L-lysine) family of dendrimers, which are the most commonly used peptide-based dendrimers.[33] It is a polypeptide dendrimer built from a benzhydrylamine core by the repetition of a sequence of protection–deprotection, steps using *N,N′*-bis(*tert*-butoxycarbonyl)-L-lysine nitrophenyl ester as reagent (Scheme 7.6). This synthesis was carried out up to the tenth generation (PLy-G10).

The majority of peptide dendrimers referred to in the literature and currently in use possess a structure based on only two of the three structural parts: branching units and surface functional groups. They are often synthesized without a core and more appropriately termed 'dendrons' rather than classical dendrimers. This kind of dendritic system is called a multiple antigen peptide (MAP) system. Interestingly, these molecules contain both α-peptide and ε-peptide.[34]

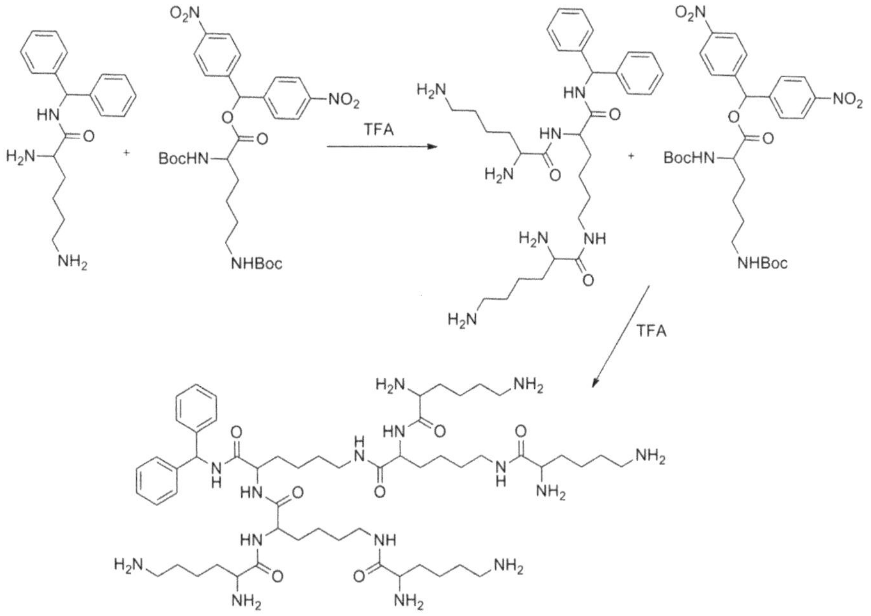

Scheme 7.6 Poly(L-lysine) dendrimer with a benzhydrylamine core.

7.2.2.4 Carbosilane Dendrimers

Fetters and co-workers reported the use of a G_1 carbosilane dendrimer with 12 end-groups for the synthesis of a star polymer as early as 1978.[35] However, the first syntheses of carbosilane dendrimers of various generations was reported independently by the groups of van der Made,[36,37] Roovers,[38,39] Muzafarov[40] and Seyferth.[41] To date, all reported carbosilane dendrimers have been synthesized *via* the divergent approach (Scheme 7.7). The dendrimer is built starting from a central core (G_0) possessing alkenyl groups, using repeating sequences of alternating hydrosilylations with chlorosilanes and ω-alkenylations with Grignard reagents.

Carbosilane dendrimers up to the fifth generation were prepared by van der Made and co-workers using tetraallylsilane as core, $HSiCl_3$ as hydrosilylation reagent and allylmagnesium bromide as ω-alkenylation reagent. However, the molecular weight and the structural perfection of these dendrimers were not substantiated by appropriate analytical methods. To obtain dendrimers possessing an open structure, Roovers and co-workers started from a tetravinylsilane core and used $HMeSiCl_2$ and vinylmagnesium bromide as branching units. Dendrimers with an even more open structure were obtained by Muzafarov *et al.*, who chose $HMeSiCl_2$ as hydrosilylation reagent, allylmagnesium chloride as ω-alkenylation reagent and triallylmethylsilane as a core. Another type of synthesis consists of using silicon tetrachloride as a core (Figure 7.3). Obviously, the synthetic route to carbosilane dendrimers offers high flexibility and versatility. Not only the

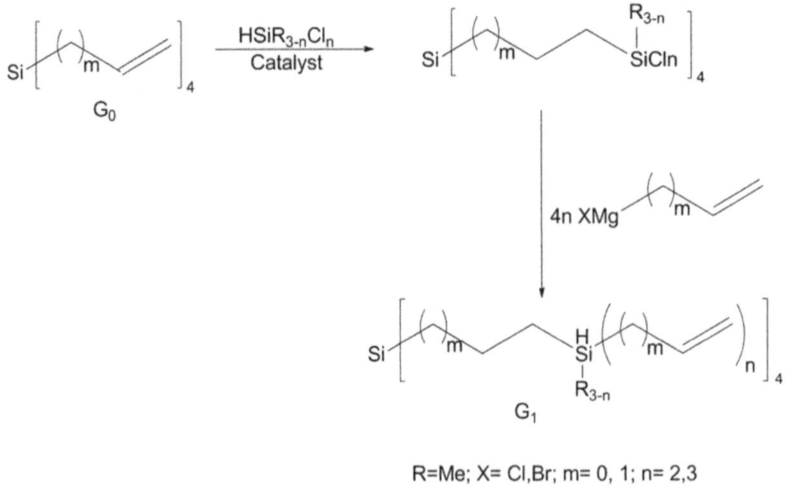

R=Me; X= Cl,Br; m= 0, 1; n= 2,3

Scheme 7.7   Divergent synthesis of a typical carbosilane dendrimer.

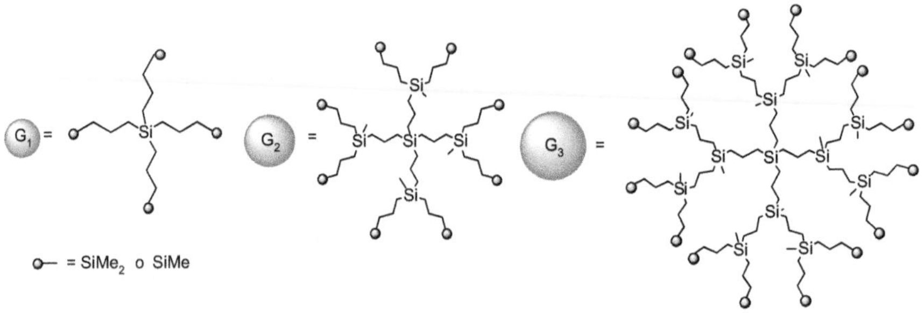

o— = SiMe$_2$ o SiMe

Figure 7.3   Carbosilane skeleton.

hydrosilylation reagent and the ω-alkenylation reagent (to some extent) but also the core molecule can be varied without drastic changes in the reaction conditions.

Anionic and cationic systems have been prepared by various reactions such as hydrosilylation and click chemistry,[42–44] among others, providing water soluble molecules that have interesting applications in biomedicine such as antibacterial agents, in drug and biomolecule delivery and as antiviral agents.

7.2.2.5   Phosphorus Dendrimers

In this type of dendrimer, the phosphorus was bonded to other heteroatoms such as N and O,[45] but compounds are also known in which phosphorus atoms are located in the core unit or in the periphery and also both in the core and at the branching points as well as the periphery.[46] Majoral and

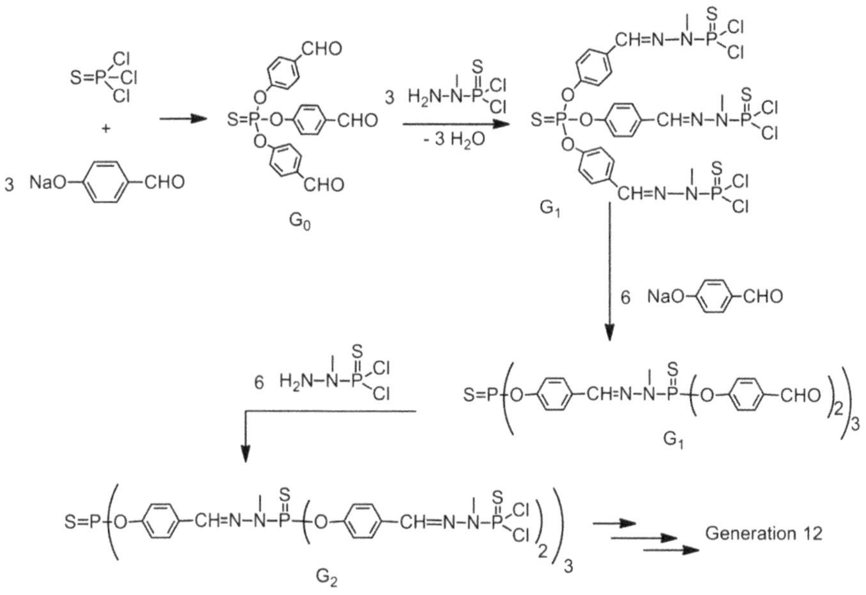

Scheme 7.8 Synthesis of phosphorus dendrimers built from the trifunctional core P(S)Cl$_3$.

co-workers described in 1994 the first neutral phosphorus-containing dendrimers and the range of compounds was subsequently substantially expanded.[45] The repetition of two consecutive stages gives rise to these dendrimers. The first conversion step is based on the reaction of a core containing n equivalents of p-hydroxybenzaldehyde and the second on condensation of the aldehyde functionality with H$_2$N–N(Me)–P(S)Cl$_2$. This process was carried out up to the twelfth generation, starting from the trifunctional core P(S)Cl$_3$ (Scheme 7.8).

This synthetic route can be applied to a large number of different cores such as a hexafunctional cyclotriphosphazene core (N$_3$P$_3$Cl$_6$).

These dendrimers contain a hydrophobic backbone functionalized with P(S)Cl$_2$ or CHO, which are also hydrophobic. However, dendritic systems acquire soluble character when the functional groups are functionalized with either positive or negative charge, which makes them interesting systems for biological uses.

7.2.2.6 Other Types of Dendrimers

7.2.2.6.1 Polyglycerol Dendrimers.
Haag *et al.* developed polyglycerol dendrimers in an iterative two-step process using tris(hydroxymethyl)ethane as core molecule (Figure 7.4).[47] The process is based on allylation of an alcohol and catalytic dihydroxylation of the allylic double bond.

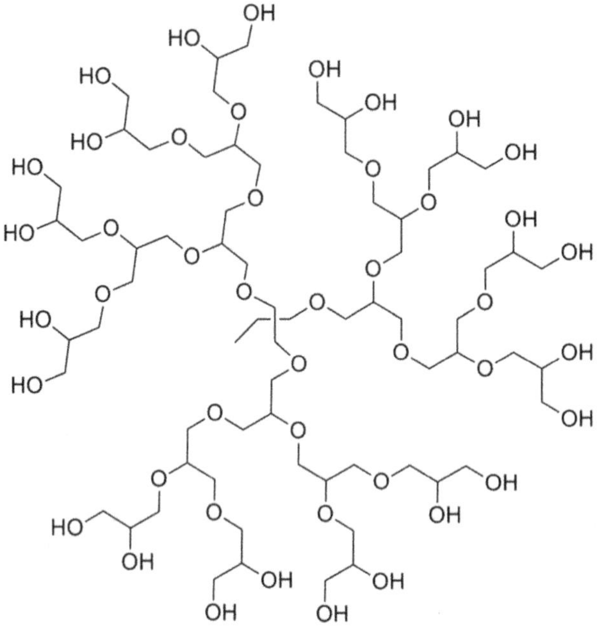

Figure 7.4   Third-generation polyglycerol dendrimer.

These dendrimers have neutral biocompatible aliphatic polyether cores and their periphery has been modified with various cationic amine terminal groups for siRNA binding and complexation.

7.2.2.6.2   Triazine Dendrimers.   In these macromolecules, the core molecule is based on a 1,3,5-triazine ring (Figure 7.5). The synthesis of this type of dendritic derivatives is based on the sequential substitution of the trichlorotriazine with amine nucleophiles; hence is it possible to prepare diverse dendrimers.[48–50] Triazine dendrimer synthesis can be escalated and therefore large generations can be prepared. To date, triazine dendrimers have been probed for a variety of medicinal applications, including drug delivery with an emphasis on cancer, non-viral DNA and RNA delivery systems, in sensing applications and as bioactive materials.[51]

7.2.2.6.3   Fréchet-Type Dendrimers.   This type of dendrimer was developed by Hawker and Fréchet.[52,53] The skeleton of these dendrimers is based on hyperbranched poly(benzyl ether). The modification of these systems resulted in molecules of biomedical interest. A particular case of these macromolecules contains carboxylic acid groups on the periphery, which provides the molecule with further functionalization sites and also enhances the solubility of these hydrophobic structures in aqueous media and in polar solvents (Figure 7.6).

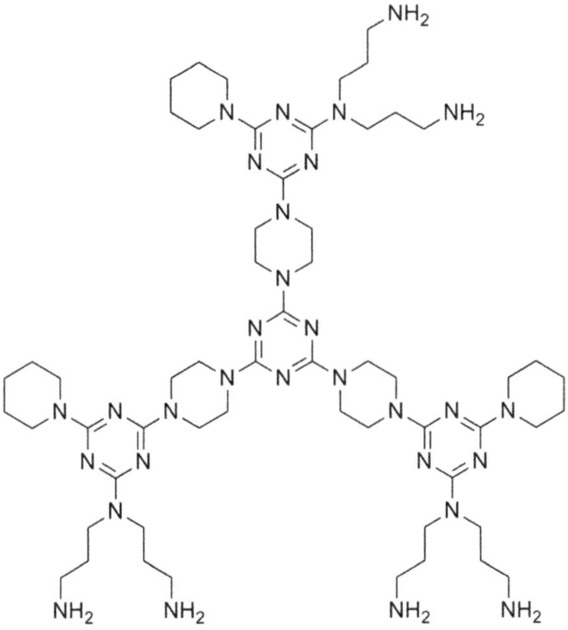

Figure 7.5 First-generation triazine dendrimer.

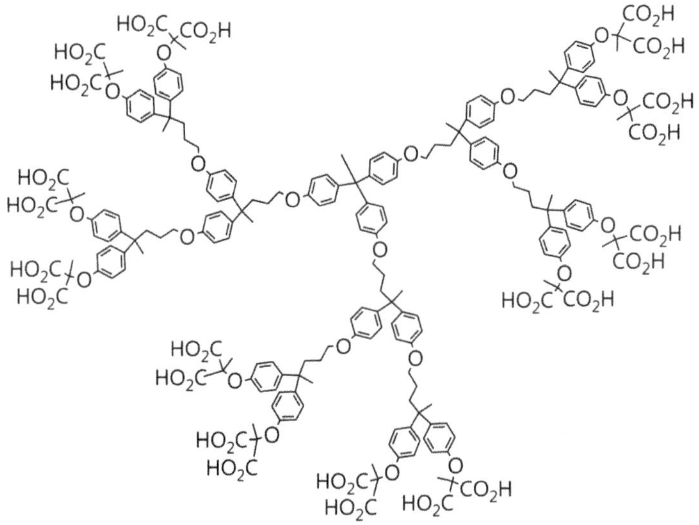

Figure 7.6 Fréchet-type dendrimer.

Other types of dendrimers include those based on hyperbranched polyesters. An example is the commercially available Boltorn dendrimers, which contain a multifunctional alcohol as building block and 2,2-bis(methylol)-propionic acid (bis-MPA) as branching units. They can be modified to obtain lipophilic and amphiphilic dendrimers.

Throughout this section, we have tried to summarize the most representative types of dendrimers. However, the number of typologies reported in the literature is more extensive.

7.2.3 Structural Characterization of Dendrimers and Dendriplexes

Dendrimers are branched polymers whose structure is formed by monomeric subunit branches diverging to all sides from a central nucleus.[54] Dendriplexes are formed by the binding of nucleic acids to cationic dendrimers (see Chapter 6). The structural characterization of dendrimers differs from that of dendriplexes because first their chemical structure should be confirmed and second the formation, organization and stability of complex has to be analysed.

7.2.3.1 Spectroscopic and Spectrometric Methods

7.2.3.1.1 Dendrimers. The chemical structures of dendrimers are usually confirmed by NMR spectroscopic, mass spectrometric and elemental analysis.[55] Concentration-dependent ^{1}H NMR and/or MALDI-TOF MS (matrix-assisted laser desorption/ionization time-of-flight mass spectrometry) investigations allow their possible aggregation in solution to be studied.[55] UV–visible absorption spectra provide the spectral properties of newly synthesized dendrimers.[55] A detailed review of the spectroscopic characterization of dendrimers by Gautam *et al.*[56] also indicated the possibility of applying infrared spectroscopy, Raman spectroscopy, fluorescence spectroscopy, X-ray photoelectron spectroscopy, electron paramagnetic resonance and X-ray scattering methods to determine their properties.[56]

7.2.3.1.2 Dendriplexes. The spectroscopic characterization of dendriplexes is directed at studying the complexation between dendrimers and nucleic acids. It includes investigations on the fluorescence polarization of labelled nucleic acids or labelled dendrimers during their binding, fluorescence intercalation assays and circular and linear dichroism.[57] The fluorescence polarization technique is based on changes in the fluorescence intensity or fluorescence polarization of fluorescein-labelled nucleic acids when a dendrimer is added (and *vice versa*).[58] A labelled nucleic acid in solution at 20–37 °C is fairly flexible. Dendriplex formation leads to a significant restriction of molecular motions of the nucleic acid and increases the molecular mass of the complex, apparent in a significant increase (up to fourfold) in its degree of fluorescence polarization.[58] The additional benefit of this technique is its possible application in complex systems when the interaction of dendriplexes with proteins or glucosaminoglycans is being studied.[59] The basis of the fluorescent dye intercalation assay is the intercalation of fluorescent dyes (ethidium bromide, PicoGreen, GelStar, *etc.*) into double-stranded DNA or RNA. The fluorescent dye occupies an effective binding site (EB) for several base pairs (bp) (2–4 bp/EB) and binding increases its fluorescence intensity significantly, causing a blue shift in its maximum emission wavelength. Compounds with higher affinity for DNA (dendrimers) displace the dye, quench its fluorescence and induce a red shift of its maximum emission wavelength.[58,60] Circular and linear dichroism are techniques for investigating the structure of DNA in solution and the conformational modifications of the double helix produced by dendrimer binding. Comparison of the experimental results with empirical spectra of representative DNA samples provides useful comparative and direct structural information.[61]

7.2.3.2 Scattering Techniques

7.2.3.2.1 Dendrimers. The application of scattering techniques to characterize dendrimers allows the study of their intermolecular structure, intramolecular cavity, radius of gyration, hydrodynamic radius, molecular weight, effective charge number of a single dendrimer molecule, water penetration into the interior of the dendrimers and the internal dynamics.[62] Specifically, the scattering techniques (Chapter 2) include small-angle neutron scattering, quasi-elastic neutron scattering, small-angle X-ray scattering and light scattering.[62]

7.2.3.2.2 Dendriplexes. As spectroscopic methods, scattering techniques are used to study the formation of complexes between dendrimers and nucleic acids and to characterize them. Two techniques are widely used: zeta potential and zeta size. Every charged particle in a solution containing ions is surrounded by an electrical double layer of ions and counterions. The potential that exists at this hydrodynamic boundary is known as the zeta potential. It is determined by electrophoresis of the sample and measuring the velocity of the particles using laser Doppler velocimetry. The zeta potential gives information on dendriplex charge ratios and its final zeta potential[57,63] (see Chapter 2). The hydrodynamic diameter of dendriplexes in solution can be determined by dynamic light scattering (DLS) and static light scattering (SLS). As is well known, colloidal particles in suspension exhibit Brownian motion, *i.e.* random movements depending on particle size, temperature and solution viscosity. DLS provides measurements of the time-dependent fluctuations in scattering intensity to determine the translational diffusion coefficient, from which the hydrodynamic diameter can be calculated using the Stokes–Einstein equation. Using DLS allows the hydrodynamic diameter of a dendriplex to be estimated.[57,63–64] Also (see Chapter 2), structural polymorphism of DNA–dendrimer complexes can be estimated by synchrotron X-ray diffraction[65] or small-angle X-ray scattering.[66]

7.2.3.3 *Microscopy*

7.2.3.3.1 Dendrimers. The structural composition of dendrimers can be determined by atomic force microscopy or electron microscopy to check the size and uniformity of the molecules.[56,67]

7.2.3.3.2 Dendriplexes. Transmission electron microscopy and scanning electron microscopy provide estimates of the size and shape of complexes and in some cases the type of packing, surface topography, composition and electrical conductivity[57,63,68] (see Chapter 2). Like electron microscopy, atomic force microscopy allows the shape and particle size distributions of dendriplexes to be studied at various charge ratios, pH values and salt concentrations. Samples in the dried, frozen or partially liquid state as indicated by planar or 3D images can be analysed.[69–71] Molecular modelling based on atomic force microscopy data allows the features and peculiarities of complexes to be determined.[71]

7.2.3.4 *Other Types of Techniques*

In addition to the techniques mentioned in Sections 7.2.3.1–7.2.3.3, other important techniques used to analyse dendrimers and dendriplexes should be mentioned.

First, gel electrophoresis and capillary electrophoresis have been widely used to characterize dendrimers and study dendriplex formation, calculate the DNA:dendrimer charge ratio, estimate the shape of the DNA (linear/supercoiled) in the dendriplex and determine the stability of the complex.[57,68,72] Second, isothermal titration calorimetry is a very promising technique that allows the heat energy of the binding process to be measured directly. It allows not only the binding constant (and thereby free energy) but also the molar and charge ratios, enthalpy and entropy of dendrimer–nucleic acid complex formation to be determined.[68] Analysis of the 'melting points' of dendriplexes gives information about their thermal stability. The melting process changes the optical properties of DNA. Thus, melting profiles can be analysed by two independent techniques: measurement of DNA absorbance at 260 nm upon heating and differential scanning calorimetry.[61,73,74] The secondary structure of the DNA component when complexed within a dendrimer can be investigated using Fourier transform infrared (FTIR) spectroscopy.[61] The fluorescence quenching technique provides information on the packing of a complex.[75] The parameters of an electron paramagnetic resonance probe (TEMPO) covalently attached to the dendrimer surface allow the interaction between DNA and dendrimer during dendriplex formation to be studied.[73]

Important techniques for the characterization of dendrimers and dendriplexes include molecular biology methods to study their cytotoxicity and transfection efficiency.[76] Cellular uptake is estimated using fluorescently labelled dendrimers or nucleic acids.[77,78] Transfection efficiency is measured by the luciferase reporter assay or β-galactosidase assay as indirect techniques or by flow cytometry or confocal microscopy for labelled proteins, such as green fluorescent protein (GFP).[76,79,80] Cytotoxicity is determined by 7AAD (7-aminoactinomycin D) labelling as a direct method or by the MTT [3-(4,5-dimethylthiazol-2-yl)-2,5-diphenyltetrazolium bromide] test or LDH (lactate dehydrogenase) assay[79–82] as indirect techniques.

Finally, it can be concluded that an essential standard set of techniques for characterizing dendriplexes is as follows: (1) analysis of the size and shape of dendriplexes in the dried/frozen state by electron or atomic force microscopy; (2) analysis of the charge/molar ratio of complexes by gel electrophoresis or ethidium bromide intercalation assay or zeta potential measurement; and (3) determination of the hydrodynamic diameter of dendriplexes in solution by DLS. For the evaluation of transfection efficiency, the essential techniques are (4) luciferase reporter assay, β-galactosidase assay, flow cytometry or confocal microscopy and (5) determination of cytotoxicity by 7AAD labelling or the MTT assay or LDH test. All of these tests allow the transfection efficiency and cytotoxicity of different kinds of dendrimers to be compared.[77]

7.3 Applications of Dendrimers in Nanomedicine

Special features of dendrimers for their individual use or for the formation of multiple protein complexes make these nanoscopic macromolecules very attractive for a number of applications.

Owing to the progress in the structural control of the synthesis process of dendrimers, the polyvalency properties, the nano-bioactivity and surface engineering, a new area of dendrimer-based nanomedicine is being developed, highlighting the excellent features of dendrimers as vectors and scaffolding for (i) gene therapy, (ii) drug delivery and (iii) therapeutic agents such as antimicrobial, antitumour and anti-inflammatory agents, and also as (iv) diagnostic/imaging agents, *etc.* The main applications of dendrimer-based drug delivery are achieving a better biodistribution throughout the body, obtaining a controlled and directed distribution compared with traditional pharmaceutical drugs and also their use for topical delivery such as antiviral agents.[83,84] However, owing to a better understanding of dendrimer surface conjugates, protein mimicry, nanoscale directed biodistribution/excretion modes, dendrimer-based unimolecular drug encapsulation and exterior complexation properties, more complex applications such as nanodevices are beginning to emerge. Advanced dendrimer-based drug delivery involves at least three approaches: (i) precise nanoscale vectors for a therapeutic approach to pharmaceuticals, targeting specific disease sites through dendrimer–prodrug conjugates, (ii) unimolecular dendrimer encapsulation of therapeutic drugs with subsequent controlled release and (iii) dendrimer complexing agents with nucleic acids or other appropriate drugs.

7.3.1 Dendrimers as Vectors for Gene Therapy

In 1990, the first gene therapy protocol was performed at the US National Institutes of Health, obtaining very good results in children with immunodeficiency due to adenosine deaminase deficiency. Gene therapy was presented as the treatment for those diseases for which conventional drugs were unsuccessful, and this led to the dedication of much effort in research in this field. As a result, in recent years there has been a breakthrough in the use of genes for therapeutic purposes and, in parallel, the necessary tools for gene transfer, such as dendrimers, have been developed (see also Chapter 6 for more information).

Gene therapy is used to treat diseases caused by missing, defective or overexpressing genes. Developing a method for introducing a therapeutic gene into target cells is the key issue in treating diseases by gene therapy. Currently two types of therapeutic approaches are under investigation: *in vivo* gene therapy, wherein the gene of interest is inserted into a vector and then the vector is transferred directly to the patient, and *ex vivo* gene therapy, in which the vector is introduced into cultured cells (which can be taken from the patient) and subsequently these genetically engineered cells are transplanted into the patient. In any case, viral or non-viral vectors (such as liposomes, cationic polymers or dendrimers) are necessary for introducing the gene of interest into cells.[85,86] Usually, nucleic acid-based therapies using plasmid DNA (pDNA), small interfering RNA (siRNA) or microRNA (miRNA) are applied with the objective of increasing efficiency, safety and therapeutic value (Chapter 6). In addition, the optimal plasmid or siRNA design is of critical importance to improve delivery strategies. Nucleic acids are highly hydrophilic, and therefore are unable to penetrate lipid cell membranes and, additionally, can be degraded by nucleases in the blood. For the successful application of pDNA, siRNA and miRNA, delivery should be achieved with the lowest toxicity and without inducing an immune response. For example, pDNA can be delivered by injection into muscle tissue or *via* hydrodynamic delivery, but in most cases requires a vector that provides for shielding, targeting and cellular uptake.[87,88]

The vehicle can be a cationic dendrimer capable of condensing genetic material containing multiple phosphate residues forming dendriplexes. Thereby, genetic material is protected from degradation and owing to an excess of positive charges, complexes are able to interact with the negatively charged cell membrane. Then, the dendriplex is internalized *via* endocytosis, and once inside the acidic endosomes, the dendrimer acts as a 'proton sponge' by buffering the pH, leading to endosomal swelling and rupture, releasing its contents to the cytosol and allowing DNA or RNA eventually to enter the nucleus.[90] For example, PEI polymer has a strong buffering capacity and can use this ability to promote endosomal release of the polymer and its siRNA cargo.[91,92] Once the target cell has been reached, proper endosomal release and the incorporation in the RNA-induced silencing complex (RISC) are necessary for a therapeutic effect [Figure 7.7(A)].[93]

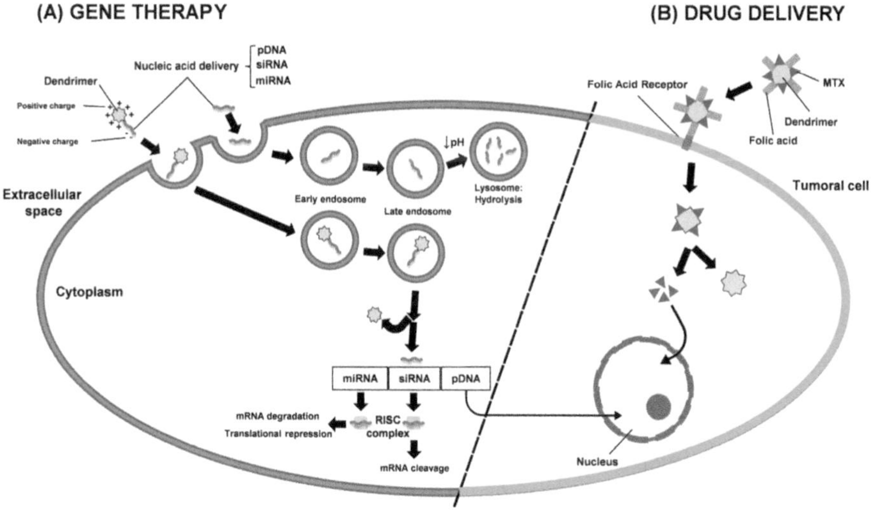

(A) GENE THERAPY (B) DRUG DELIVERY

Figure 7.7 (A) Uptake and intracellular trafficking of a targeted nucleic acid delivery vehicle. (B) Dendrimer-based nanoparticles for drug delivery for cancer therapy.

Applications of cationic dendrimers as transport vehicles have been described, with emphasis on their use as transfection agents of nucleic acids; they can transfect them into a large number of cell lines and primary culture cells.[89,94–96]

One of the best known transfection reagents of genetic material into eukaryotic cells is SuperFect, which is based on amine-terminated PAMAM dendrimers.[94,97] Interestingly, it has been shown that more flexible fractured PAMAM dendrimers exhibited better DNA and RNA transfection properties than non-fractured dendrimers. To achieve higher transfection efficiencies, recent design studies have focused on dendrimer structures incorporating higher flexibility (*e.g.* softer dendrimers).[98,99] Other transfection reagents such as carbosilane dendrimers,[89,100] phosphorus dendrimers,[95] PPI-type dendrimers[101] and triazine dendrimers,[98] are currently under detailed investigation and so far have shown interesting results.[102]

The use of RNA transfection for the knockout of specific intracellular processes has achieved great relevance as a therapeutic strategy for the treatment of different pathologies. Transfections involving low molecular weight oligonucleotides (*e.g.* siRNA or microRNA) have served not only as tools in genomic analysis but also for possible applications in gene therapy. There are different dendrimers for delivering siRNA *in vitro* in human cancer cell lines (*i.e.* liver, breast, prostate, endometrial, pancreatic, ovarian, skin and testicular cancers). For example, a PAMAM-based RNA vector was shown to down regulate the expression of heat-shock protein 27 (Hsp27) in prostate cancer cells, and also PPI dendrimers were used for knockdown of phosphoenolpyruvate carboxykinase (PEPCK) and organic cationic transporter 1

(OCT 1) in a rat cell line. These proteins are involved in the regulation of blood glucose content, so it can be an important strategy for the regulation and treatment of type 2 diabetes.[103,104] Further, gene therapy represents a promising approach for preventive therapy against HIV/AIDS. Carbosilane or phosphorus dendrimers/siRNA dendriplexes have shown high efficiency in HIV-Nef silencing, decreasing HIV replication in SupT1 T-cells and in primary peripheral blood mononuclear cells, representing a potential alternative therapy for HIV infection.[95,100,105]

7.3.2 Dendrimers as Vectors for Targeted Drug Delivery

Drugs are delivered by dendrimers by simple encapsulation, electrostatic interaction or covalent conjugation. The application of dendrimer–drug conjugates by oral, intravenous, intraperitoneal, intratumoral, transdermal, ocular, *etc.* administration has been investigated. Drug–dendrimer conjugates have two advantages over drug–dendrimer complexes and traditional drug dosing: prolonged lifetime and more stable level of the active substance. Dendrimers are also being designed to become an important tool for delivery in a tissue-specific target. The topics of drug delivery, dendrimer solubility, dendrimer toxicity and biocompatibility are closely related and influence each other.

The same passive and targeted delivery strategies as described above for genetic material delivery can also be applied to small drugs. For example, polyether dendritic compounds bearing folate residues on their surface have been prepared as model drug carriers with potential tumour cell specificity. Folic acid (FA) is required in great amounts in cancer cells; therefore, many types of cancer (lung, ovary, colon and epithelial) overexpress the high-affinity folic acid receptors (FARs). PAMAM dendrimers showed targeted toxicity towards KB cancer cell line compared with the FA alone.[106] In addition, methotrexate (MTX), a chemotherapy drug, can also be conjugated with PAMAM dendrimers which are carrying FA (G5-Ac-FITC-FA-OH-MTX). Cancer cells bind FA residues and absorb high concentrations of the drug [Figure 7.7(B)]. For research purposes, fluorescein isothiocyanate (FITC) was utilized in this multifunctional device as an imaging agent to track the location of dendrimers. These multifunction dendrimer-based conjugates, called theranostics, are generated as a tool of polymer-based nanodevices for the intracellular targeting of drugs, imaging agents and other materials.[107] The treatment of mice with theranostics showed a significant decrease in tumour growth compared with the free drug, which indicates that these theranostics delay tumour growth *in vivo*.[108] Target delivery strategies are also used for other diseases apart from cancer. There are also folate–dendrimer conjugates that are suitable as vehicles to target specific sites of inflammation and deliver an anti-arthritic drug, such as indomethacin.[109]

On the other hand, for the development of vaccines, it is of interest to design dendrimers as vectors that can specifically deliver well-defined antigenic domains to antigen-presenting cells (APCs), such as dendritic cells (DCs). In spite of improvements in this area, strategies employing *ex vivo*-generated DCs have shown limited efficacy in clinical trials. Dendrimers have been proposed as new carriers for antigen delivery with the aim of ameliorating DC antigen loading that is a pivotal point in DC approaches. Immature DCs (iDCs) and mature DCs (mDCs) loaded with cationic carbosilane dendrimers or with dendriplex do not show fundamental changes in its functions. *In vitro*-derived DCs show a high level of internal peptide signal without deregulating cellular functions; this is necessary to activate immune systems developing an efficient anti-HIV DC-based system.[110,111] Otherwise, maltose-functionalized dendrimer–peptide complexes are able to activate the immune system by way of DC stimulation.[112] These glycodendrimers could be new tools in the search for effective anti-HIV DC-based immunotherapies. In addition, biocompatible Ni(II) nitrilotriacetic acid-modified polyethylenimine–maltose (Ni-NTA–DG) is a promising carrier system for antigen delivery in immunotherapy, because advancing release of His-tagged peptides from a polyplex can be induced at pH 6 when using the long-spacered derivative Ni-NTA–PEG12–DG.[113,114] Future *in vivo* work could provide some essential information that cannot be obtained with *in vitro* experiments, especially if it achieves the presentation of MHC class I or II epitopes to antigen peptides, which are loaded to carbosilane dendrimers or Ni-NTA–PEG12–DG systems. Moreover, mannose surface-modified PAMAM dendrimers facilitate the delivery *via* C-type mannose receptors such as CD209 (DC-SIGN), which are located at the surface of APC. Furthermore, when the mannosylated dendrimer is conjugated with ovalbumin (OVA), it shows an increase in immune responses compared with OVA alone.[115] This prototype combines large nanoscale dendrimer capabilities to bind antigens thanks to the strong binding capacity of the polyvalent dendrimer surface. The use of different dendrimers as targeting vectors can be a promising approach in targeted delivery of vaccines or immunotherapy antigens. There are at least three potential therapeutic applications for phosphorus dendrimers, namely topical anti-inflammatory agents, compounds for allograft rejection or autoimmune diseases and agents inducing specific tolerance with antigen-loaded DCs against allergy reaction.

7.3.3   Dendrimers as Therapeutic Agents

There are dendrimers that have pharmaceutical effects in their native state without involving any association with traditional drugs. The degree of dendrimer modification to produce biological effects may be viewed as a continuum process, which begins with unmodified dendrimers, using them as drugs, and extends to specifically tailor multifunctional dendrimers aimed at specific targets (*e.g.* cellular receptors, tissue of microorganism). On the one hand, complexes formed by covalent attachments of drugs to dendrimers have greater beneficial pharmacological properties than the single drug has, such as (i) higher solubility of the drug–dendrimer complex, (ii) reduction of non-specific toxicity of the drug, increasing efficacy and sustained release, and (iii) drug conjugation to the scaffolding, increasing biodistribution properties. On the other hand, non-covalent attachments of drugs to dendrimers also present more beneficial pharmacological properties than the free drug. For example, dendrimer complexation with an anti-inflammatory agent such as indomethacin increases the circulation time compared with pure indomethacin.[116]

Recently, interest has been focused on the design of dendrimers that have shown promising properties related to transport of drugs through various biological barriers, *e.g.* intestinal barrier, blood–brain barrier (BBB). Crossing the BBB to achieve dendrimer entry into the central nervous system (CNS) is an extremely complex process, but fortunately some approximations can be made to achieve this, such as coating surfactants and poly(ethylene glycol)s (PEGs) to modify the biodistribution or developing nanocarriers to exploit the endogenous transport systems at the level of the BBB.[117–120] Although crossing other biological barriers also is not an easy process to achieve,[121] there is a great deal of information about the modifications needed to dendrimers to facilitate the transepithelial transport and microvascular extravasation;[122] as an example, PAMAM dendrimers can permeate across intestinal epithelial barriers, suggesting their potential use as oral drug carriers.[123]

Below various dendrimers with cationic surfaces that exhibit intrinsic pharmaceutical effects are first considered. Prions are proteins that have pathogenic properties and infectious features. They cause neurodegenerative disease by aggregating within the CNS to form amyloid plaques, disrupting the normal tissue structure. All known prion diseases affect the structure of the brain or other neural tissue and are currently untreatable and universally fatal. Aggregates are found in different neurodegenerative protein misfolding diseases such as Alzheimer's disease, Huntington's disease and Creutzfeldt–Jakob disease, among others. In this context, there is considerable interest in the properties that cationic dendrimers have, in their ability not only to solubilize prion protein (PrP) aggregates but also to dissolve mature fibrils.[124–126] Phosphorus-containing dendrimers with tertiary amine surface groups produced a prion curing effect *in vitro* and inhibited PrP aggregate formation in a mouse model. On the other hand, a direct interaction exists between PPI cationic dendrimers and misfolded PrP, which solubilizes aggregates.[125] Furthermore, guanidinium groups on the surface of PPI dendrimers have shown solubilizing effects on amyloid fibril forming prion peptide sequences. All of these data support the concept that charged surface groups can play a key role in the PrP aggregate solubilizing effect of the dendrimer.[127] Some polylysine dendrimers can be proposed as compounds for developing antiamyloidogenic drugs.[124]

Cationic dendrimers are also effective as bacteriostatic and bactericidal compounds, owing to cationic surface groups that mediate the binding to the negatively charged bacterial surface. Thus, increasing dendrimer generation and consequently the number of surface cations might be expected to increase the antibacterial activity of dendrimers. However, it may also result in less membrane penetration. PAMAM, PPI and carbosilane dendrimers with quaternary ammonium structures at their surfaces cause disruption of the bacterial membrane, producing 'lethal pore formation' thereon.[128–130] Quaternary ammonium-functionalized carbosilane dendrimers act as potent biocides owing to the multivalency together with the biopermeability of the carbosilane dendritic skeleton.[131] Quaternary ammonium-functionalized PPI dendrimers are very potent biocides, and the antimicrobial properties of these novel biocides with bromide anions are more potent than those with chloride anions.[128] In summary, cationic dendrimers can also be used for the design of highly effective eukaryotic/prokaryotic differentiating drugs.

Interestingly, changes in gene expression induced by carbosilane dendrimers in CD4 T lymphocytes, CD8 T lymphocytes and macrophages have shown their potential for application in medication. Therefore, the strong repression of IL17A, IL17F, IL23R and IL23A (proteins involved in autoimmune diseases) *in vitro* and in a mouse model by cationic carbosilane dendrimers suggests a useful potential pharmacological application that could be adopted to treat inflammatory processes driven by Tc17 cells.[132–134] PAMAM dendrimers with amine or hydroxyl terminal groups showed high anti-inflammatory effects in an arthritis model in the rat.[135] Finally, synthetically engineered macromolecules such as polyvalent dendrimer–glucosamine conjugates can be tailored to have defined immunomodulatory and antiangiogenic properties.

Second, dendrimers with anionic surfaces with pharmaceutical effects are considered. First events leading to viral infection involve ionic interactions between the virus capsid and the cellular membrane, between the positively charged haemaglutinin at the viral surface and sialic acids/phosphates at the cellular surface. With the purpose of interrupting these events, highly charged agonistic drugs based on the polyvalency features of dendrimers have been created. Polyanionic dendrimers inhibit respiratory syncitial virus (RSV),[136] and also in the case of herpes simplex virus (HSV) and HIV they show antimicrobial properties,[137–139] and further prevent the sexual transmission of HIV in animal models such as the macaque.[140] Moreover, polyanionic dendrimers with anti-HIV properties such as sulfonated naphthalene or carbosilane dendrimers are being developed as topical microbicides for human use.[83,84,141,142] These dendrimers not only have a partial capacity to block the entry of HIV-1 inside epithelial cells, but also protect the epithelial monolayer from cell disruption and reduce HIV-1 infection of activated peripheral blood cells. In addition, treatment of epithelial cells with these dendrimers did not produce changes in the proinflammatory cytokine profile, proliferation of peripheral blood cells, microbiota and sperm survival. Finally, no irritation or vaginal lesions were detected in female rabbits and CD1 (ICR) mice after vaginal administration of dendrimers. These dendrimers could be effective in inhibiting HIV infection and transmission within genital mucosa and also the spread of HIV transmission.[83,143,144] SPL7013 dendrimer (Vivagel)[145,146] is a candidate dendrimer–microbicide for the prevention of HIV and HSV infection and has demonstrated great potency.[147,148] However, Vivagel received US Food and Drug Administration (FDA) phase III clinical trial approval for use as a topical nano-pharmaceutical against bacterial vaginosis (http://www.starpharma.com/news, 2012).

Polyanionic carbosilane dendrimers are attractive microbicide candidates as they target a broad spectrum of HIV-1 strains such as laboratory or primary strains or HIV-2, HSV and SIV strains.[83,84] Moreover, combinations of dendrimers with other anti-HIV drugs acting on different steps of the viral cycle could be interesting in order to develop new potent microbicides. Synergy may allow the use of lower doses of each component, permitting a reduction in adverse effects and limiting local injuries and inflammation caused previously to sexual intercourse. This decrease in the doses also impacts the final cost of production, facilitating access for consumers, including those living in developing countries. Polyanionic carbosilane dendrimers and antiretrovirals (ARVs) are safe and effective compounds against HIV, with great potential as topical microbicides.[149–153] This supports further clinical research on these combinations as potential microbicides in the context of blocking sexual transmission of HIV-1.[143]

7.3.4   Dendrimers in Diagnostics

There are two areas of medical diagnostics, *in vitro* and *in vivo* diagnostics. The former is normally an off-line approach and covers analytical methods for biological samples such as blood and urine samples and deals with long-known methodologies such as radioimmunoassays, enzyme immunoassays and microarrays. The latter is usually performed on-line and covers the detection and characterization of diseases in patients or animals using different imaging methodologies.

Immunoassays employed for the quantitative determination of analytes are divided into two types called 'homogeneous' and 'heterogeneous.'[154] All reactions in a homogeneous bioassay are performed in an aqueous monophasic solution and do not require the separation of the free analyte from the analyte–antibody complex in a solid-phase immunoassay. However, in the heterogeneous system, the antibody–antigen complex is separated on a solid-phase system and dendrimers are used as structural components to increase the number of detector molecules, increasing the binding of target analytes in a biological sample, and therefore increasing the sensitivity. Dendrimers are good candidates as scaffolds for fluorophore groups in the monitoring of biological events.

Dendrimers can be used to enhance the covalent and non-covalent binding capacities of surfaces used for heterogeneous assays. Therefore, they are being considered in enzyme-linked immunosorbent assays (ELISAs) (polystyrene), microarrays (glass) and certain biosensor protocols involving gold surfaces and plasmon surface resonance spectroscopy. These protocols involve non-covalent binding of detector molecules to the surface, but covalent binding may be utilized to enhance the stability and selectivity of the bioassay. Dendrimers can be used for enhancing the binding efficacy and to increase the uniformity of detector molecules to provide optimum recognition activity and sensitivity.[153–157]

There is a particular novel type of dendrimer composed of PAMAM segments, named Starburst, that consists of water-soluble polymeric materials with a well-defined composition and structure. They are used for coupling specific antibodies to developing immunoassays. The dendrimer-based radial partition immunoassay (RPIA) approach combines the best attributes of homogeneous and heterogeneous immunoassay formats, provides the same advantages of a homogeneous immunoassay system such as speed, convenience and simplified automation, but also allows for high sensitivity by incorporating a partitioning step to remove any unbound labelled reagents and interfering materials, as is characteristic of a heterogeneous immunoassay system.[157] Starburst dendrimers are produced commercially by NanoSynthons (Mount Pleasant, MI, USA). Fréchet/Ihre/Hult-type dendripolyesters are offered by Polymer Factory (Nacka, Sweden), Newkome-type dendri-polyamides and Simanek-type dendri-polytriazines by Frontier Scientific (Logan, UT, USA), Majoral/Caminade-type phosphorus dendrimers by Dendris (Toulouse, France) and carbosilane dendrimers by Ambiox Biotech (Madrid, Spain).

Dendrimers can also be used for modulating the polarity and adhesion of biomolecules to solid phases by passive adsorption. Prepared bioreactive surfaces with a high density of amine groups are useful for immobilizing biological macromolecules for various biosensor applications, such as the fabrication of DNA microarrays and protein chips.[158,159] Also, an electrochemical immunosensor on a gold surface based on dendrimers has been reported, where dendrimers are used as a high-capacity scaffolding for antigen presentation involved in biotin interaction.[160] Further, attachment of DNA dendritic molecules carrying oligonucleotide cassettes to antibodies or to other ligands that mediate molecular recognition is very useful for detecting target molecules at extremely low concentrations. This is used in diagnostic assays and dendrimers are used as amplified signal-generating structures for diagnostic applications.[161,162]

Biosensors such as DNA microarrays and microchips are based on hybridizations that occur between complementary oligonucleotides, one linked to a solid surface (the probe) and the other to be analysed (the target). Two main types of uses of dendrimers for arrays can be distinguished: either the dendrimer is connected to the slide and is used as a support for the probe or the dendrimer is used as a multiply labelled entity connected to the target for easier detection. These two topics constitute the two most important fields of this chapter; the first represents approximately two-thirds of the total number of publications.

To summarize, the use of dendrimers (or dendrons) for improving the sensitivity of microarrays is still in its infancy; however, promising leads are already being identified. When used as linkers between the solid surface and the oligonucleotide probe, dendrimers generally allow a higher loading of the oligonucleotide and their remoteness from the solid favours hybridization, which occurs as easily as in solution. Furthermore, systems created using dendrimers are generally very robust and so will not be damaged after several hybridization–dehybridization cycles. Their reusability has been shown in several cases, involving dozens or hundreds of experiments. When used as fluorescent or radioactive multiply labelled entities linked to the target, dendrimers (or dendrons) are always much more easily detected than monomeric labels, inducing important increases in sensitivity. This is particularly useful when very small quantities of biological materials are analysed. Such methodology has been applied to 'suspension microarrays' using beads (microspheres) instead of planar microarrays.[163]

Microarrays are used for screening large numbers of genes (DNA arrays) and for the analysis of protein–protein interactions (protein–peptide arrays). The use of highly polyvalent dendrimer surfaces increases the 'load' of biomolecules on glass-based microarrays and the advantage is that the use of DNA microarrays based on dendrimer-coated surfaces shows a doubling in signal intensity and the slides can be reused. The increase in signal-to-noise ratio was shown in a DNA microarray designed to detect HSV.[164] Applying the same system to proteins achieved a 10-fold increase in signal with this dendrimer-based protocol compared with conventional coated surfaces.[165,166] Sensitive biosensors based on electrochemical impedance spectroscopy were developed with Tomalia-type PAMAM dendrimers and antibodies or DNA.[167,168] Finally, an optical fibre-based assay for the quantification of anomalous reflection on gold chips where amine-terminated PAMAM dendrimers are coupled to the gold surface has been described.[169,170]

A number of automated heterogeneous immunoassays based on dendrimer nanotechnology are commercially available for assaying changes in blood levels of different clinically significant markers related to a variety of pathogenic conditions. For example, the Stratus (Siemens, Erlangen, Germany) assay protocol is used for the analysis of different cardiac markers, and reduced the heart infarction diagnosis time from more than 1 h to a few minutes.[157] This dendrimer–biomolecule concept has been implemented in the formation of dendrimer–nucleotide conjugates and used in sensitive assays suitable for detecting chlamydia.[171] The Genisphere technology based on Nilsen-type poly-DNA dendrimers is useful in microarray applications,[172,173] and the NEESA system has achieved remarkably high signal intensities with a single dendrimer molecule.[161]

Regarding *in vivo* medical diagnostics, dendrimers are under investigation as polymeric carriers of contrast agents for magnetic resonance imaging (MRI), scintigraphy and X-ray techniques, *e.g.* computed tomography (CT). The objective of synthesizing high molecular weight contrast agents is to modify the pharmacokinetic behaviour of available small-sized compounds from a broad extracellular to an intravascular distribution. Major target indications include angiography, tissue perfusion determination and tumour detection and differentiation. In principle, imaging moieties, *e.g.* metal chelates for MRI and scintigraphy and triiodobenzene derivatives for CT, are coupled to a dendrimeric carrier characterized by a defined molecular weight.

The images obtained by MRI are the result of magnetically induced energy absorption accompanying the relaxation of proton nuclei associated with tissue water exchange rates. Well-defined polyvalent dendrimer surfaces can be functionalized to control specific properties such as circulation time, toxicity and targeting of specific cell or tissue receptors, being one of the principal properties that make dendrimer-based contrast agents ideal blood pool agents for imaging the vascular system. In addition, contrast agents can be designed by engineering a large number of dendrimer surface functional groups in concert with precise nanoscale generational size. This enables higher gadolinium payloads to be introduced for higher relaxation efficacy, increasing the contrast signal and, very importantly, providing control over the nanoscale size of the MRI agent that establishes the mode of excretion of the agent.

In a general way, we can divide the sites specifically targeted by dendrimer-based MRI contrast agents by two principal strategies: (i) passive targeting that involves a critical nanoscale design parameter such as nanoscale size, that can be tailored to reside longer in the bloodstream and be used as a contrast agent directed to enter into normal or tumour tissues,[174,175] or (ii) active targeting, which is receptor mediated and dependent on surface chemistry.[176] The objective is to target specific tumour receptors, for example, the use of folic acid-surfaced dendrimers that may target malignant melanoma cells,[177–179] dendrimers using pyridoxamine as surface groups for selectively targeting the liver,[180] dendrimers targeted to VEGFR-2 receptor[181] and conjugated PAMAM dendrimer which has a surface chemistry combination of (i) metalloprotease-sensitive cell-penetrating peptide moieties, (ii) fluorescent (Cy5) functional groups, (iii) DOTA-chelated gadolinium and (iv) appropriate surface PEGylation to provide an adequate circulatory residence time for tagging the active tumour sites. This is an effective strategy for guiding the surgical removal of primary and metastatic tumours *in vivo* using real-time imaging techniques. Dendrimer-based 'chemical exchange-dependent saturation transfer' (CEST) shows high pH sensitivity to the surroundings, providing a useful method for visualizing tumour tissues.[182,183] Although no dendrimer has reached the status of clinical application, Gadomer-17 is the first commercial dendrimer-based contrast agent that is in a preclinical phase and it is available for research. Gadomer-17 is a poly(L-lysine) dendrimer possessing 24 Gd–DOTA complexes on its surface that have important value and usefulness for tumour characterization and therapy monitoring.[184,185] However, major efforts are still necessary before a dendrimeric contrast agent will finally be available for widespread use in patients.

References

1. E. Buhleier, W. Wehner and F. Vögtle, *Synthesis*, 1978, **2**, 155.
2. D. A. Tomalia, H. Baker, J. Dewald, M. Hall, G. Kallos, S. Martin, *et al.*, *Polym. J.*, 1985, **17**, 117.
3. A. M. Naylor, W. A. Goddard, G. E. Kiefer and D. A. Tomalia, *J. Am. Chem. Soc.*, 1989, **111**, 2339.
4. D. A. Tomalia, A. M. Naylor and W. A. Goddard, *Angew. Chem. Int. Ed. Engl.*, 1990, **29**, 138.
5. U. Boas and P. M. H. Heegaard, *Chem. Soc. Rev.*, 2004, **33**, 43.
6. V. Sujitha, B. Sayani and P. Kalyani, *Int. J. Pharm. Res.*, 2011, **2**, 25.
7. S. Fuchs, A. Pla-Quintana, S. Mazères, A. M. Caminade and J. P. Majoral, *Org. Lett.*, 2008, **10**, 4751.
8. X. Feng, D. Taton, E. Ibarboure, E. L. Chaikof and Y. Gnanou, *J. Am. Chem. Soc.*, 2008, **130**, 11662.
9. J. Liu, Y. Feng, Y. He, N. Yang and Q. Fan, *New J. Chem.*, 2012, **36**, 380.
10. J. I. Paez, M. Martinelli and V. Brunetti, *Polymers*, 2012, **4**, 355.
11. M. Kroeger, O. Peleg and A. Halperin, *Macromolecules*, 2010, **43**, 6213.
12. E. Buhleier, W. Wehner and F. Vögtle, *Synthesis*, 1978, **55**, 155.
13. D. A. Tomalia, H. Baker, J. Dewald, M. Hall, G. Kallos, S. Martin, *et al.*, *Br. Polym. J.*, 1985, **17**, 117.
14. D. A. Tomalia, A. M. Naylor and A. W. Goddard III, *Angew. Chem. Int. Ed. Engl.*, 1990, **9**, 138.
15. G. R. Newkome, *J. Org. Chem.*, 1985, **50**, 2003.
16. C. J. Hawker and J. M. J. Fréchet, *J. Am. Chem. Soc.*, 1990, **112**, 7638.
17. T. Kawaguchi, K. L. Walker, C. L. Wilkins and J. S. Moore, *J. Am. Chem. Soc.*, 1995, **117**, 2159.
18. G. Labbe, B. Forier and W. Dehaen, *Chem. Commun.*, 1996, 2143.
19. F. Zeng and S. C. Zimmerman, *J. Am. Chem. Soc.*, 1996, **118**, 5326.
20. V. Maraval, J. Pyzowski, A. M. Caminade and J. P. Majoral, *J. Org. Chem.*, 2003, **68**, 6043.
21. P. Antoni, M. J. Robb, L. Campos, M. Montanez, A. Hult, E. Malmström, M. Malkoch and C. J. Hawker, *Macromolecules*, 2010, **43**, 6625.
22. X.-Q. Xiong, *Aust. J. Chem.*, 2009, **62**, 1371.
23. D. A. Tomalia, A. M. Naylor and W. A. Goddard III, *Angew. Chem. Int. Ed. Engl.*, 1990, **29**, 138.
24. G. J. Kirkpatrick, J. A. Plumb, O. B. Sutcliffe, D. J. Flint and N. J. Wheate, *J. Inorg. Biochem.*, 2011, **105**, 1115.
25. D. M. Sweet, R. B. Kolhatkar, A. Ray, P. Swaan and H. Ghandehari, *J. Control. Release*, 2009, **138**, 78.
26. Y. E. Kurtoglu, R. S. Navath, B. Wang, S. Kannan, R. Romero and R. M. Kannan, *Biomaterials*, 2009, **30**, 2112.
27. H. Wang, H. Shi and S. Yin, *Exp. Ther. Med.*, 2011, **2**, 777.
28. J. H. Lee, Y. B. Lim, J. S. Choi, Y. Lee, T. I. Kim, H. J. Kim, *et al.*, *Bioconjug. Chem.*, 2003, **14**, 1214.
29. E. Buhleier, W. Wehner and F. Vögtle, *Synthesis*, 1978, **55**, 155.

30. R. Moors and F. Vögtle, *Chem. Ber.*, 1993, **126**, 2133.
31. M. de Brabander-van den Berg and E. W. Meijer, *Angew. Chem.*, 1993, **105**, 1370; *Angew. Chem. Int. Ed. Engl.*, 1993, **32**, 1308; C. Wörner and R. Mülhaupt, *Angew. Chem.*, 1993, **105**, 1367; *Angew. Chem. Int. Ed. Engl.*, 1993, **32**, 1306.
32. R. G. Denkewalter, J. F. Kolc and W. J. Lukasavage, Macromolecular highly branched homogeneous compound based on lysine units, *U.S. Pat.*, 4 289 872, 1981; R. G. Denkewalter, J. F. Kolc and W. J. Lukasavage, Macromolecular highly branched homogeneous compound, *U.S. Pat.*, 4 410 688, 1983.
33. K. Luo, C. Li, G. Wang, Y. Nie, B. He, Y. Wu, *et al.*, *J. Control. Release*, 2011, **155**, 77.
34. J. P. Tam, *Proc. Natl. Acad. Sci. U. S. A.*, 1988, **85**, 5409; J. P. Tam, *J. Immunol. Methods*, 1996, **196**, 17.
35. N. Hadjichristidis, A. Guyot and L. J. Fetters, *Macromolecules*, 1978, **11**, 668.
36. A. W. van der Made and P. W. N. M. van Leeuwen, *J. Chem. Soc., Chem. Commun.*, 1992, 1400.
37. A. W. van der Made, P. W. N. M. van Leeuwen, J. C. de Wilde and R. A. C. Brandes, *Adv. Mater. Process.*, 1993, **5**, 466.
38. J. Roovers, P. M. Toporowski and L. L. Zhou, *Polym. Prepr. Am. Chem. Soc. Div. Polym. Chem.*, 1992, **33**, 182.
39. L. L. Zhou and J. Roovers, *Macromolecules*, 1993, **26**, 963.
40. A. M. Muzafarov, O. B. Gorbatsevich, E. A. Rebrov, G. M. Ignat'eva, T. B. Chenskaya, V. D. Myakushev, *et al.*, *Polym. Sci., Ser. A*, 1993, **35**, 1575.
41. D. Seyferth, D. Y. Son, A. L. Rheingold and R. L. Ostrander, *Organometallics*, 1994, **13**, 2682.
42. J. F. Bermejo, P. Ortega, L. Chonco, R. Eritja, R. Samaniego, M. Müllner, *et al.*, *Chem. Eur. J.*, 2007, **13**, 483.
43. E. Arnaiz, L. Doucede, S. Garcia-Gallego, K. Urbiola, R. Gómez, C. T. de Ilarduya, *et al.*, *Mol. Pharm.*, 2012, **9**, 433.
44. B. Rasines, J. Sanchez-Nieves, M. Maiolo, M. Maly, L. Chonco, J. L. Jiménez, *et al.*, *Dalton Trans.*, 2012, **41**, 12733.
45. N. Launay, A. M. Caminade, R. Lahana and J. P. Majoral, *Angew. Chem.*, 1994, **106**, 1682; N. Launay, A. M. Caminade and J. P. Majoral, *J. Am. Chem. Soc.*, 1995, **117**, 3282; A. M. Caminade, C. O. Turrin, R. Laurent, A. Maraval and J. P. Majoral, *Curr. Org. Chem.*, 2006, **10**, 2333.
46. J. P. Majoral and A. M. Caminade, *Chem. Rev.*, 1999, **99**, 845; J. P. Majoral and A. M. Caminade, *Top. Curr. Chem.*, 1998, **197**, 79; J. P. Majoral, A. M. Caminade, R. Laurent and P. Sutra, *Heteroat. Chem.*, 2002, **13**, 474.
47. R. Haag, A. Sunder and J. F. Stumbe, *J. Am. Chem. Soc.*, 2000, **122**, 2954.
48. O. M. Merkel, M. A. Mintzer, D Librizzi, O. Samsonova, T. Dicke, B. Sproat, *et al.*, *Mol. Pharm.*, 2010, 7, 969.
49. J. Lim, A. Chouai, S. T. Lo, W. Liu, X. Sun and E. E. Simanek, *Bioconjug. Chem.*, 2009, **20**, 2154.

50. K. K. Bansal, D. Kakde, U. Gupta and N. K. Jain, *J. Nanosci. Nanotechnol.*, 2010, **10**, 8395.
51. J. Lim and E. E. Simanek, *Adv. Drug Deliv. Rev.*, 2012, **64**, 826.
52. C. Hawker and J. M. J. Fréchet, *J. Chem. Soc., Chem. Commun.*, 1990, 1010–1012.
53. C. Hawker, K. L. Wooley and J. M. J. Fréchet, *J. Chem. Soc., Perkin. Trans. 1*, 1993, 1287–1289.
54. D. A. Tomalia, A. M. Naylor and W. A. Goddard III, *Angew. Chem. Int. Ed. Engl.*, 1990, **29**, 138.
55. C.-Q. Ma, E. Mena-Osteritz, M. Wunderlin, G. Schulz and P. Bäuerle, *Chem. Eur. J.*, 2012, **18**, 12880.
56. S. P. Gautam, A. K. Gupta, S. Agrawal and S. Sureka, *Int. J. Pharm. Pharm. Sci.*, 2012, **4**, 77.
57. D. Shcharbin, E. Pedziwiatr and M. Bryszewska, *J. Control. Release*, 2009, **135**, 186.
58. D. Shcharbin, E. Pedziwiatr, L. Chonco, J. F. Bermejo-Martín, P. Ortega, F. J. de la Mata, *et al.*, *Biomacromolecules*, 2007, **8**, 2059.
59. M. Szewczyk, J. Drzewinska, V. Dzmitruk, D. Shcharbin, B. Klajnert, D. Appelhans, *et al.*, *J. Phys. Chem. B*, 2012, **116**, 14525.
60. M. L. Ölrberg, K. Schille and T. Nylander, *Dendrimers Biomed. Appl.*, 2007, **8**, 1557.
61. C. S. Braun, M. T. Fisher, D. A. Tomalia, G. S. Koe, J. G. Koe and C. R. Middaugh, *Biophys. J.*, 2005, **88**, 4146.
62. X. Wang, L. Guerrand, B. Wu, X. Li, L. Boldon, W. R. Chen, *et al.*, *Polymers*, 2012, **4**, 600.
63. M. X. Tang and F. C. Szoka, *Gene Ther.*, 1997, **4**, 823.
64. S. Ribeiro, N. Hussain and A. T. Florence, *Int. J. Pharm.*, 2005, **298**, 354.
65. H. M. Evans, A. Ahmad, K. Ewert, T. Pfohl, A. Martin-Herranz, R. F. Bruinsma, *et al.*, *Phys. Rev. Lett.*, 2003, **91**, 075501–4.
66. K. K. Ewert, H. M. Evans, A. Zidovska, N. F. Bouxsein, A. Ahmad and C. R. Safinya, *J. Am. Chem. Soc.*, 2006, **128**, 3998.
67. I. J. Majoros, Ch. R. Williams, D. A. Tomalia and J. R. Baker Jr, *Macromolecules*, 2008, **41**, 8372.
68. D. J. Coles, S. Yang, A. Esposito, D. Mitchell, R. F. Minchin and I. Toth, *Tetrahedron*, 2007, **63**, 12207.
69. A. G. Schätzlein, B. H. Zinselmeyer, A. Elouzi, C. Dufés, Y. T. A. Chim, C. J. Roberts, *et al.*, *J. Control. Release*, 2005, **101**, 247.
70. H. G. Abdelhady, S. Allen, M. C. Davies, C. J. Roberts, S. J. B. Tendler and P. M. Williams, *Nucleic Acids Res.*, 2003, **31**, 4001.
71. E. Pedziwiatr-Werbicka, D. Shcharbin, M. Bryszewska, J. Maly, M. Maly, M. Zaborski, *et al.*, *J. Biomed. Nanotechnol.*, 2012, **8**, 57.
72. D. S. Shah, T. Sakthivel, I. Toth, A. T. Florence and A. F. Wilderspin, *Int. J. Pharm.*, 2000, **208**, 41.
73. M. F. Ottaviani, F. Furini, A. Casini, N. J. Turro, S. Jockusch, D. A. Tomalia, *et al.*, *Macromolecules*, 2000, **33**, 7842.
74. M. S. Shchepinov, K. U. Mir, J. K. Elder, M. D. Frank-Kamenetskii and E. M. Southern, *Nucleic Acids Res.*, 1999, **27**, 3035.

75. E. Pedziwiatr, D. Shcharbin, L. Chonco, P. Ortega, F. J. de la Mata, R. Gómez, *et al.*, *J. Fluoresc.*, 2009, **19**, 267.
76. D. Shcharbin, E. Pedziwiatr, J. Blasiak and M. Bryszewska, *J. Control. Release*, 2010, **141**, 110.
77. F. Kihara, H. Arima, T. Tsutsumi, F. Hirayama and K. Uekama, *Bioconjug. Chem.*, 2002, **13**, 1211.
78. L. M. Santhakumaran, T. Thomas and T. J. Thomas, *Nucleic Acids Res.*, 2004, **32**, 2102.
79. J. Haensler and F. C. Szoka, *Bioconjug. Chem.*, 1993, **1093**, 372.
80. J. F. Kukowska-Latallo, A. U. Bielinska, J. Johnson, R. Spindler, D. A. Tomalia and J. R. Baker, *Proc. Natl. Acad. Sci. U. S. A.*, 1996, **93**, 4897.
81. L. Chonco, J. F. Bermejo-Martın, P. Ortega, D. Shcharbin, E. Pedziwiatr, B. Klajnert, *et al.*, *Org. Biomol. Chem.*, 2007, **5**, 1886.
82. N. Weber, P. Ortega, M. I. Clemente, D. Shcharbin, M. Bryszewska, F. J. de la Mata, *et al.*, *J. Control. Release*, 2008, **132**, 55.
83. L. Chonco, M. Pion, E. Vacas, B. Rasines, M. Maly, M. J. Serramia, *et al.*, *J. Control. Release*, 2012, **161**, 949.
84. E. V. Cordoba, E. Arnaiz, M. Relloso, C. Sanchez-Torres, F. Garcia, L. Perez-Alvarez, *et al.*, *AIDS*, 2013, **27**, 1219.
85. S. Li and L. Huang, *Gene Ther*, 2000, 7, 31.
86. M. Nishikawa and L. Huang, *Hum. Gene Ther.*, 2001, **12**, 861.
87. K. Kawabata, Y. Takakura and M. Hashida, *Pharm. Res.*, 1995, **12**, 825.
88. J. F. Bermejo, P. Ortega, L. Chonco, R. Eritja, R. Samaniego, M. Mullner, *et al.*, *Chem. Eur. J.*, 2007, **13**, 483.
89. T. Gonzalo, M. I. Clemente, L. Chonco, N. D. Weber, L. Diaz, M. J. Serramia, *et al.*, *ChemMedChem*, 2010, **5**, 921.
90. M. Dominska and D. M. Dykxhoorn, *J. Cell Sci.*, 2010, **123**, 1183.
91. O. Boussif, F. Lezoualc'h, M. A. Zanta, M. D. Mergny, D. Scherman, B. Demeneix, *et al.*, *Proc. Natl. Acad. Sci. U. S. A.*, 1995, **92**, 7297.
92. Y. W. Cho, J. D. Kim and K. Park, *J. Pharm. Pharmacol.*, 2003, **55**, 721.
93. C. Padié, M. Maszewska, K. Majchrzak, B. Nawrot, A. Caminade and J. Majoral, *New J. Chem.*, 2009, **33**, 318.
94. J. F. Kukowska-Latallo, A. U. Bielinska, J. Johnson, R. Spindler, D. A. Tomalia and J. R. Baker Jr, *Proc. Natl. Acad. Sci. U. S. A.*, 1996, **93**, 4897.
95. V. Briz, M. J. Serramia, R. Madrid, A. Hameau, A. M. Caminade, J. P. Majoral, *et al.*, *Curr. Med. Chem.*, 2012, **19**, 5044.
96. J. Lim and E. E. Simanek, *Adv. Drug Deliv. Rev.*, 2012, **64**, 826.
97. J. Haensler and F. C. Szoka Jr, *Bioconjug. Chem.*, 1993, **4**, 372.
98. M. A. Mintzer and E. E. Simanek, *Chem. Rev.*, 2009, **109**, 259.
99. Y. N. Xue, M. Liu, L. Peng, S. W. Huang and R. X. Zhuo, *Macromol. Biosci.*, 2010, **10**, 404.
100. N. Weber, P. Ortega, M. I. Clemente, D. Shcharbin, M. Bryszewska, F. J. de la Mata, *et al.*, *J. Control. Release*, 2008, **132**, 55.

101. Y. Inoue, R. Kurihara, A. Tsuchida, M. Hasegawa, T. Nagashima, T. Mori, *et al.*, *J. Control. Release*, 2008, **126**, 59.
102. J. L. Jimenez Fuentes, P. Ortega, S. Ferrando-Martínez, R. Gomez, M. Leal, J. de la Mata, *et al.*, Dendrimers in RNAi delivery, in *Advanced Delivery and Therapeutic Applications of RNAi*, ed. K. Cheng and R. I. Kahato, Wiley, Chichester, 2013, pp. 163–186.
103. M. L. Patil, M. Zhang, O. Taratula, O. B. Garbuzenko, H. He and T. Minko, *Biomacromolecules*, 2009, **10**, 258.
104. M. Mei, Y. Ren, X. Zhou, X. B. Yuan, L. Han, G. X. Wang, *et al.*, *Technol. Cancer Res. Treat.*, 2010, **9**, 77.
105. E. Pedziwiatr-Werbicka, E. Fuentes, V. Dzmitruk, J. Sanchez-Nieves, M. Sudas, E. Drozd, *et al.*, *Colloids Surf. B*, 2013, **109**, 183.
106. K. Kono, M. Liu and J. M. J. Fréchet, *Bioconjug. Chem.*, 1999, **10**, 1115.
107. J. R. Baker, A. Quintana, L. Piehler, M. Banaszak-Holl, D. A. Tomalia and E. Raczka, *Biomed. Microdevices*, 2001, **3**, 61.
108. I. J. Majoros, C. R. Williams, A. Becker and J. R. Baker Jr, *Wiley Interdiscip. Rev. Nanomed. Nanobiotechnol.*, 2009, **1**, 502.
109. D. Chandrasekar, R. Sistla, F. J. Ahmad, R. K. Khar and P. V. Diwan, *Biomaterials*, 2007, **28**, 504.
110. M. Ionov, K. Ciepluch, B. Klajnert, S. Glinska, R. Gomez-Ramirez, F. J. de la Mata, *et al.*, *Colloids Surf. B*, 2013, **101**, 236.
111. M. Pion, M. J. Serramia, L. Diaz, M. Bryszewska, T. Gallart, F. Garcia, *et al.*, *Biomaterials*, 2010, **31**, 8749.
112. E. V. Cordoba, M. Pion, B. Rasines, D. Filippini, H. Komber, M. Ionov, *et al.*, *Nanomedicine*, 2013, **9**, 972.
113. N. Hauptmann, M. Pion, M. A. Munoz-Fernandez, H. Komber, C. Werner, B. Voit, *et al.*, *Macromol. Biosci.*, 2013, **13**, 531.
114. M. Ionov, K. Ciepluch, B. R. Moreno, D. Appelhans, J. Sanchez-Nieves, R. Gomez, *et al.*, *Curr. Med. Chem.*, 2013, **20**, 3935.
115. K. C. Sheng, M. Kalkanidis, D. S. Pouniotis, S. Esparon, C. K. Tang, V. Apostolopoulos, *et al.*, *Eur. J. Immunol.*, 2008, **38**, 424.
116. A. S. Chauhan, S. Sridevi, K. B. Chalasani, A. K. Jain, S. K. Jain, N. K. Jain, *et al.*, *J. Control. Release*, 2003, **90**, 335.
117. Y. Chen and L. Liu, *Adv. Drug Deliv. Rev.*, 2012, **64**, 640.
118. N. Denora, A. Trapani, V. Laquintana, A. Lopedota and G. Trapani, *Curr. Top. Med. Chem.*, 2009, **9**, 182.
119. V. Laquintana, A. Trapani, N. Denora, F. Wang, J. M. Gallo and G. Trapani, *Expert Opin. Drug Deliv.*, 2009, **6**, 1017.
120. A. Misra, S. Ganesh, A. Shahiwala and S. P. Shah, *J. Pharm. Pharm. Sci.*, 2003, **6**, 252.
121. A. T. Florence and N. Hussain, *Adv. Drug Deliv. Rev.*, 2001, **50**(Suppl. 1), S69.
122. K. M. Kitchens, M. E. El-Sayed and H. Ghandehari, *Adv. Drug Deliv. Rev.*, 2005, **57**, 2163.
123. W. Ke, Y. Zhao, R. Huang, C. Jiang and Y. Pei, *J. Pharm. Sci.*, 2008, **97**, 2208.

124. I. M. Neelov, A. Janaszewska, B. Klajnert, M. Bryszewska, N. Z. Makova, D. Hicks, *et al.*, *Curr. Med. Chem.*, 2013, **20**, 134.
125. J. Solassol, C. Crozet, V. Perrier, J. Leclaire, F. Beranger, A. M. Caminade, *et al.*, *J. Gen. Virol.*, 2004, **85**, 1791.
126. S. Supattapone, H. O. Nguyen, F. E. Cohen, S. B. Prusiner and M. R. Scott, *Proc. Natl. Acad. Sci. U. S. A.*, 1999, **96**, 14529.
127. P. M. Heegaard, H. G. Pedersen, J. Flink and U. Boas, *FEBS Lett.*, 2004, **577**, 127.
128. C. Z. Chen, N. C. Beck-Tan, P. Dhurjati, T. K. van Dyk, R. A. LaRossa and S. L. Cooper, *Biomacromolecules*, 2000, **1**, 473.
129. C. Z. Chen and S. L. Cooper, *Biomaterials*, 2002, **23**, 3359.
130. P. Ortega, J. L. Copa-Patino, M. A. Munoz-Fernandez, J. Soliveri, R. Gomez and F. J. de la Mata, *Org. Biomol. Chem.*, 2008, **6**, 3264.
131. B. Rasines, J. M. Hernandez-Ros, N. de las Cuevas, J. L. Copa-Patino, J. Soliveri, M. A. Munoz-Fernandez, *et al.*, *Dalton Trans*, 2009, **6**, 8704.
132. R. Gras, L. Almonacid, P. Ortega, M. J. Serramia, R. Gomez, F. J. de la Mata, *et al.*, *Pharm. Res.*, 2009, **26**, 577.
133. R. Gras, M. Relloso, M. I. Garcia, F. J. de la Mata, R. Gomez, L. A. Lopez-Fernandez, *et al.*, *Biomaterials*, 2012, **33**, 4002.
134. R. Gras, M. I. Garcia, R. Gomez, F. J. de la Mata, M. A. Munoz-Fernandez and L. A. Lopez-Fernandez, *Mol. Pharm.*, 2012, **9**, 102.
135. A. S. Chauhan, P. V. Diwan, N. K. Jain and D. A. Tomalia, *Biomacromolecules*, 2009, **10**, 1195.
136. D. L. Barnard, C. L. Hill, T. Gage, J. E. Matheson, J. H. Huffman, R. W. Sidwell, *et al.*, *Antiviral Res.*, 1997, **34**, 27.
137. N. Bourne, L. R. Stanberry, E. R. Kern, G. Holan, B. Matthews and D. I. Bernstein, *Antimicrob. Agents Chemother.*, 2000, **44**, 2471.
138. K. Ciepluch, N. Katir, A. El Kadib, A. Felczak, K. Zawadzka, M. Weber, *et al.*, *Mol. Pharm.*, 2012, **9**, 448.
139. P. Ortega, B. Macarena Cobaleda, J. M. Hernandez-Ros, E. Fuentes-Paniagua, J. Sanchez-Nieves, M. P. Tarazona, *et al.*, *Org. Biomol. Chem.*, 2011, **9**, 5238.
140. Y. H. Jiang, P. Emau, J. S. Cairns, L. Flanary, W. R. Morton, T. D. McCarthy, *et al.*, *AIDS Res. Hum. Retroviruses*, 2005, **21**, 207.
141. M. Galan, J. Sanchez-Rodriguez, M. Cangiotti, S. Garcia-Gallego, J. L. Jimenez, R. Gomez, *et al.*, *Curr. Med. Chem.*, 2012, **19**, 4984.
142. D. Tyssen, S. A. Henderson, A. Johnson, J. Sterjovski, K. Moore, J. La, *et al.*, *PLoS One*, 2010, **5**, e12309.
143. E. V. Cordoba, E. Arnaiz, F. J. de la Mata, R. Gomez, M. Leal, M. Pion, *et al.*, *AIDS*, 2013, **27**, 2053.
144. M. Witvrouw, V. Fikkert, W. Pluymers, B. Matthews, K. Mardel, D. Schols, *et al.*, *Mol. Pharmacol.*, 2000, **58**, 1100.
145. R. Rupp, S. L. Rosenthal and L. R. Stanberry, *Int. J. Nanomed.*, 2007, **2**, 561.
146. S. Telwatte, K. Moore, A. Johnson, D. Tyssen, J. Sterjovski, M. Aldunate, *et al.*, *Antiviral Res.*, 2011, **90**, 195.

147. I. McGowan, K. Gomez, K. Bruder, I. Febo, B. A. Chen, B. A. Richardson, *et al.*, *AIDS*, 2011, **25**, 1057.
148. C. F. Price, D. Tyssen, S. Sonza, A. Davie, S. Evans, G. R. Lewis, *et al.*, *PLoS One*, 2011, **6**, e24095.
149. P. J. Klasse, R. Shattock and J. P. Moore, *Annu. Rev. Med.*, 2008, **59**, 455.
150. M. M. Lederman, R. S. Veazey, R. Offord, D. E. Mosier, J. Dufour, M. Mefford, *et al.*, *Science*, 2004, **306**, 485.
151. C. Herrera and R. J. Shattock, *Curr. HIV Res.*, 2012, **10**, 42.
152. Q. Abdool Karim, S. S. Abdool Karim, J. A. Frohlich, A. C. Grobler, C. Baxter, L. E. Mansoor, *et al.*, *Science*, 2010, **329**, 1168.
153. R. E. Berger, *J. Urol.*, 2011, **185**, 1729.
154. D. W. Chan, in *Immunoassay Automation: a Practical Guide*, ed. D. W. Chan, Academic Press, San Diego, CA, 1992, pp. 11–13.
155. M. I. Montanez, C. Mayorga, M. J. Torres, M. Blanca and E. Perez-Inestrosa, *Nanomedicine*, 2011, **7**, 682.
156. A. J. Ruiz-Sanchez, M. I. Montanez, C. Mayorga, M. J. Torres, N. S. Kehr, Y. Vida, *et al.*, *Curr. Med. Chem.*, 2012, **19**, 4942.
157. P. Singh, F. Moll III, S. H. Lin, C. Ferzli, K. S. Yu, R. K. Koski, *et al.*, *Clin. Chem.*, 1994, **40**, 1845.
158. S. S. Mark, N. Sandhyarani, C. Zhu, C. Campagnolo and C. A. Batt, *Langmuir*, 2004, **20**, 6808.
159. M. Yang, E. M. Tsang, Y. A. Wang, X. Peng and H. Z. Yu, *Langmuir*, 2005, **21**, 1858.
160. H. C. Yoon, M. Y. Hong and H. S. Kim, *Anal. Biochem.*, 2000, **282**, 121.
161. S. Capaldi, R. C. Getts and S. D. Jayasena, *Nucleic Acids Res.*, 2000, **28**, E21.
162. T. W. Nilsen, J. Grayzel and W. Prensky, *J. Theor. Biol.*, 1997, **187**, 273.
163. H. Kobayashi, S. Kawamoto, S. K. Jo, H. L. Bryant Jr, M. W. Brechbiel and R. A. Star, *Bioconjug. Chem.*, 2003, **14**, 388.
164. H. M. Striebel, E. Birch-Hirschfeld, R. Egerer, Z. Foldes-Papp, G. P. Tilz and A. Stelzner, *Exp. Mol. Pathol.*, 2004, **77**, 89.
165. R. Benters, C. M. Niemeyer, D. Drutschmann, D. Blohm and D. Wohrle, *Nucleic Acids Res.*, 2002, **30**, E10.
166. R. Benters, C. M. Niemeyer and D. Wohrle, *ChemBioChem*, 2001, **2**, 686.
167. T. Selvaraju, J. Das, S. W. Han and H. Yang, *Biosens. Bioelectron.*, 2008, **23**, 932.
168. N. Zhu, H. Gao, Y. Gu, Q. Xu, P. He and Y. Fang, *Analyst*, 2009, **134**, 860.
169. A. Bosnjakovic, M. K. Mishra, H. J. Han, R. Romero and R. M. Kannan, *Anal. Chim. Acta*, 2012, **720**, 118.
170. A. Syahir, K. Y. Tomizaki, K. Kajikawa and H. Mihara, *Langmuir*, 2009, **25**, 3667.
171. J. M. J. Fréchet and D. A. Tomalia (eds), *Dendrimers and Other Dendritic Polymers*, Wiley, New York, 2001.
172. M. K. Borucki, J. Reynolds, D. R. Call, T. J. Ward, B. Page and J. Kadushin, *J. Clin. Microbiol.*, 2005, **43**, 3255.
173. Y. Li, Y. T. Cu and D. Luo, *Nat. Biotechnol.*, 2005, **23**, 885.

174. H. Maeda, G. Y. Bharate and J. Daruwalla, *Eur. J. Pharm. Biopharm.*, 2009, **71**, 409.
175. H. Sarin, A. S. Kanevsky, H. Wu, A. A. Sousa, C. M. Wilson, M. A. Aronova, *et al.*, *J. Transl. Med.*, 2009, 7, 51.
176. D. A. Tomalia, L. A. Reyna and S. Svenson, *Biochem. Soc. Trans.*, 2007, **35**, 61.
177. S. D. Konda, M. Aref, S. Wang, M. Brechbiel and E. C. Wiener, *MAGMA*, 2001, **12**, 104.
178. A. Quintana, E. Raczka, L. Piehler, I. Lee, A. Myc, I. Majoros, *et al.*, *Pharm. Res.*, 2002, **19**, 1310.
179. S. D. Swanson, J. F. Kukowska-Latallo, A. K. Patri, C. Chen, S. Ge, Z. Cao, *et al.*, *Int. J. Nanomed.*, 2008, **3**, 201.
180. G. P. Yan, B. Hu, M. L. Liu and L. Y. Li, *J. Pharm. Pharmacol.*, 2005, **57**, 351.
181. D. G. Udugamasooriya, S. P. Dineen, R. A. Brekken and T. Kodadek, *J. Am. Chem. Soc.*, 2008, **130**, 5744.
182. S. Langereis, J. Keupp, J. L. van Velthoven, I. H. de Roos, D. Burdinski, J. A. Pikkemaat, *et al.*, *J. Am. Chem. Soc.*, 2009, **131**, 1380.
183. J. A. Pikkemaat, R. T. Wegh, R. Lamerichs, R. A. van de Molengraaf, S. Langereis, D. Burdinski, *et al.*, *Contrast Media Mol. Imaging*, 2007, **2**, 229.
184. H. Kobayashi, S. Kawamoto, M. Bernardo, M. W. Brechbiel, M. V. Knopp and P. L. Choyke, *J. Control. Release*, 2006, **111**, 343.
185. R. C. Orth, J. Bankson, R. Price and E. F. Jackson, *Magn. Reson. Med.*, 2007, **58**, 705.

CHAPTER 8

Bicellar Systems: Characterization and Skin Applications

GELEN RODRÍGUEZ,[a] LUCYANNA BARBOSA-BARROS,[a] MERCEDES CÓCERA,[a] LAIA RUBIO,[a] CARMEN LÓPEZ-IGLESIAS,[b] ALFONS DE LA MAZA*[a] AND OLGA LÓPEZ[a]

[a] Departament de Tecnologia Química i de Tensioactius, Institut de Química Avançada de Catalunya (IQAC), Consejo Superior de Investigaciones Científicas (CSIC), C/Jordi Girona 18-26, 08034 Barcelona, Spain; [b] Centres Científics i Tecnològics, Universitat de Barcelona, Parc Científic de Barcelona, C/Josep Samitier 1–5, 08028 Barcelona, Spain
*Email: amresl@cid.csic.es

8.1 Introduction

The structure of bilayered discoidal micelles – also known as 'bicelles' – has been proposed in aqueous solutions containing mixtures of long- and short-chain lipids. This fascinating category of lipid assemblies has been increasingly utilized in several research fields. The fact that they form disk-shaped aggregates makes them suitable model systems for high-resolution NMR studies of membrane-interacting peptides. Bicelles are formed when long-chain lipids are brought in contact with surfactant molecules. Long-chain lipids alone form lipid bilayers, whereas surfactant molecules on their own form micelles. When they are mixed, lyotropic mesophases are observed

RSC Nanoscience & Nanotechnology No. 34
Soft Nanoparticles for Biomedical Applications
Edited by José Callejas-Fernández, Joan Estelrich, Manuel Quesada-Pérez and Jacqueline Forcada
Published by the Royal Society of Chemistry, www.rsc.org

that combine the properties of both bilayers and micelles. In the simplest case, such a bicellar phase is made up of disk-like aggregates where a central bilayer patch is enclosed by a 'rim' of surfactant molecules. However, this simple scheme does not apply to all bicellar phases. The presence of natural lipids and the fact that the surface of the central lipid region is more-or-less flat eliminate some of the limitations associated with micelles.

When preparing bicelles, three main traits should be taken into account. First, they contain lipid bilayers with very similar properties to those found in biological membranes. Second, these bilayers form flat patches rather than having a more or less pronounced curvature found in vesicles. Third, they can potentially be macroscopically aligned by an external magnetic field. This last trait is especially relevant for NMR studies.

An important contribution in the design and application of bicellar structures was made in 1992 by Sanders and Schwonek,[1] who demonstrated that the short-chain phospholipid dihexanoylphosphatidylcholine (DHPC) could be used to replace the surfactant CHAPSO, 3-[(3-cholamidopropyl)di-methylammonio]-2-hydroxy-1-propanesulfonate, eliminating the need for a potentially disadvantageous non-lipid surfactant. In the same contribution, it was shown that long-chain phospholipids and DHPC in bicelles are spatially separated, presumably into bilayered patches and surfactant rims. A first comprehensive review on bicelles, especially with regard to their potential in the study of membrane proteins, was published in 1994[2] and has since been cited by most studies employing bicelles. Remarkably, the term 'bicelles' was not introduced until a year later[3] to denote 'binary, bilayered, mixed micelles bearing a resemblance to the classical model for bile-salt phosphatidylcholine aggregates.'[4]

Considering the above definition, the distinguishing structural feature of a bicelle is a central planar bilayer formed by the long-chain phospholipid, surrounded by a rim of short-chain phospholipid or surfactant that shields the long-chain lipid tails from water. A scheme of this structure is shown in Figure 8.1.

For the long-chain phospholipid component, dimyristoylpho-sphatidylcholine (DMPC) is often used and it can be combined with phospholipids that have identical chain lengths but different headgroups [*e.g.* dimyristoylphosphatidylserine (DMPS) or dimyristoylphosphatidylglycerol (DMPG)] to alter the charge characteristics of the interface and to provide versatility in phospholipid composition.[5] Bicelles can also be prepared with dipalmitoylphosphatidylcholine (DPPC) or dilaurylphosphatidylcholine (DLPC) to vary the total bilayer thickness. As mentioned, the rim can be composed of either a bile-salt derivative surfactant such as CHAPSO or a short-chain phospholipid such as DHPC.[6,7] Since the long-chain phospholipids in bicelles are sequestered into the planar core region, which is devoid of short-chain phospholipids or surfactants, the core region of the bicelles mimics a section of natural membrane much better than standard micelles. Micelles have very little planar surface area and the surfactant is usually uniformly dispersed throughout the phospholipid. Hence the bicelles may

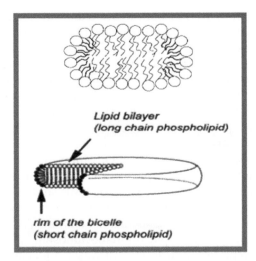

Figure 8.1 Schematic representation of a bicelle, showing the flat bilayer region composed of long-chain phospholipid, surrounded by a rim of short-chain phospholipid.

provide a better membrane model to facilitate NMR structural studies of many membrane-associated biomolecules. In this chapter, we discuss some of the most exciting applications of bicelles in structural and dynamic studies of membrane proteins, paying special attention to their suitability for skin-related applications

8.2   Bicelles and Bicellar Systems: Key Parameters

The size and physicochemical properties of bicelles can be varied as a function of the long-chain to short-chain phospholipid molar ratio (q) and the total phospholipid concentration expressed as weight/weight (c_L) to provide a variety of similar bicelle phases that are amenable to different types of biological studies. At high ratios ($q > 3$) and high total phospholipid concentration ($c_L = 15$–25%), bicelles form discoidal lipid aggregates of approximately 4080 kDa and 50 nm diameter.[8] These phospholipid mixtures spontaneously align in magnetic fields of >2 T [Figure 8.2(A)] at temperatures above the transition temperature (T_m) of the long-chain phospholipid. The transition temperature is the temperature at which phospholipids undergo a transition from the gel to liquid phase.

As mentioned, this property has made it possible to use solid-state NMR techniques to examine both the orientation of a peptide associated with the bicelle with respect to the bicelle surface and the effects of the peptide on lipid packing within the bicelle.[9–11] In this phase (high q), DMPC bicelles have been shown to form stacks of perforated lamellae, where the bicelles form edge-to-edge contacts, whereas, below T_m, these phospholipid mixtures have been shown to contain the standard bicelle morphology.[12,13]

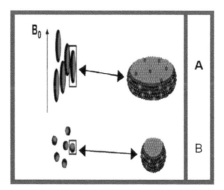

Figure 8.2   Schematic representation of bicelle alignment in a magnetic field. Large bicelles ($q > 3$ and $c_L = 15$–25% w/w) align with their normals oriented perpendicular to the field (A) while small bicelles ($q < 1$ and $c_L = 10$–15% w/w) are unaligned (B).

As the amount of short-chain phospholipid is increased ($q < 1$ and $c_L = 5$–15%), bicelles form an unaligned phase [Figure 8.2(B)] that is suitable for high-resolution NMR studies.[14,15] Several complementary techniques have been used to show that these bicelles, which are much smaller than their alignable counterparts (252 kDa and 8 nm diameter), remain discoidal with segregated lipid pools.[16,17] This is of the same morphology as that observed for bicelles in the aligned phase. These isotropic bicelles exist over a wide range of sample conditions and have a viscosity amenable to solution-state studies of peptides and proteins.[14] Their isotropic nature makes them suitable for high-resolution NMR studies of protein structure. These isotropic bicelles were found to be stable over a wide range of phospholipid ratios ($q = 0.05$–0.5) and temperatures (5–37 °C) and maintained a free (not bicelle-associated) short-chain phospholipid concentration of 7 mM for DHPC. If the total phospholipid concentration is between 1 and 5% w/w, the short-chain phospholipid dissociates from the bicelle to maintain the free short-chain phospholipid concentration at 7 mM. This increases q, which, in turn, leads to a doubling of the bicelle size over this concentration range. At $c_L < 1\%$, the bicelles undergo a drastic change in morphology, resulting in all of the short-chain phospholipids leaving the bicelle, causing the long-chain phospholipids to form a large, soluble lipid aggregate, presumably vesicular in structure.[17]

Hence bicelles offer a unique system in which different techniques can be employed to examine the interaction of proteins and model membranes. The aligned-phase bicelles, $q > 3$, allow the use of solid-state NMR techniques to determine the orientation of the protein and to determine the orientation of the phospholipids. The isotropic bicelles can be used to conduct NMR structural studies on the enzyme. In both cases, the protein sees the identical structure and environment so the results should be directly comparable.

The last requirement for a membrane model is that enzymes that act on native membranes are able to act on the bicelles in the same fashion. Previous studies of diacylglycerol kinase (DAGK) activity in various large bicelle systems ($q > 2.5$) have shown that the enzyme activity was comparable to those found in mixed micelles and vesicles. We present data below that indicate that small isotropic bicelles ($q = 0.5$) also provide an excellent substrate for phospholipase A2, a lipolytic enzyme. Hence the isotropic bicelles appear to be a good membrane model for at least this enzyme. The aligned phase can be used to determine the orientation of the enzyme at the surface. Moreover, the isotropic phase can be used to analyse the structure of the enzyme and to measure the enzyme activity.

8.3   Morphological Transformations

This section includes studies of the morphological transformations of two specific bicellar systems due to the changes in their lipid composition, temperature and time after preparation. These systems are the classic one formed by DMPC and DHPC and another formed by DPPC and also by DHPC. In general, the alignment of the systems has been studied using ^{31}P NMR spectroscopy. The characterization of the structural changes produced in bicelles by the effect of different factors is focused upon, combining different techniques such as: small-angle X-ray scattering (SAXS), dynamic light scattering (DLS) and freeze fracture electron microscopy high-pressure freezing (FFEM-HPF) (see Chapter 2).

8.3.1   DMPC–DHPC Bicelles

8.3.1.1   *Effect of Temperature on the Physicochemical Characteristics of Bicelles*

In recent years, diverse studies have shown the morphology and phase diagram of DMPC–DHPC bicelles to be very complex and they are subject to some debate.[18–21] Nieh and co-workers proposed a model for bicelles with and without lanthanide cations (Ln^{3+}).[12,19] According to this model, at temperatures below the T_m of the DMPC, the resulting bicelles were disk shaped.[12] As the temperature increased (from temperatures at which bicelles are in the gel phase to those for a liquid crystalline phase), the bicelles fused together in an end-to-end manner to form lamellar sheets with perforated holes that were lined with DHPC. Further increases in temperature caused phase separation with the formation of DHPC-rich mixed micelles and DMPC-rich oriented lamellae, and the DHPC-rich mixed micelles became incorporated into the oriented bilayers at even higher temperatures.[22] In the absence of Ln^{3+}, bicelles were disk shaped in the gel phase and chiral nematic in the liquid crystalline phase, which were described as worm-like micelles, and at higher temperatures were multilamellar vesicles.[19] Additionally, recent studies have demonstrated that the inclusion of a small amount of charged lipids eliminated the appearance of the worm-like or ribbon phase.[23] Hereafter, the use of the term 'bicelle' in most studies refers only to the sample composition and not to the disk-shaped morphology. A comparative study of alignable and non-alignable bicelles associated with the temperature changes of the system could be helpful in clarifying the morphological transitions involved in this process.

Barbosa-Barros and co-workers published studies[24,25] in which high-pressure freeze fixation and freeze fracture electron microscopy techniques[26,27] were combined with ^{31}P NMR spectroscopy (see Chapter 2) to correlate the magnetic alignment of DMPC–DHPC bicelles, morphological transitions and the changes in the temperature of the system. Two lipid molar ratios ($q = 2$ and 3.5) were investigated. Direct observations showed that at $q = 2$ the bicelles were very fluid and transparent at 20 °C, but as the temperature increased, the sample became viscous with a gel-like appearance, maintaining the transparency until 60 °C. At $q = 3.5$, the system became fluid and transparent only at temperatures below the T_m of DMPC (23.5 °C).[28] At 25 °C the sample became viscous, at 30 °C it was viscous and slightly milky and from 40 to 60 °C it was milky and fluid. Figure 8.3 plots the ^{31}P NMR spectra, in which the alignment of $q = 2$ and $q = 3.5$ bicelles is shown in the temperature range 20–60 °C[24] (see Chapter 2 for the interpretation of the NMR spectra).

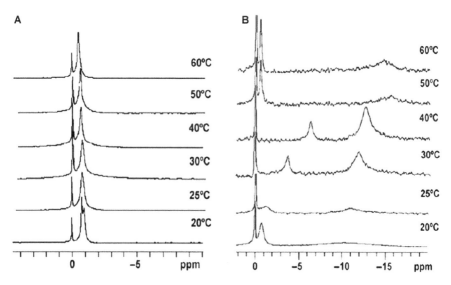

Figure 8.3   ^{31}P NMR spectra under ^{1}H decoupling as a function of temperature of the $q = 2$ system (A) and the $q = 3.5$ system (B). The temperature is indicated and was increased smoothly starting from 15 °C with an equilibration time of 15 min between different temperatures – the reference resonance at 0 ppm arises from 85% phosphoric acid. The water content in both systems was 80%.
Reprinted with permission from Ref. 24. Copyright 2009 Royal Microscopical Society.

The reference peak is shown at 0 ppm. The spectra of the $q = 2$ bicelles [Figure 8.3(A)] showed isotropic behaviour at all temperatures. At 20 °C, the spectrum was a doublet near −1 ppm, reflecting the slightly different environments of the DMPC sites (bilayer) and DHPC sites (rim) on the bicelle structure.[29] At 25 °C, these peaks overlapped. The single peak shifted slightly downfield with increase in temperature until 60 °C, showing a characteristic behaviour of non-alignable phospholipid mixtures.[2,8] The spectra of $q = 3.5$ bicelles (B) showed considerable variations depending on the temperature. Thus, at 20 °C a single resonance peak was observed at −0.77 ppm and a broad protuberance was seen at around −10 ppm, suggesting that the sample had some orientation with respect to the magnetic field. The presence of two well-differentiated resonances in bicellar samples was reported by Bax and Gaemers to be a result of the alignment occurring with the bilayer normal oriented perpendicular to the magnetic field.[13] The high-field resonance would correspond to DMPC oriented bilayers and the low-field resonance to DHPC forming disks or hole edges.[30] Just above the T_m of DMPC, at 25 °C, the spectrum showed two broad resonances at −1.19 and −10.82 ppm. From this temperature, the observed resonances move towards the right and the lines become more intense, indicating an improvement of the magnetic alignment, which reaches a maximum at 40 °C. The integration of the high- and low-field peaks gave a relative intensity of 3.3. This value was close to the q ratio of this sample (3.5), confirming that the downfield peak corresponded to the DHPC and the upfield peak to the DMPC. The gradual upfield shift of the DMPC peak reflected an increase in the bilayer order with respect to the magnetic field and the upfield shift of the DHPC peak was related to the increasing presence of DHPC in the DMPC bilayer. That is, the increase in temperature promoted the miscibility of the DHPC molecules in the DMPC bilayer, which resulted in the migration of the DHPC molecules from the edges of the bicelles to the bilayer area. Hence the absence of sufficient DHPC molecules to fill the bicelle edges led to the fusion of bilayers, which increased the bicellar diameters and improved the alignment until a certain point. From 50 °C, the sample orientation was lost and a typical isotropic low-field peak at around −0.54 ppm and a broad protuberance at around −15 ppm were present. According to Triba *et al.*,[29] this change in the spectra was characteristic of a phase transition from bicellar aggregates to larger structures with slow motion. To elucidate the morphological characteristics of these bicellar systems, electron microscopy experiments were performed. To this end, alignable $q = 3.5$ and non-alignable $q = 2$ samples, below the T_m of DMPC and above this temperature, were analysed. Bicelles were high-pressure frozen (HPF) at −150 °C from sample temperatures of 20 and 40 °C. These temperatures were chosen to compare the morphologies of these alignable and non-alignable samples, to correlate further with the corresponding ^{31}P NMR data.

The HPF method was developed to allow for the freezing of hydrated material from a defined state.[27] The great advantage of this method is that sample pretreatment, which frequently leads to the formation of artefacts, is not required. The freeze fracture replicas were visualized mainly by a cryo-scanning electron microscopy (SEM) technique.[25] Replica cleaning for transmission electron microscopy (TEM) is tedious and sometimes difficult, large replicas tend to disintegrate during cleaning and the field of view is restricted by the grid bars. These problems are circumvented by cryo-SEM,[31] in which no replica cleaning is required. In SEM, the sample does not have to be transparent to the electron beam and bulk samples can be analysed. Using the combination of HPF and cryo-SEM, samples difficult to clean for TEM were imaged and large areas were investigated.

All investigated samples were free of visible ice crystal artefacts. The cryo-SEM images for $q = 2$ and 3.5 bicelles are shown in Figures 8.4 and 8.5, respectively. At 20 °C, a temperature below the corresponding T_m, small aggregates were observed in both cases

Figure 8.4 shows small rounded aggregates with diameters of around 20 nm (arrows). In Figure 8.5, the aggregate diameters of $q = 3.5$ bicelles were about 40 nm (slightly larger than those of $q = 2$ bicelles), also being visualized as a discoidal structure shape. Two different zones in the discoidal structures are observed. The arrows indicate aggregates viewed face-on and the arrow heads point to aggregates viewed edge-on.

Comparison of these images with the corresponding ^{31}P NMR spectra revealed a relation between the aggregate structure and its magnetic orientation. The rounded shape aggregates of the $q = 2$ bicelles image did not allow for magnetic alignment, whereas the two well-differentiated areas of the disks visualized in the $q = 3.5$ bicelles micrograph accounted for the onset of orientation detected in Figure 8.1(B) by the presence of the two separated resonances.

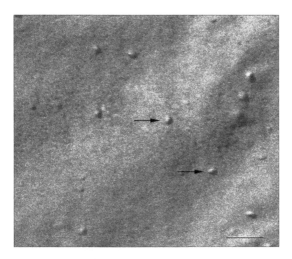

Figure 8.4   Electron micrograph of the $q = 2$ system at 20 °C. The arrows indicate small bicelles of about 20 nm. Bar = 100 nm.
Reprinted with permission from Ref. 24. Copyright 2009 Royal Microscopical Society.

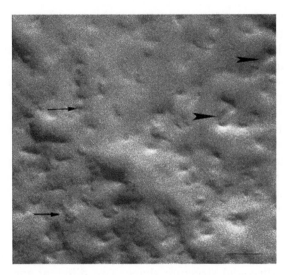

Figure 8.5   Electron micrograph of the $q = 3.5$ system at 20 °C. The arrows denote
disks viewed face-on and the arrowheads denote disks viewed edge-on.
Bar $= 50$ nm.
Reprinted with permission from Ref. 24. Copyright 2009 Royal Microscopical Society.

At a temperature of 40 °C, large aggregates were observed for both the
$q = 2$ and 3.5 systems. Figure 8.6 shows an image for $q = 2$ with elongated
aggregates of about 2000 nm. Similar results were reported by van Dam *et al.*
for DMPC–DHPC systems at $q = 2$, low-concentration samples ($c_L = 3\%$), at
temperatures of 36 and 46 °C using a cryo-TEM technique.[32]

In Figure 8.7, a cryo-SEM image of the sample with $q = 3.5$ at 40 °C is
displayed, in which extended areas of stacked lamellar sheets (arrows) were
observed. The presence of multilamellar vesicles was also detected.

Figure 8.8 depicts a detail of one of these multilamellar vesicles with a size
of around 1000 nm. The large structures observed in this micrograph account for the milky appearance of this sample. This lamellar phase probably
represents the most ordered phase of this sample since the best alignment
in ^{31}P NMR experiments was obtained at this temperature.

The current discussion of the morphology of the DMPC–DHPC aligned
aggregates may be summarized in two hypotheses. Both coincide in the
morphological transitions of the aggregates from disks to elongated micelles, perforated lamellar sheets and mixed multilamellar vesicles. Also,
both hypotheses take into account the initial increase in sample viscosity
and the subsequent decrease in this viscosity caused by the continuous increase in temperature. Nevertheless, they diverge in the morphology of the
structures when the sample shows magnetic alignment. One of the theories
proposes that elongated aggregates correspond to the morphology of the
magnetic aligned phase and are present when the sample viscosity increases.[33] The other claims that the perforated lamellar sheets correspond to

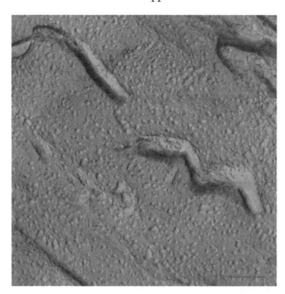

Figure 8.6   Electron micrograph image of the $q = 2$ system at 40 °C showing isolated elongated aggregates. Bar = 1.00 μm.
Reprinted with permission from Ref. 24. Copyright 2009 Royal Microscopical Society.

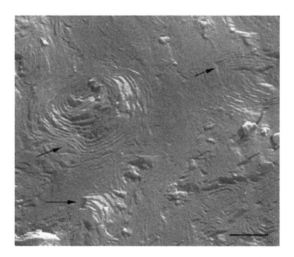

Figure 8.7   Electron micrograph image of the $q = 3.5$ system at 40 °C showing stacked bilayer sheets (arrows). Bar = 500 nm.
Reprinted with permission from Ref. 24. Copyright 2009 Royal Microscopical Society.

the morphology of the magnetic aligned phase. The formation of these lamellar structures would take place when the sample viscosity decreases and the sample appearance is milky.[30] Moreover, in a detailed ^{31}P NMR study,

Triba *et al.* reported the compatibility of both elongated and perforated lamellar structures with the NMR aligned spectra, although they did not disprove the discoidal structure as possible in the aligned phase.[29]

The results obtained by Barbosa-Barros *et al.*[24] are in accordance with some points in the mentioned theories. In the experiments with $q = 2$ bicelles, we observed an increase in sample viscosity when the sample was heated. The cryo-SEM images of this sample at 40 °C showed elongated structures, as described by Harroun *et al.*[33] (Figure 8.6). Nevertheless, this sample did not align in magnetic fields [Figure 8.3(A)]. At this q ratio, the proximity of the DMPC and DHPC molecules in the elongated structures formed at 40 °C and also in the spherical objects formed at 20 °C did not allow for an anisotropic environment. Hence the two phosphorus signals in the ^{31}P NMR spectrum overlap and a single resonance is seen. In this way, although the increase in sample viscosity can be attributed to the formation of elongated aggregates, these structures did not correspond to the morphology of the magnetic aligned phase.

On the other hand, evidence for the morphology of aligned aggregates was obtained by the analysis of the $q = 3.5$ bicelle micrographs. This sample also showed an increase in viscosity when the temperature increased, but when the best alignment was detected (40 °C), the sample had a milky appearance. The cryo-SEM images at this temperature (Figures 8.7 and 8.8) showed areas with plenty of stacked bilayer sheets and some multilamellar liposomes.

The closing up of the bilayer sheets in vesicles is the result of a spontaneous process probably caused by the reduction of the energetic cost of bending by the effect of temperature. The existence of a mixed system

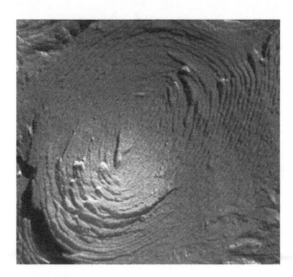

Figure 8.8   Electron micrograph showing a multilamellar vesicle present in the $q = 3.5$ system at 40 °C. Bar = 250 nm.
Reprinted with permission from Ref. 24. Copyright 2009 Royal Microscopical Society.

accounts for the broad peaks obtained in the spectrum at 40 °C. The multilamellar vesicles probably interfere with the perfect alignment of the structures which would produce narrow resonance peaks. In any case, the segregation of DMPC and DHPC in the $q = 3.5$ aggregates accounts for the orientation of this system in the magnetic field. Even below T_m, the discoidal shape of these aggregates (Figure 8.5) caused the onset of orientation reflected in the appearance of the broad resonance at -10 ppm in the NMR spectra at 20 °C [Figure 8.3(B)]. An improvement in the alignment of this sample was observed with increase in temperature, reaching a maximum at 40 °C, the temperature at which stacked lamellar sheets were mainly observed. From 50 °C, the lost magnetic alignment and the resonance signal at -15 ppm probably indicated the complete closing up of the aligned bilayer sheets into isotropic vesicles. From these data, it may be assumed that the combination of NMR, HPF and FFEM techniques was appropriate for evaluating the relation between the magnetic alignment and the morphology of the two bicellar systems.

Concerning the isotropy of the non-alignable $q = 2$ bicelles, morphological evolutions from spherical to elongated aggregates were reported, verifying that despite their growth, these elongated aggregates were not able to orient in magnetic fields. For $q = 3.5$ bicelles, these samples presented anisotropy for magnetic orientation even below T_m, given that the morphology of the aggregates allows distinct phosphorus environments. As the temperature increased, the sample achieved best alignment and the morphology of the aggregates evolved from disks to lamellae. At the temperature of the best magnetic alignment, the aggregates were formed by stacked bilayer sheets and multilamellar liposomes.

The present data could be very useful for modulating the alignment of systems in magnetic fields and therefore optimizing the use of bicellar systems as membrane mimics.

8.3.1.2 Stability with Time

This section considers studies on the temporal stability of DMPC–DHPC bicelles based on the aforementioned techniques.[34–36] Figure 8.9 shows the ^{31}P NMR spectra for bicelles (A) 24 h and (B) 14 days after preparation.[37]

A doublet was observed with peaks at around -0.754 and -0.927 ppm [Figure 8.9(A)] and -0.764 and -0.935 ppm [Figure 8.9(B)], in addition to the reference peak at 0.000 ppm. As mentioned, this spectral appearance was reported previously by Triba *et al.* for small disks in rapid rotation.[29] The reproducibility of the bicelle characteristic resonances after 24 h and 14 days indicated good structural stability of the sample at least for 2 weeks. Figure 8.10 plots the SAXS curve of bicelles 24 h after preparation.[37]

The fact that no differences were detected between this scattering curve and that obtained 14 days after preparation (data not shown) also corroborates the stability of the system. The lamellar repeated distance d was estimated from the analysis of the peak by Bragg's law and was attributed to

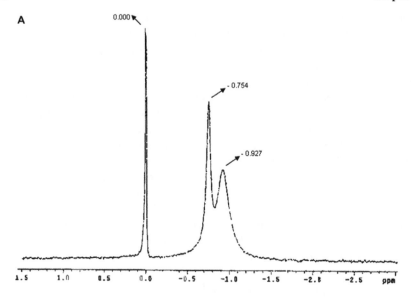

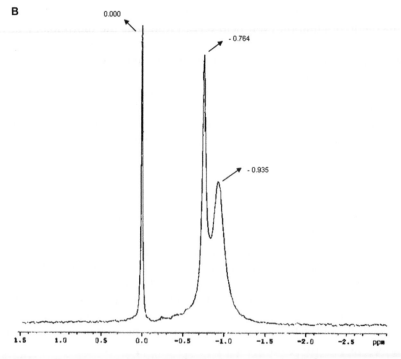

Figure 8.9 ^{31}P NMR spectra at 20 °C for DMPC–DHPC bicelles ($q = 2$ and 80% water content), (A) 24 h and (B) 14 days after preparation. Reprinted with permission from Ref. 37. Copyright 2010 Elsevier.

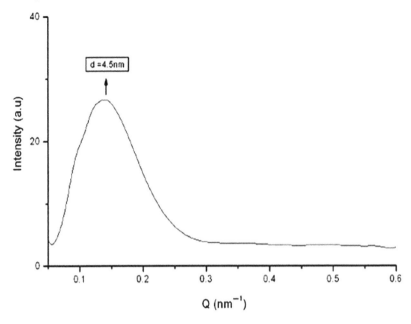

Figure 8.10 SAXS curve at 20 °C for bicelles ($q = 2$, 80% water content) 24 h after preparation.
Reprinted with permission from Ref. 37. Copyright 2010 Elsevier.

the bilayer thickness in a similar way to studies with liposomes and other bilayer models.[38] A repeat distance value of $d = 4.5$ nm was obtained in both spectra (24 h and 14 days after preparation), indicating that no alteration in the bilayer occurred during the 14 days of the experiment.

The size distribution curves obtained by DLS exhibited a monomodal distribution with a hydrodynamic diameter (HD) of 15.9 nm and a poly-dispersity index (PI) of 0.211, 24 h after preparation.[37] Values for the same sample 14 days after preparation were very similar (HD = 16.3 nm and PI = 0.303). The HD obtained by DLS is that of a hypothetical hard sphere that diffuses with the same speed as the particle under experiment. As the bicelles present a flat appearance, the values of particle size obtained may be considered as an estimation of the structure's dimension. The particle size values obtained showed fairly good agreement with the structure sizes visualized by SEM. Figure 8.11 shows an SEM image of the $q = 2$ bicelles at 20 °C, 14 days after preparation.

The arrows indicate the small rounded bicelles of about 20 nm (bar = 100 nm) similar to those shown in an SEM image of $q = 2$ bicelles after preparation also at 20 °C (Figure 8.4). Similar structures obtained using other electron microscopy techniques have been reported.[24,26,28,39]

To summarize, the physicochemical characterization of the bicellar system with $q = 2$ indicated structural stability for at least 14 days after preparation.

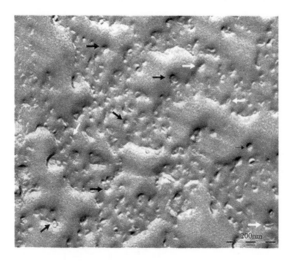

Figure 8.11   Electron micrograph of DMPC–DHPC $q = 2$ bicelles. The image de-
                   picts a zone full of bicelles. Black arrows denote disks viewed face-on
                   and white arrows denote disks viewed edge-on.
                   Reprinted with permission from Ref. 37. Copyright 2010 Elsevier.

8.3.2   DPPC–DHPC Bicelles

In this section, the characterization and the physicochemical properties of
bicellar systems formed by DPPC and the conventional DHPC are considered. In these systems, the q ratio was also adjusted to 3.5 in order to
decrease the concentration of DHPC and c_L was adjusted to 20%. Some
reasons are presented for introducing these changes. One reason is based
on the difference in the T_m values (23.5 °C for DMPC and 41.4 °C for
DPPC),[40,41] despite the increase of only two carbons in the long-chain lipid
(from 14 to 16). This results in significantly different polarities above and
below their T_m.[42] Furthermore, although bicelles formed by both lipids
exhibit similar dynamic behaviour, the DPPC lipid acyl chain displays a
lower degree of mobility, as evidenced by higher generalized order parameters throughout the acyl chain.[43] On the other hand, thinking of the
application of bicellar systems as colloidal carriers, the preservation of the
shape and size of bicelles would be a requirement. Hence it is necessary
that the T_m of lipids building bicelles exhibit higher values than the experimental temperature (for *in vivo* applications about 37 °C). The increase
in the q value to 3.5 (decrease in the relative proportion of DHPC) was
chosen in order to reduce the surfactant character of the lipid mixture,
making the model more appropriate for *in vivo* studies. Therefore, here we
modified the dimensions and morphology of DPPC–DHPC bicelles and the
bicelle–vesicle transition that takes place by the effect of dilution. Dilution
of the systems would be desirable in order to use bicelles as colloidal delivery systems.

8.3.2.1 Dimensions and Morphology

The characterization of the DPPC–DHPC bicelles is in general performed following the same methodology and techniques as used to characterize the DMPC–DHPC bicelles (Section 8.3.1).[24,44]

Applying the form factor squared of the simplified Gaussian model to SAXS, a bilayer thickness value of 5.4 nm was obtained.[24,44] The size distribution curve obtained by DLS shows an HD value of 11.3 nm and a PI of 0.072. This particle size provides a relative measurement of the structural dimensions. Therefore, these data should be interpreted taking into consideration also the electron microscopy images, in which direct visualization of the bicellar structures is obtained.

Figure 8.12 shows a cryo-SEM image at 37 °C, in which small discoidal aggregates about 15 nm in diameter are observed. In this image, the face (black arrows) and the edges (white arrows) of the disks are visualized. As a consequence, this result shows fairly good agreement with the DLS data, despite the different resolution of each technique.

8.3.2.2 Bicelle to Vesicle Transition States Obtained by Dilution

To monitor the effect of dilution on the structure of the bicelles, 2 mL of a bicelle sample ($q = 3.5$ and $c_L = 20\%$) was sequentially diluted with deionized water in seven steps (D1–D7) to obtain concentrations from 5 to 0.07%, and

Figure 8.12 Electron micrograph of DPPC–DHPC bicelles at 37 °C. The black arrows denote the disks in a face-on disposition and the white arrows correspond to the disks viewed edge-on.
Reprinted with permission from Ref. 44. Copyright 2008 American Chemical Society.

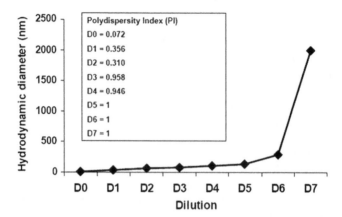

Figure 8.13   Evolution of the average particle size of the DPPC–DHPC bicelles with $q = 3.5$ by the effect of dilution as measured by DLS. The polydispersity indices are given in the inset.
Reprinted with permission from Ref. 44. Copyright 2008 American Chemical Society.

24 h after the dilutions the HD and PI of each diluted sample were measured by DLS at 37 °C. An overview of this process is plotted in Figure 8.13. The DLS curves of the diluted samples showed that the HDs of the structures increase upon dilution from about 11.3 nm (assigned to bicelle disks) to large-sized aggregates with HD >1 μm.

For a better understanding of the results, aside from the evolution of the average particle sizes, a detailed analysis of the scattering intensity was performed. The particle sizes obtained were separated into three groups: P1 included particle sizes in the range 10–100 nm, P2 included intermediate particle sizes ranging from 101 to 500 nm and P3 included particle sizes >500 nm. The HDs and the percentage of light scattered for each group are plotted separately as a function of the sequential sample dilution in Figure 8.14.

It is noteworthy that the intensity for small bicelles (P1 in Figure 8.14) diminished with dilution, whereas the HD increased. Intermediate aggregates (P2 in Figure 8.14) were present from the first dilution on and display a maximum of intensity and size at D3. From D4, the intensity of these aggregates diminishes and larger structures are detected (P3 in Figure 8.14), coexisting with P1 and P2 until D6. The high PI values (shown in Figure 8.13 from D3 onwards) are explained by the wide variety of aggregates present in these samples.

It is noteworthy that the percentage on the intensity curves does not represent the percentage of the structures present in the systems. In DLS, the HD is calculated from the intensity of the scattered light and gives information about the different particles present in the sample. From the Rayleigh approximation, it follows that the intensity is proportional to d^6 (d being the particle diameter); therefore, the contribution of the light

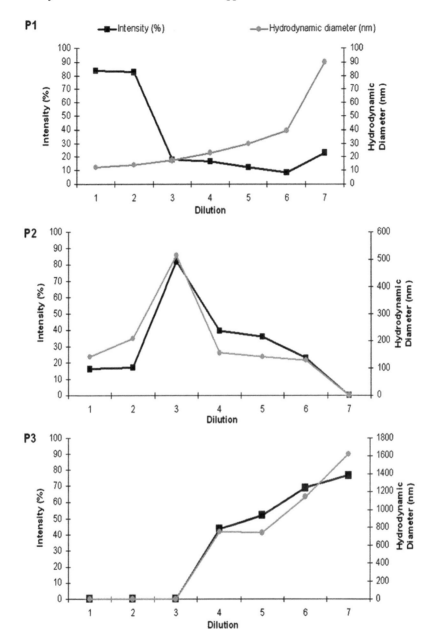

Figure 8.14 Evolution of the DLS scattered intensity and hydrodynamic diameter of the DPPC–DHPC bicelles upon dilution. P1 denotes the evolution of particles with sizes in the range 10–100 nm. P2 denotes intermediate-sized particles with diameters around 100–500 nm and P3 represents the larger particles of >500 nm.
Reprinted with permission from Ref. 44. Copyright 2008 American Chemical Society.

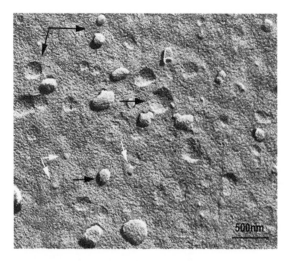

Figure 8.15 Micrograph of the diluted DPPC–DHPC bicelles (D5) at 37 °C. The black arrows denote vesicle structures of about 200–500 nm and the white arrows point to bicellar structures.
Reprinted with permission from Ref. 44. Copyright 2008 American Chemical Society.

scattered from small particles is relatively small compared with that of large particles that scatter much more light. Hence the intensity curves obtained indicate the appearance of larger aggregates by dilution but do not accurately quantify them. In general, the percentage intensity of these structures and their HD increase in each dilution (Figure 8.14). The coexistence of these large aggregates with the small bicelles is detected until the last dilution performed (D7).

In order to investigate the morphology of these systems, the diluted bicellar solutions were analysed by electron microscopy. A representative cryo-SEM image of sample D5 is shown in Figure 8.15.

This image reveals the presence of vesicles of about 200–500 nm (black arrows) together with small bicelles (white arrows). Comparing this micrograph with that of the original system (Figure 8.12), the increase in structures size and the variety of larger aggregates in the sample are noteworthy. This result is fully consistent with DLS data corroborating the transition of bicelles from disks to vesicles by the effect of dilution. This transition took place by progressive steps that implied the coexistence of different aggregate structures in the medium.

In general terms, although the classical bicelle model formed by DMPC and DHPC has been successfully used, another bicelle systems made up of lipids with different acyl chains, backbones or head-groups have offered alternatives for different studies.[5,45–48] As aforementioned, the DPPC is one of the most studied lipids for bilayer models due to its longer acyl chains and the bilayer thickness[49] and has the additional advantage of having a higher value of T_m.[50–52]

The phase transitions that occurred at T_m involve morphological changes in the bicellar structures, that is, from disks to cylindrical micelles to perforated lamellar sheets and mixed multilamellar vesicles.[33,53] Thus, a system of small size at physiological temperature that could be used, for instance, for *in vivo* applications would be the DPPC–DHPC $q = 3.5$ bicelles. At 37 °C, below the T_m of DPPC, these structures have dimensions of about 15 nm in diameter and 5.4 nm in thickness. These values are consistent with previously reported data for DPPC bilayer thickness[16,54] and bicellar disk dimensions.[55,56] The passage of these small bicelles through tissues such as the skin stratum corneum (SC) seems reasonable, considering that this region is formed by lipid lamellae with narrow interlamellar spaces (between 6 and 10 nm).[57]

The DLS and FFEM analyses of the bicellar diluted samples demonstrated the tendency of the bicellar aggregates to grow and form vesicles by the effect of dilution (Figures 8.14 and 8.15). It is noteworthy that the phase transitions that the bicelles underwent due to the variation of lipid concentration, temperature and phospholipid molar ratio[58] are very similar to those involved in the reconstitution of the surfactant–lipid micellar systems.[59–61] The transformation of these structures in vesicles has been widely discussed.[62–65] In earlier studies, we investigated kinetic aspects of this process.[66] A model for the micelle-to-vesicle transition proposed by Leng *et al.* described the rapid formation of disk-like aggregates and their growth and closure to form vesicles.[67] This model took into account the line tension dominating the bending energy. Certainly, the resemblance of surfactant-lipid micelles and phospholipid bicelles justifies their similar behaviour. DHPC solubilizes the DPPC bilayer, forming the bicellar structures in a similar way that a surfactant solubilizes lipid vesicles forming micelles. In addition, the high water solubility of DHPC accounts for the structural changes that bicelles underwent. When DHPC is removed from the bicelles upon dilution (q increases), the bilayers tend to fuse and the bicellar diameter increases. Morphological changes in the bicelle structure then take place. When sufficient DHPC is removed from the bicelles, the precipitation of large aggregates is visible, that is, phase separation occurs.[68]

8.4   Bicelle Suitability for Skin-Related Applications

In the last decade, liposomes have been intensively studied as drug carriers for topical delivery because of their capacity to enhance drug penetration into the skin,[69] their ability to improve therapeutic effectiveness[70] and a decrease in their drug side effects.[71] However, although several mechanisms of vesicle–skin interactions have been described, no evidence of intact vesicle penetration into the deeper skin layers has been found.[72] It was suggested that vesicles likely disintegrate in the skin's surface and their components disperse into the intercellular lipid matrix, where they mix with the SC lipids, modifying the lipid lamellae or inducing new vesicle-like structures.[73] In 2009, Karande and Mitragotri published a review of the enhancement of transdermal drug delivery *via* synergistic chemical action, employing solvent mixtures, microemulsions, eutectic mixtures, complex self-assembled vesicles and the inclusion of complexes.[74] Currently, new delivery systems are being developed to improve aggregate surfaces and their effects on skin. Dragicevic-Curic *et al.*[75] reported that loaded liposomal hydrogels were able to deliver active compounds in efficient doses into the SC and deeper skin layers. Alves *et al.*[76] reported a new drug model for *in vitro* skin penetration that consisted of semisolid topical formulations containing nanocarriers (nanospheres, nanocapsules and nanoemulsions). They demonstrated the influence of polymers and different types of nano-carriers (matrix, vesicular or emulsion) on drug penetration through human skin. Küchler *et al.*[77] similarly reported that nanoparticulate carrier systems, solid lipid nanoparticles and dendritic core–multishell nanotransporters were suitable skin drug delivery systems for hydrophilic agents.

The motivation for using bicelles in the skin arises from the resemblance of these structures to micelles and liposomes. Bicelles combine some of the most attractive characteristics of these systems and therefore present several advantages for skin applications compared with either micelles or lipo-somes. Bicelle structures contain a bilayer that allows for the incorporation of different molecules, but these structures are much smaller (~ 15–40 nm) than a regular liposome (~ 200 nm). This is due the presence of DHPC molecules on the edges of the structures, which control the diameter of the assembly. These molecules are responsible for discoidal-shaped bicelles and the formation of other structures, such as small vesicles. In fact, systems formed by long alkyl chain phospholipids and DHPC have been reported to produce reasonably monodisperse unilamellar vesicles that are thermo-dynamically stable, with radii ranging from 10 to 40 nm.[78–80] Several kinetic studies have shown that discoidal bicelle morphology is a precursor to small-vesicle morphology.[80] DHPC molecules solubilize the DMPC bilayer simi-larly to a surfactant; however, DHPC is a phospholipid with the same polar headgroup as DMPC. These two lipids differ only in the lengths of their hydrophobic chains. The use of systems composed only of lipids avoids damage to the skin barrier function caused by surfactants, which is char-acterized by breaking of the corneocyte envelopes and disorganization of the intercellular lipid structures.

Bicelle compositions have been adapted for use in the skin. As already mentioned, above the T_m of DMPC, DMPC–DHPC bicelles undergo phase transitions, changing from small aggregates of ~ 15–20 nm in size to structures larger than 500 nm. Because skin intercellular spaces lie in the range 6–10 nm and physiological skin temperatures are near 37 °C, this phase transition would present a handicap for skin penetration. DMPC–DHPC bicelles are probably not able to penetrate into SC intercellular spaces and their effects are limited to the skin's surface.

A different scenario is obtained if the long-chain phospholipid DMPC is replaced by another phospholipid with higher T_m, such as DPPC. This phospholipid has two additional carbons in its hydrophobic chain and a T_m

of 41.4 °C. At physiological temperatures, bicelles composed of DPPC and DHPC are still small aggregates. Some studies have found that at 37 °C, these structures have dimensions of ∼15 nm in diameter and ∼5.4 nm in thickness. These structures have been shown to penetrate through the intercellular spaces of the skin SC, affecting deeper internal layers.[44]

By varying bicelle lipid compositions and/or lipid ratios, novel systems with unique physicochemical characteristics can be obtained. The morphologies and sizes of bicellar structures may be varied to obtain a given effect, which may be superficial or profound depending on the particular application, allowing the modulation of bicelle penetration into the SC. These features provide bicelles with the potential to become new multi-functional, skin-compatible platforms.

8.4.1 Application of Bicelles to the Skin

The structure of the SC is very specific in composition and behaviour. The lipid matrix is formed by bilayers composed mainly of ceramides, choles-terol, fatty acids and small amounts of cholesterol sulfate and cholesterol esters. This structure ensures cohesiveness between the corneocytes and accounts for the permeability of the SC.[81,82] This exceptional composition and organization are not observed in other biological membranes, which in general consist predominantly of phospholipids as bilayers. Because the SC is remarkably devoid of phospholipids, its ability to form bilayers is sur-prising. The interactions of the systems with the SC are key factors in de-termining their potential for skin delivery and other skin-related applications

8.4.1.1 In Vitro *Interactions*

8.4.1.1.1 Microstructural Studies. Experiments have been reported in which human and pig SC samples were incubated with bicelles to study the effects of these systems on the SC microstructure. In one of these studies, fresh human SC samples were incubated with DMPC–DHPC bicelles with $c_L = 20\%$ and $q = 2$ for 18 h at 25 °C. Treated and untreated SC samples were cryofixed, cryosubstituted and visualized with TEM. No differences, including microstructural alterations and/or apparent damage, were observed in images of the treated SC [Figure 8.16(B) and (C)] com-pared with untreated samples [Figure 8.16(A)].

Given the small size of the bicelles (20 nm in diameter and 4.5 nm in thickness) and their bilayered structures, lipid dispersion through the SC lamellae and the reinforcement of the SC bilayer area would be expected. However, this was not observed, suggesting that these bicelles were not able to penetrate or disperse through the SC lipid area.[37] This result was clarified by Rodríguez *et al.*[83] using attenuated total reflectance Fourier transform infrared (ATR-FTIR) spectroscopy. They reported that the application of DMPC–DHPC $q = 2$ bicelles caused phase transitions in the SC lipid

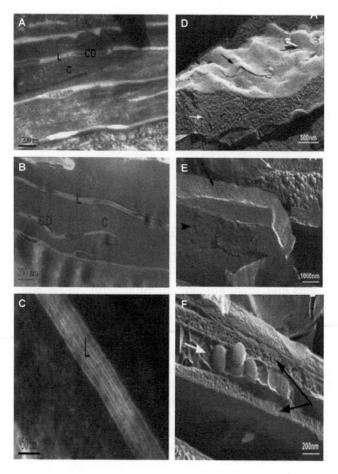

Figure 8.16   Freeze substitution transmission electron microscopy (FSTEM) images of (A) native SC and (B) and (C) SC treated with DMPC–DHPC
bicelles. In all cases the images show regular areas of corneocytes (C),
lipid intercellular spaces (L) and corneodesmosomes (CD). This figure
also shows cryo-SEM images of (D) native SC and (E, F) SC treated with
DPPC–DHPC bicelles. In (D), the white arrow indicates the corneocyte
area and the black arrow indicates the intercellular lipid area. In (E),
the white arrow shows vesicle structures with sizes of ~200 nm and
the arrowhead indicates the lamellar-like structures in the intercellular lipid areas. The black arrow shows a corneocyte area. (F) displays a
magnification of vesicles in the intercellular lipid area (white arrows)
between two cross-fractured corneocytes (black arrows).
Reprinted with permission from Ref. 44. Copyright 2008 American
Chemical Society.

conformation from the gel state to the liquid-crystalline state. This transition would be promoted by the incorporation of phospholipids from bicelles
in the SC lipid lamellar structure. This process involves an increase in the
fluidity and/or disorder of the lipids. An analysis of phosphate vibrations

only detected effects from DMPC–DHPC bicelles on lipids of the outer layer of the SC, suggesting that the majority of DMPC–DHPC bicelles remained in the outermost part of the tissue. This is likely explained because Barbosa-Barros et al.[37] found no microstructural differences between untreated and treated SC. Transitions occurring mainly on the SC surface do not imply structural modifications that can be visualized in electron microscopy experiments.

In another study, DPPC–DHPC bicelles with $c_L = 10\%$ and $q = 3.5$ were incubated with fresh pig skin SC samples for 18 h at 37 °C. As mentioned in Section 8.3, this DPPC–DHPC system was specially developed to obtain improved effects on the SC microstructure and better skin penetration under physiological conditions. Because the T_m of DPPC is 41.4 °C, the structures do not undergo phase transitions at physiological temperatures (~ 37 °C) and small bicelle structures are favoured during the incubation process. After incubation, both treated and untreated SC samples were HPF treated from an initial temperature of 37 °C, freeze fractured and observed using cryo-SEM.[44] These bicelles produced different results in the SC than in the DMPC–DHPC $q = 2$ system.

Figure 8.16 also shows cryo-SEM images of native and treated SC at two magnifications [images (D), (E) and (F), respectively]. Comparison of these images reveals that the DPPC–DHPC system penetrated and interacted with the SC, forming lipid vesicles and new lamellar-like structures. To study this phenomenon, Rodriguez et al.[84] studied the effects of DPPC–DHPC on SC lipids using ATR-FTIR spectroscopy coupled with a tape-stripping methodology. Analysis of the lipid organization in terms of chain conformational order and lateral packing showed that bicelles hampered the temperature-dependent fluidization of SC lipids in the most superficial layers of the SC and led to a lateral packing corresponding to a stable hexagonal phase. CH_2 stretching and phosphate vibrations in the ATR-FTIR spectra of subsequent stripping indicated that DPPC–DHPC bicelles penetrated into and were widely distributed in deep layers of the SC. These results corroborate the presence of different DPPC–DHPC structures inside the SC layers and bicellar reinforcement of the SC structures was observed.

The appearance of vesicles inside the SC lipid layers was investigated. Various groups have reported studies of bicellar state transitions from disks to vesicles that were induced by dilution or temperature changes. As mentioned in Section 8.2, there is a general consensus that these transitions occur in progressive steps, implying the coexistence of different aggregates in the medium.[5,30,85] To study the bicelle-to-vesicle transition, DLS studies were undertaken by Barbosa-Barros et al.[44] A DPPC–DHPC sample with $q = 3.5$ and $c_L = 10\%$ was sequentially diluted with water in seven steps and each diluted sample was measured using DLS at 37 °C to imitate physiological conditions. The DLS curves indicated that the hydrodynamic diameter (HD) of the structures increased upon dilution from 11.3 nm (assigned to bicelle disks) to aggregates larger than 1 μm. The morphologies of the aggregates were analysed using electron microscopy, which confirmed the DLS results, indicating a dilution-induced transition from disks to vesicles.

The transitions that disk-shaped bicelles undergo on variation of the lipid concentration, temperature or lipid molar ratio[79] are very similar to those involved in the reconstitution of surfactant–lipid mixed micelles in vesicles. This phenomenon, which also occurs *via* dilution, has been discussed in a number of studies.[86,87] A resemblance between surfactant–lipid micelles and phospholipid bicelles explains their similar behaviour. In phospholipid bicelles, the DHPC molecules solubilize the DPPC bilayer, forming disk-shaped structures similar to surfactant-solubilized liposomes forming surfactant–lipid micelles. In bicellar systems, the DHPC molecules are found mainly on the edges of the disk structures and in the water (as monomers). With increasing dilution, the DHPC concentration in water decreases and DHPC is transferred from the bicelle edges into solution, maintaining monomeric equilibrium. Hence disk diameters increase and high dilutions lead to the fusion and closure of large bilayered disks, forming vesicles.

This bicelle-to-vesicle transition explains the presence of vesicles in the SC intercellular spaces treated with DPPC–DHPC.[44] This process would have been promoted by the dilution of aggregates because the SC pieces were washed with water after incubation. Bicelles that have been transformed into vesicles presumably follow a process similar to that observed in the DLS observations of diluted bicellar solutions outside the skin. A similar process was reported by López *et al.*, who applied octylglucoside–phosphatidylcholine mixed micelles to the SC.[88]

8.4.1.1.2   Percutaneous Penetration Studies.   Rubio *et al.*[89] performed percutaneous penetration studies with DMPC–DHPC and DPPC–DHPC systems with $c_L = 10\%$ and $q = 2$ to evaluate their effects on the skin penetration of diclofenac diethylamide (DDEA). They reported that the incorporation of DDEA in the bicelles led to markedly decreased bicelle sizes, indicating that DDEA tends to be located at the bicelle edges, similarly to DHPC.

Figure 8.17 shows cryo-TEM images of DPPC–DHPC bicelles with and without DDEA.

Both systems decreased the percutaneous absorption of the DDEA compared with an aqueous solution of DDEA, suggesting a penetration retarder effect after treatment with bicelles. This effect was more marked for the DMPC–DHPC bicelles. This could be related to the different T_m values for the two systems. At 37 °C, DMPC–DHPC ($T_m = 23.5$ °C) bicelles were in the liquid-crystalline phase, in contrast to the gel phase of SC lipids ($T_m \approx 60$ °C). However, DPPC–DHPC bicelles ($T_m = 41.4$ °C) were in the gel phase, similarly to the SC lipids. A different mixing behaviour of lipids from bicelles with SC lipids could induce different effects on the retention of DDEA in the upper layers of the skin. This retarder effect was ascribed in part to rigidity in the headgroups of bicelle phospholipids caused by the carboxyl groups of DDEA. This rigidity would hinder the penetration of DDEA through the skin. However, DDEA was likely unable to diffuse out of the bicellar systems because of its high affinity for its vehicle.

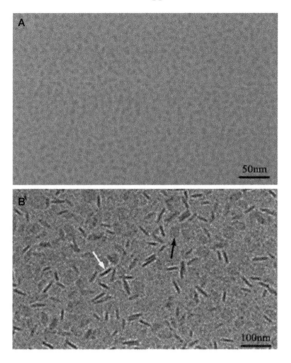

Figure 8.17 Cryo-TEM images of DPPC–DHPC bicellar systems with reduced sizes due to (A) diclofenac diethylamide (DDEA) or (B) bicellar structures of regular size without DDEA.
Reprinted with permission from Ref. 89. Copyright 2010 Elsevier.

A second percutaneous penetration assay was also performed by Rubio *et al.*[89] to evaluate the potential of bicelles as penetration enhancers. In this experiment, the skin was pretreated with the bicellar systems (DMPC–DHPC and DPPC–DHPC systems with $c_L = 10\%$ and $q = 2$) before the application of aqueous DDEA solution. The overall results obtained showed that pretreatment of the skin with bicelles promotes the percutaneous absorption of DDEA, with no significant differences between DMPC–DHPC and DPPC–DHPC systems. These studies suggest that treatment with these bicelle systems prior to drug treatment can enhance drug absorption. This enhancement is ascribed to an initial interaction of bicelles with the SC that causes some disorganization of the intercellular lipids, which are responsible for the SC barrier functionality. These results suggest a route to aid the absorption of DDEA through the skin.

8.4.1.2 In Vivo *Interaction of Bicelles with the Skin*

To estimate the effects of bicelles on the skin *in vivo*, non-invasive biophysical studies were performed with healthy volunteers. These studies reported mainly on measurements of skin hydration, elasticity, erythema and transepidermal water loss (TEWL). Skin hydration and elasticity are useful measures of skin water content, distensibility, extensibility and tonicity.[90,91] Erythema indicates skin tolerance[92] and TEWL indicates barrier function integrity.[93,94]

The application of DMPC–DHPC bicelles with $c_L = 20\%$ and $q = 2$ to the skin of healthy volunteers has been reported.[37] In this experiment, intra-individual comparisons of three test areas on the volar forearms of six healthy Caucasian female volunteers (aged 25–38 years) with no visible skin abnormalities were performed. In the test areas, bicelles, deionized water and control (non-treated) areas were randomized regarding the test sites on each subject. The solutions were applied daily over a period of 10 days and the skin properties were measured each day before application.

It was observed that successive bicelle applications led to an increase in TEWL from day 0 to 11. This increase was moderate and did not reach pathological levels, which are from 25 to 40 g m^{-2} h^{-1}.[95,96] Decreasing skin hydration was also observed. Elasticity, in turn, showed improvement with the application of bicelles. The changes in the erythema of the skin, considering inter- and intra-individual variability, were not indicative of an irritation process.[97] The bicelles were found to promote increases in TEWL and skin elasticity and harmless decreases in skin hydration.

Bicelles act to enhance penetration, causing phase transformations in lipid domains that may be relevant to skin permeation.[98] As reported by Rodriguez *et al.*,[83] the phase transition of the SC lipid conformation from the gel state to the liquid-crystalline state, which causes the fluidity of these lipids, explains the increase in TEWL *in vivo*. Compared with other enhancers, bicelles would have the additional advantage of not causing skin irritation.[99]

Another study reported the effects of the application of DPPC–DHPC bicelles with $c_L = 20\%$ and $q = 2$ on the skin of healthy volunteers. The results were similar to those obtained with the DMPC–DHPC system, although they were more discrete. In a similar way, this system led to an increase in TEWL and a decrease in skin hydration. However, the effects were ~75% and ~50% less intense, respectively, than those obtained with the DMPC–DHPC system.[100] This result is not unexpected because the DPPC–DHPC system contains less DHPC, so the aggregates formed are slightly larger and have longer bilayer areas, which would exert a protective effect on the SC. In addition, as observed in the *in vitro* studies, this system penetrates into the skin SC and reconstitutes its lipids in lipid vesicles inside the skin lamellae, which reinforce the lipid structure of the tissue.

8.5 Conclusion

Bicelles have quickly emerged as another amazing possibility in the host of lipid morphologies. The application of bicelles in structural studies of membrane proteins has a number of unique advantages. Bicelles with small q values can be used for high-throughput solution-state NMR studies whereas those with large q values are ideal for solid-state NMR studies. At the same time, research that is focused on bicelles in their own right has elucidated most of their basic structural and dynamic properties.

One of the most interesting activities is in the limits of the field, where new imaginative applications of bicelles are constantly emerging. Thus, the use of bicelles in skin is a promising scientific novelty. This relatively new lipid system represents a unique versatile structure that has different effects on the skin depending on the self-assembly adopted. Control over bicelle physicochemical properties is required for their use in biomedical applications. Hydration and temperature determine their self-assembly parameters (size, morphology and structure), while selection of an appropriate bicelle composition, for instance, the use of lipids with appropriate transition temperatures to yield a specific particle size, is a key factor to render the use of these lipid nanostructures more efficient.

The application of bicelles to the skin modifies its biophysical parameters without affecting the SC lipid microstructure or promoting irritation. The penetration and growth of DPPC–DHPC bicelles inside the SC open up new avenues for the treatment of these systems. Bicelles are an effective skin carrier owing to their size, structure and composition. Although bicelles have no aqueous internal compartment for encapsulating drugs, their bilayered structure allows for the encapsulation of lipophilic and amphiphilic compounds. Because of their ability to increase the permeability of the SC, these structures enhance the penetration of hydrophilic components dissolved in an aqueous medium. Further, the conversion of bicelles into vesicles inside the SC hinders their migration outside the tissue and allows a lipid reinforcement effect on the skin. This property could be very useful for intensifying the effects of specific compounds carried by bicelles into the SC layers. By modulating their physical and chemical characteristics, bicelles may be useful for a wide range of applications. Bicelles are therefore promising nanostructures that represent new platforms for skin-related applications.

References

1. C. R. Sanders and J. P. Schwonek, *Biochemistry*, 1992, **31**, 8898.
2. C. R. Sanders, B. J. Hare, K. P. Howard and J. H. Prestegard, *Prog. Nucl. Magn. Reson. Spectrosc.*, 1994, **26**, 421.
3. C. R. Sanders and G. C. Landis, *Biochemistry*, 1995, **34**, 4030.
4. K. Müller, *Biochemistry*, 1981, **20**, 404.
5. J. O. Struppe, J. A. Whiles and R. R. Vold, *Biophys. J.*, 2000, **78**, 281.
6. C. R. Sanders, J. E. Schaff and J. H. Prestegard, *Biophys. J.*, 1993, **64**, 1069.
7. C. R. Sanders and J. H. Prestegard, *Biophys. J.*, 1990, **58**, 447.
8. R. R. Vold and R. S. Prosser, *J. Magn. Reson. B*, 1996, **113**, 267.
9. K. P. Howard and S. J. Opella, *J. Magn. Reson.*, 1996, **12**, 91.
10. J. A. Losonczi and J. H. Prestegard, *Biochemistry*, 1998, **37**, 706.

11. J. Struppe, E. A. Komives, S. S. Taylor and R. R. Vold, *Biochemistry*, 1998, **37**, 5523.
12. M. Nieh, C. J. Glinka, S. Krueger, R. S. Prosser and J. Katsaras, *Langmuir*, 2001, **17**, 2629.
13. A. Bax and S. Gaemers, *J. Am. Chem. Soc.*, 2001, **123**, 12343.
14. R. R. Vold, R. S. Prosser and A. J. Deese, *J. Biomol. NMR*, 1997, **9**, 329.
15. K. J. Glover, J. A. Whiles, R. R. Vold and G. Melacini, *J. Biomol. NMR*, 2002, **22**, 57.
16. P. A. Luchette, T. N. Vetman, R. S. Prosser, R. E. Hancock, M. P. Nieh, C. J. Glinka, S. Krueger and J. Katsaras, *Biochim. Biophys. Acta*, 2001, **1513**, 83.
17. K. J. Glover, J. A. Whiles, G. Wu, N. Yu, R. Deems, J. O. Struppe, R. E. Stark, E. A. Komives and R. R Vold, *Biophys. J.*, 2001, **81**, 2163.
18. B. A. Rowe and S. L. Neal, *Langmuir*, 2003, **19**, 2039.
19. M. P. Nieh, V. A. Raghunathan, C. J. Glinka, T. A. Harroun, G. Pabst and K. Katsaras, *Langmuir*, 2004, **20**, 7893.
20. H Wang, M. P. Nieh, E. K. Hobbie, C. J. Glinka and J. Katsaras, *J. Phys. Rev. E*, 2003, **67**, 060902.
21. S. Faham and J. U. Bowie, *J. Mol. Biol.*, 2002, **316**, 1.
22. R. S. Prosser, J. S. Hwang and R. R. Vold, *Biophys. J.*, 1998, **74**, 2405.
23. S. Mahabir, W. Wan, J. Katsaras and M. P. Nieh, *J. Phys. Chem. B*, 2010, **114**, 5729.
24. L. Barbosa-Barros, A. de la Maza, P. Walther, A. M. Linares, M. Feliz, J. Estelrich and O. López, *J. Microsc.*, 2009, **233**, 35.
25. L. Barbosa-Barros, A. de la Maza, P. Walther, J. Estelrich and O. López, *J. Microsc.*, 2008, **230**, 16.
26. P. Walther, *J. Microsc.*, 2003, **212**, 34.
27. P. Walther and A. Ziegler, *J. Microsc.*, 2002, **208**, 3.
28. E. Sternin, D. Nizza and K. Gawrisch, *Langmuir*, 2001, **17**, 2610.
29. M. N. Triba, D. E. Warschawski and P. F. Deveaux, *Biophys. J.*, 2005, **88**, 1887.
30. L. van Dam, G. Karlsson and K. Edwards, *Langmuir*, 2006, **22**, 3280.
31. P. Echlin, *Low-Temperature Microscopy and Analysis*, Springer, New York, 1992.
32. L. van Dam, G. Karlsson and K. Edwards, *Biochim. Biophys. Acta*, 2004, **1664**, 241.
33. T. A. Harroun, M. Koslowsky, M. P. Nieh, C. F. de Lannoy, V. A. Raghunathan and J. Katsaras, *Langmuir*, 2005, **21**, 5356.
34. M. Plaza, T. F. Tadros, C. S. Solans and R. Pons, *Langmuir*, 2002, **18**, 5673.
35. G. Pabst, R. Koschuch, B. Pozo-Navas, M. Rappolt, K. Lohner and P. Laggner, *J. Appl. Crystallogr.*, 2003, **36**, 1378.
36. G. Pabst, *Biophys. Rev. Lett.*, 2006, **1**, 57.
37. L. Barbosa-Barros, C. Barba, M. Cócera, L. Coderch, C. López-Iglesias, A. de la Maza and O. López, *Int. J. Pharm.*, 2008, **352**, 263.
38. M. Singh, S. Ghosh and R. A. Shannon, *J. Appl. Crystallogr.*, 1993, **26**, 787.

39. A. Arnold, T. Labrot, R. Oda and E. J. Dufourc, *Biophys. J.*, 2002, **83**, 2667.
40. M. R. Vist and J. H. Davis, *Biochemistry*, 1990, **29**, 451.
41. P. F. Almeida, W. L. Vaz and T. E. Thompson, *Biochemistry*, 1992, **31**, 6739.
42. C. J. Wohl, M. A. Helms, J. O. Chung and D. Kuciauskas, *J. Phys. Chem. B*, 2006, **110**, 22796.
43. J. Lind, J. Nordin and L. Maler, *Biochim. Biophys. Acta*, 2008, **1778**, 2526.
44. L. Barbosa-Barros, A. de la Maza, J. Estelrich, A. M. Linares, M. Feliz, P. Walther, R. Pons and O. López, *Langmuir*, 2008, **24**, 5700.
45. F. Aussenac, B. Lavigne and E. J. Dufourc, *Langmuir*, 2005, **21**, 7129.
46. M. Ottiger and A. Bax, *J. Biomol. NMR*, 1999, **13**, 187.
47. S. Cavagnero, H. J. Dyson and P. E. Wright, *J. Biomol. NMR*, 1999, **13**, 387.
48. J. A. Whiles, K. J. Glover, R. R. Vold and E. A. Komives, *J. Magn. Reson.*, 2002, **158**, 149.
49. E. K. Tiburu, D. M. Moton and G. A. Lorigan, *Biochim. Biophys. Acta*, 2001, **1512**, 206.
50. R. N. Lewis, N. Mak and R. N. McElhaney, *Biochemistry*, 1987, **26**, 6118.
51. R. G. Laughlin, R. L. Munyon, Y.-C. Fu and A. J. Fehl, *J. Phys. Chem.*, 1990, **94**, 2546.
52. D. Marsh, *CRC Handbook of Lipid Bilayers*, CRC Press, Boca Raton, FL, 1990.
53. S. H. Park, A. A. De Angelis, A. A. Nevzorov, C. H. Wu and S. J. Opella, *Biophys. J.*, 2006, **91**, 3032.
54. J. F. Nagle and S. Tristram-Nagle, *Biochim. Biophys. Acta*, 2000, **1469**, 159.
55. M. Ottiger and A. Bax, *J. Biomol. NMR*, 1998, **12**, 361.
56. D. Uhrikova, P. Rybar, T. Hianik and P. Balgavy, *Chem. Phys. Lipids*, 2007, **145**, 97.
57. L. Coderch, O. López, A. de la Maza and J. L. Parra, *Am. J. Clin. Dermatol.*, 2003, **4**, 107.
58. L. Barbosa-Barros, A. de la Maza, J. Estelrich, A. M. Linares, M. Feliz and O. López, *Biophys. J.*, 2007, **92**, 234a.
59. R. M. Pinaki and T. Blume, *J. Phys. Chem. B*, 2002, **106**, 10753.
60. O. López, M. Cocera, E. Wehrli, J. L. Parra and A. de la Maza, *Arch. Biochem. Biophys.*, 1999, **367**, 153.
61. O. López, A. de la Maza, L. Coderch, C. Lopez-Iglesias, E. Wehrli and J. L. Parra, *FEBS Lett.*, 1998, **426**, 314.
62. A. N. Goltsov and L. I. Barsukov, *J. Biol. Phys.*, 2000, **26**, 27; *Pharmaceutics*, 2011, **3**, 663.
63. H. Kashiwagi and M. Ueno, *Yakugaku Zasshi*, 2008, **128**, 669.
64. K. Dwiecki, P. Gornas, A. Wilk, M. Nogala-Kalucka and K. Polewski, *Cell. Mol. Biol. Lett.*, 2007, **12**, 51.
65. M. Almgren, *Biochim. Biophys. Acta*, 2000, **1508**, 146.
66. O. López, M. Cócera, L. Coderch, J. L. Parra, L. I. Barsukov and A. de la Maza, *J. Phys. Chem. B*, 2001, **105**, 9879.

67. J. Leng, S. U. Egelhaaf and M. E. Cates, *Biophys. J.*, 2003, **85**, 1624.
68. J. Struppe and R. R. Vold, *J. Magn. Reson.*, 1998, **135**, 541.
69. D. D. Verma, S. Verma, G. Blume and A. Fahr, *Eur. J. Pharm. Biopharm.*, 2003, **55**, 271.
70. P. Mura, F. Maestrelli, M. L. Gonzalez-Rodriguez, I. Michelacci, C. Ghelardini and A. M. Rabasco, *Eur. J. Pharm. Biopharm.*, 2007, **67**, 86.
71. A. K. Seth, A. Misra and D. Umrigar, *Pharm. Dev. Technol.*, 2004, **9**, 277.
72. M. E. van Kuijk-Meuwissen, H. E. Junginger and J. A. Bouwstra, *Biochim. Biophys. Acta*, 1998, **1371**, 31.
73. M. Kirjavainen, A. Urtti, I. Jaaskelainen, T. M. Suhonen, P. Paronen, R. Valjakka-Koskela, J. Kiesvaara and J. Monkkonen, *Biochim. Biophys. Acta*, 1996, **1304**, 179.
74. P. Karande and S. Mitragotri, *Biochim. Biophys. Acta*, 2009, **1788**, 2362.
75. N. Dragicevic-Curic, S. Winter, M. Stupar, J. Milic, D. Krajisnik, B. Gitter and A. Fahr, *Int. J. Pharm.*, 2009, **373**, 77.
76. M. P. Alves, A. L. Scarrone, M. Santos, A. R. Pohlmann and S. S. Guterres, *Int. J. Pharm.*, 2007, **341**, 215.
77. S. Küchler, M. Abdel-Mottaleb, A. Lamprecht, M. R. Radowski, R. Haag and M. Schäfer-Korting, *Int. J. Pharm.*, 2009, 377, 169.
78. M. P. Nieh, T. A. Harroun, V. A. Raghunathan, C. J. Glinka and J. Katsaras, *Biophys. J.*, 2004, **86**, 2615.
79. B. Yue, C. Y. Huang, M. P. Nieh, C. J. Glinka and J. Katsaras, *J. Phys. Chem. B*, 2005, **109**, 609.
80. M. P. Nieh, V. A. Raghunathan, S. R. Kline, T. A. Harroun, C. Y. Huang, J. Pencer and J. Katsaras, *Langmuir*, 2005, **21**, 6656.
81. J. R. Hill and P. W. Wertz, *Lipids*, 2009, **44**, 291.
82. D. Kessner, A. Ruettinger, M. A. Kiselev, S. Wartewig and R. H. Neubert, *Skin Pharmacol. Physiol.*, 2008, **21**, 58.
83. G. Rodriguez, L. Barbosa-Barros, L. Rubio, M. Cocera, A. Diez, J. Estelrich, R. Pons, J. Caelles, A. de la Maza and O. López, *Langmuir*, 2009, **25**, 10595.
84. G. Rodriguez, L. Rubio, M. Cocera, J. Estelrich, R. Pons, A. de la Maza and O. López, *Langmuir*, 2010, **26**, 10578.
85. J. Bolze, T. Fujisawa, T. Nagao, K. Norisada, H. Saitô and A. Naito, *Chem. Phys. Lett.*, 2000, **329**, 215.
86. M. Ollivon, S. Lesieur, C. Grabielle-Madelmont and M. Paternostre, *Biochim. Biophys. Acta*, 2000, **1508**, 34.
87. J. L. Rigaud and D. Levy, *Methods Enzymol.*, 2003, **372**, 65.
88. O. López, M. Cócera, C. López-Iglesias, P. Walter, L. Coderch, J. L. Parra and A. de la Maza, *Langmuir*, 2002, **18**, 7002.
89. L. Rubio, C. Alonso, G. Rodriguez, L. Barbosa-Barros, L. Coderch, A. de la Maza, J. L. Parra and O. Lopez, *Int. J. Pharm.*, 2010, **386**, 108.
90. S. Astner, E. Gonzalez, A. C. Cheung, F. Rius-Diaz, A. G. Doukas, F. William and S. Gonzalez, *J. Invest. Dermatol.*, 2005, **124**, 351.
91. R. Bazin and C. Fanchon, *Int. J. Cosmet. Sci.*, 2006, **28**, 453.

92. J. Reuter, C. Huyke, H. Scheuvens, M. Ploch, K. Neumann, T. Jakob and C. M. Schempp, *Skin Pharmacol. Physiol.*, 2008, **21**, 306.
93. J. W. Fluhr, K. R. Feingold and P. M. Elias, *Exp. Dermatol.*, 2006, **15**, 483.
94. M. Miteva, S. Richter, P. Elsner and J. W. Fluhr, *Exp. Dermatol.*, 2006, **15**, 904.
95. N. Branco, I. Lee, H. Zhai and H. I. Maibach, *Contact Dermatitis*, 2005, **53**, 278.
96. D. W. Kim, J. Y. Park, G. Y. Na, S. J. Lee and W. J. Lee, *Int. J. Dermatol.*, 2006, **45**, 698.
97. E. A. Holm, H. C. Wulf, L. Thomassen and G. B. Jemec, *J. Am. Acad. Dermatol.*, 2006, **55**, 772.
98. G. M. El Maghraby, M. Campbell and B. C. Finnin, *Int. J. Pharm.*, 2005, **305**, 90.
99. N. Kanikkannan and M. Singh, *Int. J. Pharm.*, 2002, **248**, 219.
100. L. Barbosa-Barros, C. Barba, G. Rodríguez, M. Cócera, L. Coderch, C. López-Iglesias, A. de la Maza and O. López, *Mol. Pharm.*, 2009, **6**, 1237.

CHAPTER 9

Soft Hybrid Nanoparticles: from Preparation to Biomedical Applications

TALHA JAMSHAID,[a,b] MOHAMED EISSA,[a,c] NADIA ZINE,[b] ABDELHAMID ERRACHID EL-SALHI,[b] NASIR M. AHMAD[d] AND ABDELHAMID ELAISSARI*[a]

[a] University of Lyon, 69622 Lyon, France; University of Lyon-1, Villeurbanne, CNRS, UMR-5007, LAGEP-CPE; 43 boulevard 11 Novembre 1918, 69622 Villeurbanne, France; [b] Institut des Sciences Analytiques (ISA), Université Lyon, Université Claude Bernard Lyon-1, UMR-5180, 5 rue de la Doua, 69100 Villeurbanne, France; [c] Polymers and Pigments Department, National Resaerch Centre, Dokki, Giza 12622, Egypt; [d] Polymer and Surface Engineering Laboratory, Department of Materials Engineering, School of Chemical and Materials Engineering (SCME), National University of Sciences and Technology (NUST), Islamabad-44000, Pakistan
*Email: elaissari@lagep.univ-lyon1.fr

9.1 Introduction

Colloidal particles provide highly useful diagnostic solutions and standardized results through efficient screening processes for several diseases.[1–6] Furthermore, newer diagnostic techniques require minute samples and also need to be less invasive, safe and quick and with the highest possible accuracy in terms of detection.[7] In this context, various types of functional

RSC Nanoscience & Nanotechnology No. 34
Soft Nanoparticles for Biomedical Applications
Edited by José Callejas-Fernández, Joan Estelrich, Manuel Quesada-Pérez and Jacqueline Forcada
© The Royal Society of Chemistry 2014
Published by the Royal Society of Chemistry, www.rsc.org

colloidal particles have been developed and explored for their biomedical applications. Among the key features of such functional colloidal particles are their soft hybrid core-shell structures, which are engineered in such a way that the core consists of an inorganic material whereas the shell is composed of a soft organic substance.[8] In other words, hybrid particles are a class of materials that include both organic and inorganic moieties at the same time and also possess interesting magnetic, optical and mechanical properties.[9,10] These particles are used for optical investigations of biological interactions and their other significant properties such as super-paramagnetism, biodegradability and fluorescence.[11] These particles also have the ability to occupy space much more efficiently, allowing the dispersed phase to reach much higher volume fractions than random close packing of spheres.[12] These materials have potential applications in areas such as gas separation, catalysis, storage and, most importantly, bio-materials for diverse biomedical applications. Their unique characteristics make them suitable materials for developing desirable patterns or functions by exploiting nanoscale phenomena through build-up *via* atom-by-atom or molecule-by-molecule methods (top-down) or through self-organization (bottom-up).[13]

Extensive research efforts are being made to develop and explore the biomedical applications of soft hybrid nanoparticles.[1-6] In addition, numerous types of the soft hybrid nanoparticles have been developed to target specific biomedical applications and to perform desirable diagnostics or therapeutic functions. For example, with current rapid advances in nano-medicine, colloidally engineered particles are gaining immense importance in biomedical fields such as cancer and gene therapy,[14-17] disease diagnosis[18] and bioimaging.[19-21] During the past decade, numerous colloidal particles have been created and widely investigated for medical use.[22-25] Innovations in development have been able to incorporate diverse functionalities, and colloidal particles bearing reactive groups (*e.g.* –COOH, –NH$_2$, –SH) and dendrimers have been developed for the covalent binding of bio-molecules in order to be used as a solid support for the specific capture of targets and imaging. Colloidal particles are also designed in such a way that they have suitable particle size, hydrophilic–lipophilic balance (HLB), size distribution and surface reactive groups for the target diagnosis.[26] The attachment of many macromolecules (*e.g.* polysaccharides, proteins, lecithins, antibodies, glycoproteins) can be non-covalent and generally probes such as gold acquire the properties of attached macromolecules. Their stability upon storage is also very good.[27] Furthermore, gold nanoparticles can absorb light over a wide spectral range, from the visible to near-infrared, and can be used for diagnosis as optically tuneable carriers because of their shape and size.[28] Soft hybrid nanoparticles of calcium phosphate/hydroxyapatite and meso-porous silica particles have been employed to carry drugs and DNA to intracellular organelles by interaction with cells.[29-31] Polymer-based latexes have been widely used as carriers for antigen and antibody reactions in immunoagglutination assays and were first used to detect rheumatoid factor.[32] Various other core materials have also been employed for diagnostics and bioimaging, such as quantum dots (QDs), silver nanorods to separate viruses, bacteria and microscopic components of blood samples (allowing the identification of these parameters in less than 1 h) and carbon nanotubes (CNTs), which are used in sensors to detect proteins specific to oral cancer.[33] Other very interesting types of colloidal particles are based on magnetically active inorganic particles in the core to introduce magnetic properties, which enhances the concentration of the targeted biomolecules and consequently the sensitivity of the biomedical diagnostic. Magnetically responsive hybrid iron oxide colloidal nanoparticles have superparamagnetic properties and are extensively used in bioimaging to provide high diagnostic accuracy in the detection of atherosclerosis, cancer, arthritis and many other diseases.[34-36] owing to their magnetic character, which can be controlled through the application of an external magnetic field, magnetic particles and latexes are extensively used in biomedical diagnosis such as in molecular biology, bacteria and virus isolation, immunoassays, nucleic acid extraction (NAE) and cell sorting.[37] Dual magnetic and luminescent properties which help in different types of diagnosis and the incorporation of functional groups on soft hybrid colloidal nanoparticles have also been developed and many biomolecules have been immobilized for selective organelle labelling used in the detection of polynucleotides[12,38] (DNA, RNA).

9.2 Development of Soft Hybrid Core-Shell Nanoparticles

Many methods and techniques have been developed to synthesize and characterize soft hybrid core-shell colloidal particles, as discussed elsewhere.[1,2,5,34] Each method and technique has its own merits that are mainly dependent on the nature of the particles, safety considerations, composition, size distribution, *etc*. Keeping in mind the importance of developing soft hybrid nanoparticles, a few important methods are discussed below to provide an overview of the state of the art in this important area.

9.2.1 Polymer Immobilization on Preformed Particles

In order to develop and find uses for soft hybrid core-shell colloidal particles, there is always a need to incorporate certain characteristics to meet the specific demands of desirable biomedical applications. These characteristics include a certain degree of hydrophilicity, composition, architecture, surface energy and stimuli-responsive character. For this purpose, the immobilization of polymers provides perhaps the most powerful tool for tuning the surface properties of the shell without much affecting the characteristics of the core particles. In this direction, extensive research work has been carried out and an attempt is made below to outline concisely some of these techniques.

9.2.2   Adsorption of Polymers on Colloidal Particles

Adsorption is a process that mostly occurs at the surface of an adsorbent.[39] In the case of polymers, when these interact with colloidal particles, their adsorption on the surface of particles takes place through specific interactions of functional groups present in the chains. The stabilization of colloidal particles can be controlled through various parameters at the solid/liquid interface.[40,41] A scheme of the adsorption of polymer chains on colloidal particles is illustrated in Figure 9.1. Adsorption of polymers on solid particle surfaces takes place because of either electrostatic (Coulombic) interactions or others, such as polar or non-polar forces or weaker van der Waals forces.[42] It is also possible that weaker or incomplete adsorption leads to the aggregation of particles. Colloidal dispersions are generally stable only if there is no aggregation and this stability develops only in the presence of some repulsive forces, the nature of which can be electrostatic or steric. Particles having surface charges play an important role in electrostatic stabilization. Diffuse adsorbed layers that start to overlap induce an electrostatic repulsive force and produce stable dispersions because the range of this repulsive force is greater than the van der Waals attraction.

Polymer adsorption on the particle surface plays an important role in steric stabilization because the thickness of the polymer layer protects particles from aggregation due to van der Waals forces. Adsorbed polymer chains have part of their segments in contact with the surface, but they do not constitute rigid structures and continually change their conformation. When polymer chains change the relative position of their adsorbed segments, they retain a large degree of their conformational freedom. In other words, the segments in contact with the surface continually exchange their position with non-adsorbed segments. Two surfaces with adsorbed polymers tend to confine these polymers in approaching each other, so this process is not entropically favourable and can eventually cause repulsion between

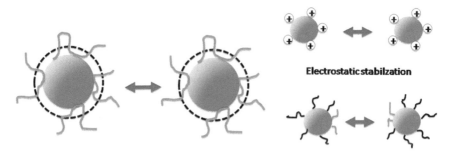

Electrostatic stabilzation

Steric stabilization

Figure 9.1   Top: electrostatic contribution to colloid stability, two like-charged particles repelling each other. Bottom: steric contribution to colloid stability, polymer chains being pushed together and confined are repelled owing to the unfavourable decrease in entropy.

them. Among the various polymers, perhaps diblock copolymers are the most widely used owing to the different advantages of steric stabilization. In fact, steric stabilization offers more advantages than charge stabilization.[43] Hence the use of adsorbed polymers either to stabilize or to flocculate plays an important role in applications such as colloidal stabilization, lubrication and adhesion.[44–46]

9.2.3 Adsorption of Polymers *via* Layer-by-Layer Self-Assembly

This method provides a versatile approach to tuning the properties of colloidal particles by adsorption of polyelectrolytes.[47] The adsorption generally takes place through electrostatic Coulombic interactions between oppositely charged polyelectrolytes and colloidal particles. Generally, irreversible electrostatic attraction helps polyelectrolyte adsorption at supersaturating bulk concentrations. There are also other types of interaction that can be utilized for assembly in layer-by-layer (LBL) systems, namely van der Waals and hydrophobic interactions and hydrogen bonding. The important features that need to be considered in the LBL method include control of the number of layers, thickness, functionalities, roughness, morphology, composition, surface energy and many more. Furthermore, desirable surface properties on the colloidal particles or templates of almost any topography (spherical or others) can be incorporated by simply varying different parameters such as pH, salt concentration, type of polyelectrolyte (weak or strong) and solvent.[3,48] The formed polyelectrolyte-coated colloidal particles can be further adsorbed with additional oppositely charged nanoparticles of various types including gold and molecular precursors to give ultrathin multilayers films of molecular precursor layer and nanoparticles,[48] as shown in Figure 9.2.[49] Similarly, by employing the LBL technique, oil-in-water submicron magnetic emulsions have also been obtained, as presented in Figure 9.2(iii).[50]

9.2.4 Adsorption of Nanoparticles on Colloidal Particles

There are many techniques that can be employed to adsorb polymer nanoparticles including (but not limited to) various polymerization methods, physiochemical LBL methods and stepwise heterocoagulation.[51–53] In the stepwise heterocoagulation method, electrostatic interactions are exploited to induce the coating of relatively small particles on the larger ones. One of the pioneering examples included the adsorption of small polystyrene (PS) particles with cationic charges particles on large PS particles with anionic charges.[51] The heterocoagulation driven by electrostatic interaction produces a monolayer of smaller particles on the larger particles through an adsorption process resulting in the formation of core-shell colloidal particles. By applying the heterocoagulation method, several types of soft core–shell colloidal particles with well-controlled morphologies have

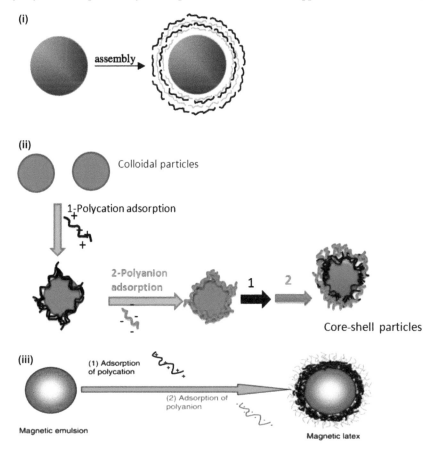

Figure 9.2   (i) Schematic illustration of LBL adsorption on a particle surface through self-assembly. (ii) Polyelectrolyte multilayers *via* LBL adsorption. (iii) LBL adsorption on a magnetic emulsion using a polycation and a polyanion.

been reported, including the adsorption of magnetic nanoparticles on sub-micron-sized PS latex particles.[51] In order to stop any release of magnetic nanoparticles, the encapsulation of the magnetic layer can be carried out *via* polymerization to form a cross-linked shell.[51] In order to overcome low sedimentation rates and specific areas for non-specific or specific binding with biomolecules for such larger sized particles, various modifications to the stepwise heterocoagulation have been proposed. For example, sub-micron highly magnetic particles and also fluorescent polymer particles were designed based on innovative stepwise heterocoagulation processes, as shown in Figure 9.3. The synthesis of polymer core–shell particles by stepwise heterocoagulation has typically used smaller coating particles with low glass transition temperatures, T_g. This low T_g was exploited and used as a model to develop the heterocoagulates, which were obtained by heating at a temperature

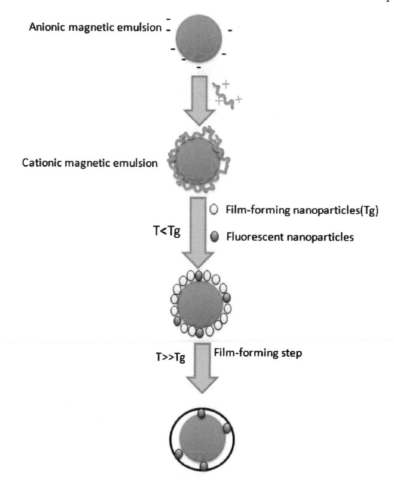

Figure 9.3   Stepwise heterocoagulation processes to form magnetic particles.
Adapted from Refs. 54 and 55.

lower than the T_g of the larger particles, but above the T_g of the smaller film-forming nanolatex particles to form a homogeneous polymer shell.[52] Therefore, such a process does not require any additional (or final) polymerization step to encapsulate the adsorbed nanoparticles *via* polymerization.[54,55]

9.2.5   Chemical Grafting of Preformed Polymers

Grafting, in the context of polymer chemistry, refers to the addition of polymer chains to a surface.[56,57] The grafting is a useful tool for altering the physical and/or chemical activity or interaction of the surface through processes of grafting 'to' and 'from'. Ultrathin (end-) grafted polymer layers are well known to affect dramatically the surface properties of substrates such as adhesion, lubrication, wettability, friction and biocompatibility. The layers are frequently used to modulate the surface properties of various materials,

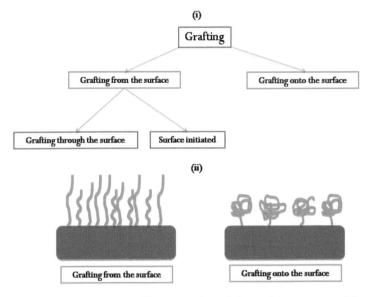

Figure 9.4   Grafting processes: (i) an overview of the grafting process; (ii) 'grafting from' and 'grafting to' a surface.

including colloidal particles, without altering their bulk performance. The chemical grafting of polymers can be accomplished by 'grafting to' or 'grafting from' methods. According to the 'grafting to' technique, end-functionalized polymer molecules react with complementary functional groups located on the surface to form tethered chains. The 'grafting from' technique utilizes the polymerization initiated from the substrate surface by initiating groups usually attached by covalent bonds. Most of the grafting methods developed – 'to' and 'from' – require the attachment of (end-)functionalized polymers or low molecular weight substances (*e.g.* initiators) to the substrate for the synthesis of the polymer brush. Usually the coupling methods are relatively complex and specific for certain substrate–(macro) molecule combinations.

A schematic overview of these two different grafting processes is presented in Figure 9.4 and discussed below.

9.2.6   Polymerization from and on to Colloidal Particles

Soft hybrid core-shell colloidal materials can be made by either the 'grafting to' or the 'grafting from' mechanism.[58] The mechanisms and advantages of these grafting processes are different and widely used to prepare colloidal particles bearing a polymer shell as shown in Figures 9.5 and 9.6. These two processes adopt two possible conformations known commonly as mushroom and polymer brush.

In the 'grafting to' mechanism, a preformed polymer chain is adsorbed on a surface from solution. Because prepolymerized chains used in this

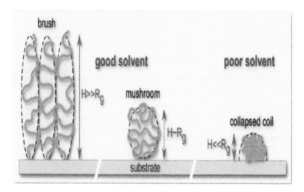

Figure 9.5   Possible conformations of 'grafted from' polymers from a surface in brush form and surface grafted in the mushroom shape in a good solvent. The conformation of 'grafted to' polymer in a poor solvent is also shown. R_g is the radius of gyration of the polymer chain and H is its thickness.[59]

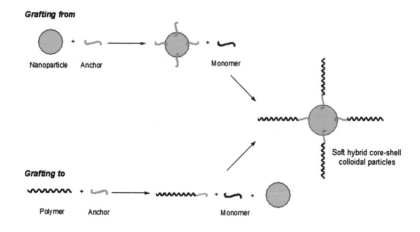

Figure 9.6   Schematic representation of the 'grafting from' and 'grafting to' systems to form a shell on the core surface to develop soft hybrid core-shell colloidal particles.[56]

mechanism have a thermodynamically favoured conformation in solution (an equilibrium hydrodynamic volume), their adsorption density is self-limiting. The radius of gyration of the coil polymer chain is therefore the limiting factor in the number of polymer chains that can reach the surface and adhere. In the 'grafting from' mechanism, an initiator is adsorbed on the surface followed by initiation of the chain and propagation of monomer *via* surface-initiated polymerization. As shown in Figure 9.7, various polymerization methods have been developed to carry out the 'grafting from' process. Specifically, these include various control radical polymerization (CRP) techniques such as atom-transfer radical polymerization (ATRP), reversible addition–fragmentation chain-transfer polymerization (RAFT) and

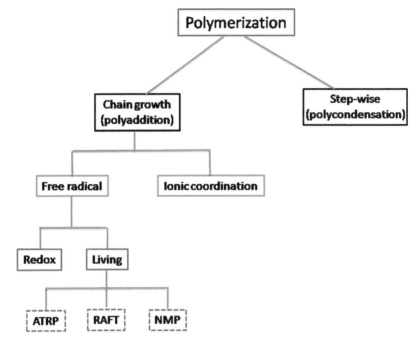

Figure 9.7 Various polymerization methods that can be used to form a shell on the core of different materials including colloidal particles to develop soft hybrid core-shell colloidal particles.

nitroxide-mediated radical polymerization (NMP). The CRP techniques have attracted much attention as they represent a facile approach to making end-grafted polymers with flexibility on the surface of various nanoparticles. Currently, CRP is regarded as the essential means to synthesize polymers with well-controlled functionality, composition, structure and architecture.

The 'grafting from' method has the advantage of allowing higher grafting densities than the 'grafting to' process. As shown in Figures 9.5 and 9.6, the 'grafting to' mechanism generally results in a 'mushroom regime,' to adhere to the surface of either a droplet or bead in solution. Owing to the larger volume of the coiled polymer and the steric hindrance that this causes, the grafting density is lower for 'grafting to' than for 'grafting from.' In contrast, the 'extended conformation' of the polymerized monomers from the surface of the bead means that the monomer must be in the solution and therefore be lyophilic in nature. This results in a polymer that has favourable inter-actions with the solution, allowing the polymer to form a more linear structure. 'Grafting from' therefore has a higher grafting density since there is greater access to chain ends. Peptide synthesis provides an example of a 'grafting from' synthetic process. In this approach, an amino acid chain is grown by a series of condensation reactions from a polymer bead surface. This grafting technique allows excellent control over the peptide com-position as the bonded chain can be washed without desorption from the

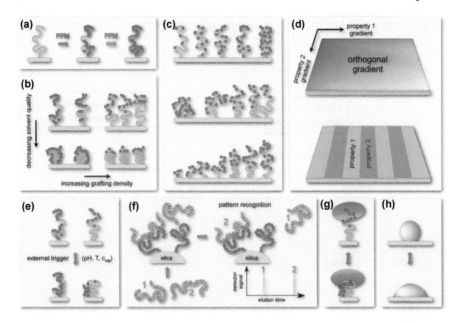

Figure 9.8   Schematic representing the formation (a–d) and function (e–h) of surface-grafted polymer systems generated by post-polymerization modification (PPM) protocols.[59]

polymer. Polymeric coatings are another area of application of grafting techniques. In the formulation of water-borne paints, latex particles are often surface modified to control particle dispersion and thus coating characteristics such as viscosity, film formation and environmental stability (UV exposure and temperature variations). Owing to such unique characteristics, grafting processes have been extensively applied to tune the surface properties of materials without much influencing the bulk characteristics. The grafting systems of both 'grafting from' and 'grafting to' have found extensive applications in various important areas, as shown in Figure 9.8.

9.2.7   Click Chemistry

Click chemistry can be defined as a concept of reactions for the rapid synthesis of new compounds through heteroatom links (C-X-C).[60] Typical examples of click chemistry include the Cu(I)-catalysed azide–alkyne-'click' (CuAAC) reaction between terminal azides and alkynes,[61–68] cycloaddition click reactions[69] and the thiol–ene click reactions.[70] Two approaches in click chemistry involve 'grafting to' and 'grafting from' mechanisms. In the 'grafting to' method, the azide- or alkyne-functionalized polymer is grafted to the corresponding functionalized nanoparticles (NPs).[71–75] For example, SiO_2 NPs have been modified with PS and polyacrylamide through the CuAAC reaction using the 'grafting to' method, as shown in Figure 9.9.[76]

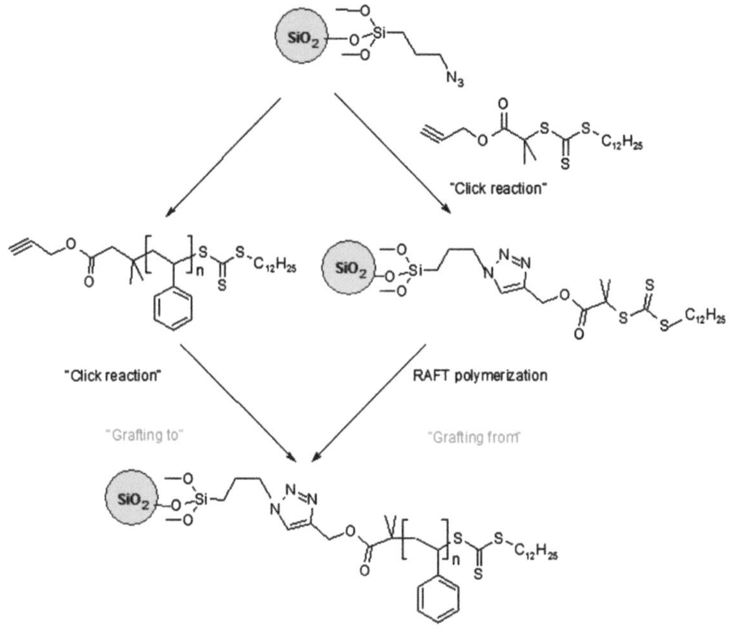

Figure 9.9   Modification of SiO$_2$ NPs with polystyrene *via* 'grafting to'[71] and 'grafting from'[80,81] methods based on click chemistry employing RAFT polymerization.
Adapted from Ref. 76.

Other NPs such as Fe$_2$O$_3$,[77] Pt[78] and Au[79] have also been grafted with various polymers. In this direction, the 'grafting from' approach has found useful applications.[80,81] For example, first a RAFT initiator clicked on the surface of SiO$_2$ NPs and polymerization was initiated, as shown in Figure 9.10. Low molecular weight monomers having high diffusibility and high grafting density are used in the 'grafting from' method, whereas, the 'grafting to' approach involves the use of the preformed high molecular weight whole polymer chain that restricts the polymer diffusibility.[76]

9.2.8   Atom-Transfer Radical Polymerization (ATRP)

Different methods have been used for grafting polymers to various substrates.[82–99] To tune the properties of core-shell colloidal particles, surface-initiated polymerization *via* the 'grafting from' approach is the most widely employed technique owing to its capability to impart good functionality control, molecular weight of polymer brushes and grafting density.[100] Furthermore, among the various polymerization methods, ATRP has been used extensively to modify the properties of various substrates.[85–99] ATRP has also been used to produce hydrophilic polymer layers having remarkable thickness and surface density.[95–99] Similarly, various functional and stimuli-responsive polymers have also been incorporated on particle surfaces using

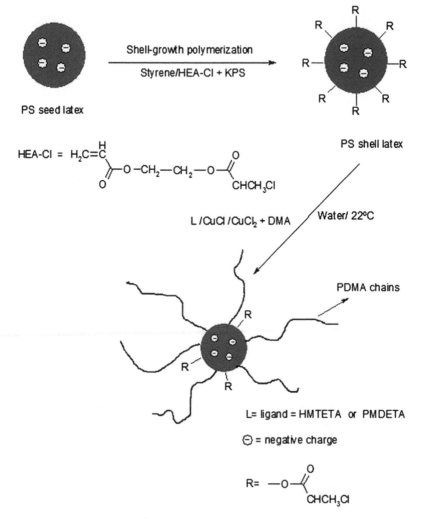

Figure 9.10 Graft polymerization of *N,N*-dimethylacrylamide (DMA) from PS particle surface based on the aqueous ATRP.
Adapted from Ref. 101.

ATRP. For example, in an interesting study, a controlled synthesis of poly(*N*-isopropylacrylamide) (PNIPAM) brushes was carried out from a PS particle surface using ATRP (see Figure 9.10).[101]

This process is based on several steps. First, PS latex particles were synthesized through surfactant-free emulsion polymerization of styrene monomer. Second, by the use of potassium persulfate (KPS) as an initiator, the polymerization of styrene and (2,2′-chloropropionato)ethyl acrylate (HEA-Cl) was performed to form the shell on the core PS seed particles. Third, aqueous ATRP polymerization of *N*-isopropylacrylamide (NIPAM) monomer from the preformed seed latex particle in the presence of

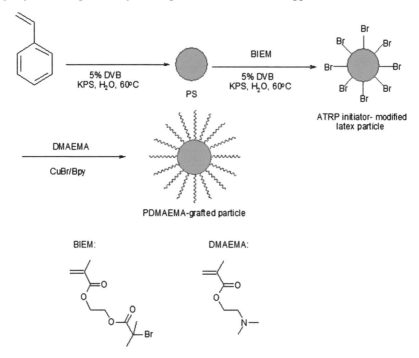

Figure 9.11 Formation of poly[2-(dimethyl amino)ethyl methacrylate] (PDMAEMA) brushes on the surface of colloid particles by employing ATRP. Adapted from Ref. 104.

1,1,4,7,10,10-hexamethyltriethylenetetramine (HMTETA)–CuCl–CuCl$_2$ powder under continuous stirring at room temperature was carried out. Similarly, different temperature-sensitive brushes such as PNIPAM, PMEA-*b*-PNIPAM, poly(*N,N*-dimethylacrylamide)[102,103] and poly[2-(dimethylamino)ethyl methacrylate] (PDMAEMA) (see Figure 9.11) have been also grafted on the surface of colloidal particles.[104] Such grafting *via* ATRP clearly demonstrates the potential of this polymerization technique to confer desirable characteristics on the hybrid soft core-shell colloidal particles.

9.2.9 Reversible Addition–Fragmentation Chain-Transfer Radical (RAFT) Polymerization

RAFT polymerization is among the most widely used controlled radical polymerization techniques. In RAFT processes, the polymerization conditions are similar to those in conventional radical polymerization except that they involve the addition of a specific chain-transfer agent (CTA).[105–109] RAFT polymerization employs a CTA in the form of a RAFT agent based on thiocarbonylthio compounds such as dithioesters, xanthates and thiocarbamates to mediate the polymerization. A typical CTA generally used in RAFT processes is shown in Figure 9.12.[110] Here the Z group serves to

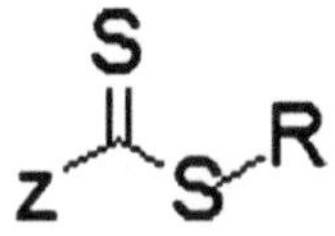

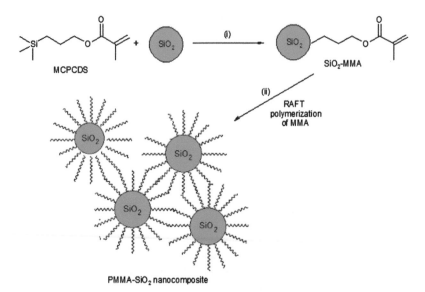

Figure 9.12 Chain-transfer agents used in RAFT polymerization. R and Z groups
are described in the text.

Figure 9.13 Grafting a polymer to the surface of silica by the 'grafting through'
method using RAFT polymerization.
Adapted from Ref. 118.

activate or deactivate the reactivity of the C=S bond towards the addition
reaction and the R group forms a stable free radical. RAFT polymerization
via a reversible chain-transfer process is capable of generating polymer
particles with low polydispersity indexes and a prechosen molecular weight
during the free-radical polymerization.

Among the distinctive features of RAFT polymerization is its tolerance to
different kinds of reaction conditions such as a large range of solvents including water in a wide temperature range. It is also suitable for different
functional monomers and does not require highly rigorous removal of oxygen and other impurities.[106,108,111–113] Owing to such advantages, RAFT
polymerizations can be used to design polymers of complex architectures,
such as linear block copolymers, comb-like and star polymers, dendrimers
and also polymer brushes. In the last case, RAFT polymerization has been
employed to graft polymers to the surface of various substrates, including
colloidal particles. For example, RAFT polymerization was used to graft
polymers to silica particles by using either 'grafting to'[114,115] or 'grafting
from'[116,117] methods. In a more recent study, a new 'grafting through'
method was used to graft polymer chains to the surface of silica by using

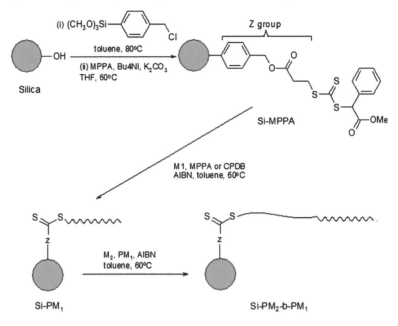

Figure 9.14   Synthetic route to polymer-grafted silica particles using the Z group by RAFT polymerization.
Adapted from Ref. 124.

RAFT polymerization as illustrated in Figure 9.13.[118] This method has also been used to graft a polymer on silica without the use of RAFT polymerization.[119–123] However, the 'grafting from' approach has been widely used to synthesize silica nanocomposites by employing different types of RAFT agents that utilized the Z group to attach the RAFT agent to the silica surface, as shown in Figure 9.14,[124] and other typical RAFT methods used the R group by attaching the RAFT agent to the surface of a substrate.[116,117]

9.2.10   Nitroxide-Mediated Polymerization (NMP)

NMP is another controlled living free radical polymerization process like ATRP and RAFT polymerization. NMP makes use of a compound known as an alkoxyamine initiator, which is viewed as an alcohol bound to a secondary amine by an N-O single bond as shown in Figure 9.15.[125]

The typical NMP reaction is controlled by reversible capping and decapping of growing (radical) polymer chains by the nitroxide radical. This results in a decrease in the concentration of growing radicals and also slows the polymerization. Consequently, the radical chain grows at a controlled rate to generate polymers of well-defined molecular weight, polydispersity index and architecture. The NMP process allows a relative lack of true termination, which in turn allows polymerization to continue as long as there is available monomer. Because of this, the NMP is said to be 'living' in

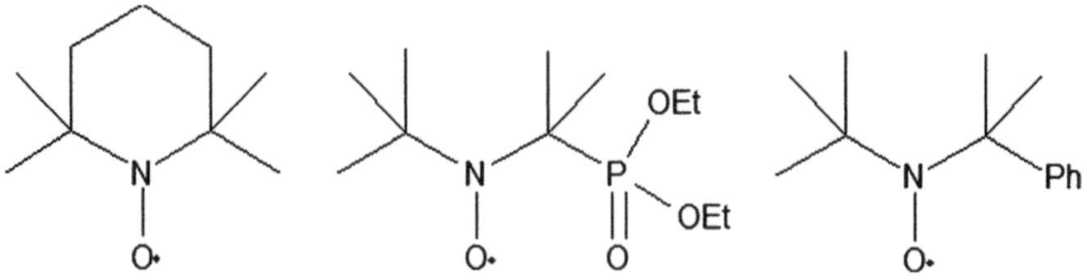

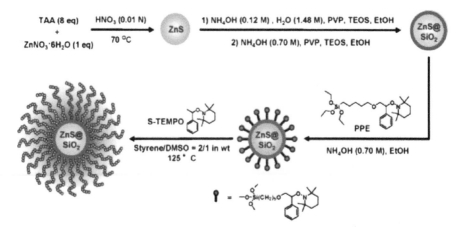

Figure 9.15 Commonly used alkoxyamine nitroxide initiating agents for NMP.[125]

Figure 9.16 Synthesis of ZnS@SiO₂–PS particles by a surface-initiated nitroxide-mediated polymerization process.[126]

nature and almost all of the growing chains are 'capped' by a mediating nitroxide. This consequently results in the dissociation and growth of chains at very similar rates to generate a largely uniform polymer chain length and structure. Because of such unique characteristics, the NMP process has been successfully employed to carry out surface-initiated polymerization (SIP) to alter the characteristics of particle surfaces. For example, a fairly recent study involved the synthesis of hybrid colloidal ZnS@SiO₂–PS particles by surface-initiated nitroxide-mediated polymerization as shown in Figure 9.16.[126] Here the procedure involved various steps including the synthesis of monodisperse ZnS particles by a homogeneous precipitation–aggregation process and then covering through uniform layer formation of SiO₂ particles. These ZnS@SiO₂ particles were subsequently functionalized with alkoxyamine moieties to initiate the surface polymerization *via* the NMP process. The SIP *via* NMP of styrene in DMSO solvent using S-TEMPO as sacrificial initiator was achieved with relatively good control over the molecular weight and molecular weight distribution of the grafted polymer, and polymer brushes of various molecular weights and high graft density were synthesized.

9.2.11 Conventional Seed Radical Polymerization

The conventional seed radical polymerization process offers another versatile approach to developing soft hybrid core-shell colloidal particles. In this method, free-radical polymerization of the desired monomer from the core or seed particles is carried out. One of the main advantages is that monomers with multiple functionalities can be polymerized from the surface of the seed particles. This is important since there has not been much work reporting the elaboration of core-shell particles grafted with functional polymers that possess hydrophilic and particularly temperature-sensitive magnetic latexes. In one of this studies, the elaboration of thermally sensitive magnetic latexes using a stepwise approach to obtain submicron composite particles was reported.[55] The inverse microemulsion polymerization process has also been studied to obtain hydrophilic nanoparticles of low magnetic content and narrow size distribution.[127–129]

In another interesting study, latex particles with different magnetic contents were prepared using conventional seed radical polymerization of functional monomers.[130] This study was focused on the elaboration of hydrophilic cationic and thermally sensitive magnetic latexes from an oil-in-water magnetic emulsion. To achieve such objective, two-stage polymerization processes were used: (i) the elaboration of seed magnetic latex particles and (ii) the functionalization of the seed by using water-soluble monomers. The cationic character was induced by using an amino-containing monomer. The encapsulation was performed using water-soluble reactants and the functionalization was induced by using the amino-containing monomer. Thermal sensitivity was induced by the use of NIPAM. The use of a large amount of functional monomer leads to high water-soluble polymer formation and highly cationic particles. Through this method, colloidal particles with different functionalization and applications can be prepared. For example, for the elaboration of hairy soft hybrid particles, temperature-sensitive highly magnetic latex particles have been reported.[130] The polymerization occurred in the presence of stabilized oil-in-water magnetic droplets by using a hydrophobic monomer as shown in Figure 9.17. This polymerization was performed in two steps. First, the

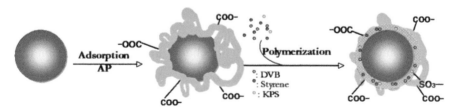

Figure 9.17   Transformation of oil-in-water ferrofluid droplet (composed of highly concentrated iron oxide nanoparticles stabilized *via* surfactant adsorption and dispersed in an aliphatic solvent) into magnetic latex particles using an amphiphilic polymer (AP).
Adapted from Ref. 130.

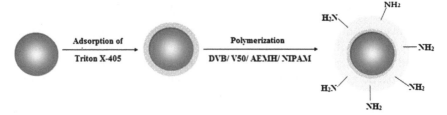

Figure 9.18   Conventional seed radical polymerization using NIPAM, MBA and AEM as a functional monomer to form a shell of cross-linked polymer on a magnetic core.
Adapted from Ref. 130.

preparation of seed magnetic latex particles was carried out. Then, functionalized magnetic latexes were obtained through seed polymerization by using water-soluble reactants. Typical reaction conditions involved the preparation of seeded magnetic latexes particles by first washing with Triton X-405. To avoid the formation of water-soluble polymer in order to incorporate NIPAM on the particle surface, two types of cross-linkers, divinylbenzene (DVB) and *N,N*-methylenebisacrylamide (MBA), were used. Encapsulation depended on the water-soluble cross-linker (MBA). Water-soluble polymer formation was reduced by increasing the amount of water-soluble cross-linker (MBA) during polymerization, which led to the formation of a polymer shell.[130] The composite latexes obtained possessed a shell of cross-linked polymer and a magnetic core as shown in Figure 9.18.

9.3   Biomedical Applications of Soft Hybrid Particles

The soft hybrid colloidal particles are not only designed to control surface functionality, morphology or chemical composition and are not only under evaluation in various biomedical applications, but are also used in practice at various biomedical levels: (i) they are used as solid supports for sample preparation and in bionanotechnology such as in rapid diagnostics; (ii) hairy-like particles bearing end chains and containing functional groups (mainly carboxylic groups in the surface) are generally used for chemical grafting of biomolecules in order to enhance the capture efficiency of a targeted analyte; (iii) charged soft hybrid colloidal particles are used for generic extraction of biomolecules such as adsorption of nucleic acids, proteins and viruses thanks to the highly charged polymer tentacles on the particles, as described in the following.

9.3.1   Extraction of Nucleic Acids

The main purpose in biomedical applications is to enhance sensitivity and specificity, but in the area of nucleic acid probes, problems such as the level of sensitivity are encountered. However, this problem can be solved by increasing the concentration of the target (RNA and/or DNA) in the considered medium. Before specific detection of the target probe, the use of appropriate hybrid particles for extraction, purification and desorption of already adsorbed nucleic acids would be of great interest.[131] The control of the adsorption and desorption of nucleic acid molecules is effected by interaction between prepared magnetic latex particles and nucleic acid extraction. The adsorption of nucleic acid molecules is performed at acidic pH and low salinity.[131] In *in vitro* biomedical diagnostics, magnetic particles are used for sample manufacture and separation of particles from the aqueous phase. Generally, specific and non-specific processes are used for nucleic acid extraction.

9.3.1.1   Non-Specific Capture of Nucleic Acids

The non-specific capture of nucleic acid molecules using particles is mainly based on electrostatic interactions. Being positively charged, the attractive electrostatic forces lead to rapid and high adsorption. Then, nucleic acid extraction is achieved by using cationic magnetic particles and controlling the pH and salinity of the medium.[37] The extracted nucleic acid molecules are subsequently amplified on magnetic beads. After desorption, the magnetic particles are removed by using a polymerase chain reaction (PCR)[132] in the case of DNA molecules or Reverse Transcriptase (RT)-PCR[133] in the case of RNA, as illustrated in Figure 9.19.[134]

9.3.1.2   Specific Capture of Nucleic Acids

The specific capture of nucleic acids[135,136] using soft hybrid magnetic latex particles is generally performed as follows. First, the captured probe of well-defined sequence is chemically immobilized on the magnetic latex particles.

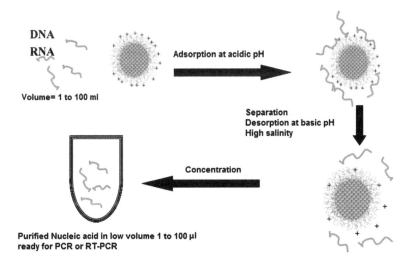

Figure 9.19   Illustration of non-specific capture, purification and concentration of nucleic acid molecules.
Adapted from Ref. 134.

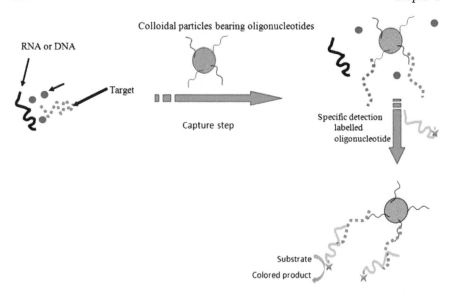

Figure 9.20 Illustration of specific capture and detection of nucleic acids (ODN, oligonucleotide; ssDNA, single-stranded DNA fragment). The intensity of the supernatant due to the presence of coloured product was determined. Specific capture is performed by addition of a substrate that reacts with enzyme in order to form a coloured medium. Adapted from Ref. 138.

Second, the biological sample is mixed with magnetic particle–ODN (ODN = oligonucleotide) conjugates. The specific capture of the target is effected by hybridization (particularly hydrogen bonding). The detection process is performed by adding a detection probe (*i.e.* oligonucleotide labelled with enzyme).[131] Then, the substrate addition is oxidized by the enzyme, which leads to a coloured supernatant as in immunoassay. This specific capture of nucleic acid molecules combined with a well-optimized detection process leads to the enhancement of this molecular biology-based diagnostic method[137] as illustrated in Figure 9.20.[138]

9.3.2 Extraction of Proteins

In biomedical applications, latex particles have been used as a particulate carrier, so substantial work has been carried out on the preparation of hydrophobic magnetic particles.[139,140] For protein denaturation and irreversible adsorption, polystyrene has been used as a solid phase when applied in the diagnostics field. The use of thermosensitive microgel particles with low critical solution temperature (LCST) is of great interest in the biomedical field, because their physicochemical properties and the adsorption of proteins can be controlled by changing the pH, the salinity of the medium and the temperature.[141,142] In addition such particles give non-denaturing support to immobilized proteins, in immunoassays and in protein purification

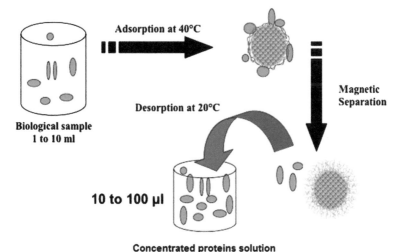

Figure 9.21   Schematic representation of protein concentration using thermosensitive latex particles.
Adapted from Ref. 143.

and concentration. Thermosensitive magnetic latex particles[143] have been used for protein concentration and purification by controlling both adsorption and desorption processes, as illustrated in Figure 9.21.

The adsorption and desorption of protein on/from thermosensitive magnetic latex particles have been studied using human serum albumin (HSA) as a protein model. Seed precipitation polymerization has been used to prepare thermosensitive core–shell magnetic latex by using magnetic polystyrene particles as a seed, NIPAM as the main monomer, MBA as a cross-linker and potassium persulfate (KPS) as an initiator. The polymerization conversion and water polymer formation have been examined. The adsorption of HSA protein was principally governed by hydrophobic interactions, since high adsorbed amounts of protein were obtained above the LCST of PNIPAM whereas negligible adsorption was evidenced below the LCST. The adsorption kinetics of HSA protein were found to be complete within 10 min at 40 °C. The amount of HSA protein adsorbed on anionic thermosensitive magnetic latex particles at 40 °C (above the LCST) decreased with increase in the pH and the salinity of the medium, reflecting the contribution of electrostatic interactions to the adsorption process. HSA desorption occurred after performing adsorption at acidic pH ($\sim$4.5). Various experimental factors, such as incubation time, pH, salinity and temperature of the medium, drastically affected the amount of protein desorbed. The maximum amount of protein desorbed was obtained at basic pH ($\sim$8.6), an ionic strength of $\sim$0.1 M and a temperature of 20 °C.[143]

The performance of such thermosensitive magnetic latex particles as soft hybrid material in the purification and concentration of HSA protein was

improved by careful control of both the adsorption and desorption conditions (pH, salinity, incubation time and temperature). The results are quite
encouraging, indicating that such particles could serve as an alternative
route to other techniques used for purifying and concentrating proteins,
e.g. precipitation using media of high salt concentration and an ion-
exchange column, such as a diethylaminoethyl-Sepharose gel column system.

9.3.3 Extraction of Viruses

The extraction of haemorrhagic fever viruses (such as Lassa and Ebola) is a
recent diagnostic method for developing countries. In recent years, only a
little work has been carried out using magnetic beads for virus detection and
concentration.[144,145] Moreover, magnetic latex particles have been used for
the capture of enveloped viruses.[146] This method is based on the use of
functionalized magnetic latex particles for capturing biological samples.
After capture, the separation of magnetic particles from supernatant occurred *via* a permanent magnet. The extraction of nucleic acids from the
captured viruses was performed by using a commercial kit. Usually, the
methodology for the concentration of viruses is used to increase the virus
concentration in the sample with no loss of viral activity and infectivity. PCR
and RT-PCR were applied in the measurement of the sensitivity of the captured assay. Satoh *et al.*[144] reported that a few RNA and DNA viruses were
concentrated more than 100- and 1000-fold, respectively, using polyethylenimine (PEI)-conjugated magnetic beads. Elaissari *et al.*[145] produced
functionalized magnetic beads for generic virus extraction and purification.
The magnetic colloidal emulsion was functionalized by surface immobilization of polymers. A polycation, PEI, was first adsorbed and then a polyanion, poly(maleic anhydride-*co*-methyl vinyl ether), was either adsorbed or

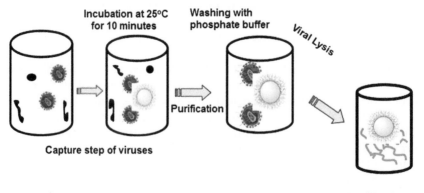

Figure 9.22   Capture of viruses using magnetic particles.
Adapted from Ref. 145.

covalently grafted. Negatively charged particles were used as a solid support for generic capture of viruses with yellow fever virus as a model; 90% of the virus could be captured and 10% released after washing. The processes of capture and detection of viral RNA were performed in the same Eppendorf tube. A short incubation of the sample was performed at room temperature and then the particle washing was performed by magnetic separation, supernatant elimination and redispersion. Finally, viral lysis and RT-PCR were carried out, as shown in the Figure 9.22.[145]

9.4   Conclusion

The preparation of soft hybrid nanoparticles has attracted considerable attention owing to the potential application of such materials in various fields. Such materials are a combination of both organic and inorganic moieties in well-defined structure and morphology. The softness of such nanoparticles is mainly due to the soft polymer shell. Such materials can be prepared using different processes starting from easy to more complex approaches.

The first and easiest method is basically the adsorption of polymer layers on hybrid nanoparticles. This adsorption can be performed in one step or multiple steps such as random sequential adsorption of oppositely charged polymers. It is interesting that the adsorption process leads to the presence of some aggregated particles due to the bridging flocculation phenomenon. Consequently, this adsorption approach was generally discarded.

In order to avoid the aggregation process, chemical grafting has been widely explored using batch grafting of reactive polymers, 'grafting from' or 'grafting to.' The chemical grafting approach involves the immobilization of preformed polymers or using chemical synthesis approaches, including mainly conventional seeded polymerization processes 'from' or 'to' the colloidal particles. However, conventional polymerization leads to layers of heterogeneous thickness and to an uncontrolled degree of polymerization of the formed brush soft layer.

In order to control the microstructure and the soft shell part, polymerization methods such as atom-transfer radical polymerization (ATRP), reversible addition–fragmentation chain-transfer (RAFT) polymerization and nitroxide-mediated radical polymerization (NMP) have been used. Another interesting polymerization technique based on click chemistry has also been investigated and gave promising results, but RAFT polymerization remains the most interesting approach and leads to a well-controlled soft layer.

Colloidal particles and especially soft hybrid nanoparticles have potential and promising applications, particularly in the biomedical diagnostics domain. The fascinating features of such colloidal particles make them good and suitable candidates for rapid conjugation and sensitive detection of various biomolecules in a very small volume of analyte.

Functionalized soft nanoparticles are used for sample preparation in which the extraction of biomolecules such as proteins, nucleic acids, bacteria and viruses are of paramount importance in order to enhance the sensitivity of *in vitro* biomedical diagnostics. Since the discovery of bionanotechnology, special attention has been dedicated to the elaboration of multifunctionalized and stimuli-responsive soft hybrid nanoparticles for use not only as solid supports for the immobilization of biomolecules, but also for transport and even for detection in microsystems combining microfluidic and lab-on-a-chip technology.

References

1. A. K. Gupta and M. Gupta, *Biomaterials*, 2005, **26**, 3995.
2. S. S. Davis, *Trends Biotechnol.*, 1997, **15**, 217.
3. C. E. Mora-Huertas, H. Fessi and A. Elaissari, *Int. J. Pharm.*, 2010, **385**, 113.
4. A.-H. Lu, E. L. Salabas and F. Schüth, *Angew. Chem. Int. Ed.*, 2007, **46**, 1222.
5. A. V. Bychkova, O. N. Sorokina, M. A. Rosenfeld and A. L. Kovarski, *Russ. Chem. Rev.*, 2012, **81**, 1026.
6. M. M. Rahman and A. Elaissari, *J. Colloid Sci. Biotechnol*, 2012, **1**, 3.
7. P. Boisseau and B. Loubaton, *C. R. Phys.*, 2011, **12**, 620.
8. E. Hübner, J. Allgaier, M. Meyer, J. Stellbrink, W. Pyckhout-Hintzen and D. Richter, *Macromolecules*, 2010, **43**, 856.
9. S. M. Gravano, R. Dumas, K. Liu and T. E. Patten, *J. Polym. Sci.: Polym. Chem.*, 2005, **43**, 3675.
10. L. Bombalski, H. Dong, J. Listak, K. Matyjaszewsk and M. R. Bockstaller, *Adv. Mater.*, 2007, **19**, 4486.
11. L. Zhou, J. Yuan, W. Yuan, M. Zhou, S. Wu, Z. Li, X. Xing and D. Shen, *Mater. Lett.*, 2009, **63**, 1567.
12. Á. Muñoz-Noval, V. Sánchez-Vaquero, V. Torres-Costa, D. Gallach, V. Ferro-Llanos, J. Javier Serrano, M. Manso-Silván, J. P. García-Ruiz, F. del Pozo and R. J. Martín-Palma, *J. Biomed. Opt.*, 2011, **16**, 025002.
13. I. W. Hamley, *Angew. Chem. Int. Ed.*, 2003, **42**, 1692.
14. J. H. Lee, H. H. Ahn, K. S. Kim, J. Y. Lee, M. S. Kim, B. Lee, G. Khang and H. B. Lee, *J. Tissue Eng. Regen. Med.*, 2008, **2**, 288.
15. G. Jiang, K. Park, J. Kim, K. S. Kim, E. J. Oh, H. Kang, S.-E. Han, Y.-K. Oh, T. G. Park and S. Kwang Hahn, *Biopolymers*, 2008, **89**, 635.
16. F. X. Gu, R. Karnik, A. Z. Wang, F. Alexis, E. Levy-Nissenbaum, S. Hong, R. S. Langer and O. C. Farokhzad, *Nano Today*, 2007, **2**, 14.
17. C. J. Sunderland, M. Steiert, J. E. Talmadge, A. M. Derfus and S. E. Barry, *Drug Dev. Res.*, 2006, **67**, 70.
18. J. H. Park, S. Lee, J.-H. Kim, K. Park, K. Kim and I. C. Kwon, *Prog. Polym. Sci.*, 2008, **33**, 113.
19. S. S. Rajan, H. Y. Liu and T. Q. Vu, *ACS Nano*, 2008, **2**, 1153.
20. C. Tekle, B. van Deurs, K. Sandvig and T.-G. Iversen, *Nano Lett.*, 2008, **8**, 1858.

21. J. Yan, M. C. Estévez, J. E. Smith, K. Wang, X. He, L. Wang and W. Tan, *Nano Today*, 2007, **2**, 44.
22. T. T. Hien Pham, C. Cao and S. J. Sim, *J. Magn. Magn. Mater.*, 2008, **320**, 2049.
23. R. Bhattacharya and P. Mukherjee, *Adv. Drug Deliv. Rev.*, 2008, **60**, 1289.
24. C. Sun, J. S. H. Lee and M. Zhang, *Adv. Drug Deliv. Rev.*, 2008, **60**, 1252.
25. X. Song, L. Huang and B. Wu, *Anal. Chem.*, 2008, **80**, 5501.
26. A. Elaissari and H. Fessi, *Braz. J. Phys.*, 2009, **39**, 146.
27. M. Horisberger, *Scanning Electron Microsc.*, 1980, 9.
28. C. Loo, A. Lowery, N. Halas, J. West and R. Drezek, *Nano Lett.*, 2005, **5**, 709.
29. V. Sokolova, A. Kovtun, O. Prymak, W. Meyer-Zaika, E. A. Kubareva, E. A. Romanova, T. S. Oretskaya, R. Heumann and M. Epple, *J. Mater. Chem.*, 2007, **17**, 721.
30. I. W. Bauer, S.-P. Li, Y.-C. Han, L. Yuan and M.-Z. Yin, *J. Mater. Sci. Mater. Med.*, 2008, **19**, 1091.
31. I. I. Slowing, J. L. Vivero-Escoto, C.-W. Wu and V. S.-Y. Lin, *Adv. Drug Deliv. Rev.*, 2008, **60**, 1278.
32. J. M. Singer and C. M. Plotz, *Am. J. Med.*, 1956, **21**, 888.
33. N. G. Portney and M. Ozkan, *Anal. Bioanal. Chem.*, 2006, **384**, 620.
34. M. G. Harisinghani, J. Barentsz, P. F. Hahn, W. M. Deserno, S. Tabatabaei, C. H. van de Kaa, J. de la Rosette and R. Weissleder, *N. Engl. J. Med.*, 2003, **348**, 2491.
35. M. E. Kooi, V. C. Cappendijk, K. B. J. M. Cleutjens, A. G. H. Kessels, P. J. E. H. M. Kitslaar, M. Borgers, P. M. Frederik, M. J. A. P. Daemen and J. M. A. van Engelshoven, *Circulation*, 2003, **107**, 2453.
36. Y. K. Kim, H. S. Kwak, C. S. Kim, G. H. Chung, Y. M. Han and J. M. Lee, *Radiology*, 2006, **238**, 531.
37. A. Elaissari, *Macromol. Symp.*, 2009, **281**, 14.
38. T. A. Taton, C. A. Mirkin and R. L. Letsinger, *Science*, 2000, **289**, 1757.
39. J. Huang, X. Wang and X. Deng, *J. Colloid Interface Sci.*, 2009, **337**, 19.
40. B. R. Sharma, N. C. Dhuldhoya and U. C. Merchant, *J. Polym. Environ.*, 2006, **14**, 195.
41. A. V. Dobrynin and M. Rubinstein, *Prog. Polym. Sci.*, 2005, **30**, 1049.
42. E. Killmann, D. Bauer, A. Fuchs, O. Portenlänger, R. Rehmet and O. Rustemeier, in *Structure, Dynamics and Properties of Disperse Colloidal Systems*, ed. H. Rehage and G. Peschel, Steinkopff, Heidelberg, 1998, pp. 135–143.
43. H. D. Bijsterbosch, *Copolymer Adsorption and the Effect on Colloidal Stability*, thesis, Wageningen Agricultural University, 1998.
44. S. Åkerman, K. Åkerman, J. Karppi, P. Koivu, A. Sundell, P. Paronen and K. Järvinen, *Eur. J. Pharm. Sci.*, 1999, **9**, 137.
45. T.-H. Young, J.-N. Lu, D.-J. Lin, C.-L. Chang, H.-H. Chang and L.-P. Cheng, *Desalination*, 2008, **234**, 134.
46. W. K. Idol and J. L. Anderson, *J. Membr. Sci.*, 1986, **28**, 269.

47. F. Caruso, in *Colloid Chemistry II*, ed. P. D. M. Antonietti, Springer, Berlin, 2003, pp. 145–168.

48. F. Caruso, *Adv. Mater.*, 2001, **13**, 11.

49. Z. Liang, C. Wang, Z. Tong, W. Ye and S. Ye, *React. Funct. Polym.*, 2005, **63**, 85.

50. R. Veyret, T. Delair and A. Elaissari, *J. Magn. Magn. Mater.*, 2005, **293**, 171.

51. M. Lansalot, M. Sabor, A. Elaissari and C. Pichot, *Colloid Polym. Sci.*, 2005, **283**, 1267.

52. M. Okubo, K. Ichikawa, M. Tsujihiro and Y. He, *Colloid Polym. Sci.*, 1990, **268**, 791.

53. R. H. Ottewill, A. B. Schofield, J. A. Waters and N. S. J. Williams, *Colloid Polym. Sci.*, 1997, **275**, 274.

54. F. Sauzedde, A. Elaissari and C. Pichot, *Colloid Polym. Sci.*, 1999, **277**, 846.

55. F. Sauzedde, A. Elaissari and C. Pichot, *Colloid Polym. Sci.*, 1999, **277**, 1041.

56. L. Hu, A. Percheron, D. Chaumont and C.-H. Brachais, in *Microwave Heating*, ed. U. Chandra, InTech, Rijeka, Croatia, 2011, Chapter 12, pp. 265–290.

57. B. Charleux, F. D'Agosto and G. Delaittre, in *Hybrid Latex Particles*, ed. A. M. van Herk and K. Landfester, Springer, Berlin, 2010, pp. 125–183.

58. J. Pyun and K. Matyjaszewski, *Chem. Mater.*, 2001, **13**, 3436.

59. C. J. Galvin and J. Genzer, *Prog. Polym. Sci.*, 2012, **37**, 871.

60. H. C. Kolb, M. G. Finn and K. B. Sharpless, *Angew. Chem. Int. Ed.*, 2001, **40**, 2004.

61. M. Meldal, *Macromol. Rapid Commun.*, 2008, **29**, 1016.

62. M. Meldal and C. W. Tornøe, *Chem. Rev.*, 2008, **108**, 2952.

63. C. W. Tornøe, C. Christensen and M. Meldal, *J. Org. Chem.*, 2002, **67**, 3057.

64. V. V. Rostovtsev, L. G. Green, V. V. Fokin and K. B. Sharpless, *Angew. Chem. Int. Ed.*, 2002, **41**, 2596.

65. W. H. Binder and R. Sachsenhofer, *Macromol. Rapid Commun.*, 2008, **29**, 952.

66. W. H. Binder and R. Sachsenhofer, *Macromol. Rapid Commun.*, 2007, **28**, 15.

67. W. H. Binder and C. Kluger, *Curr. Org. Chem.*, 2006, **10**, 1791.

68. W. H. Binder and R. Sachsenhofer, in *Click Chemistry for Biotechnology and Materials Science*, ed. J. Lahann, Wiley, Chichester, 2009, pp. 119–175.

69. H. D. B. Gacal, *Macromolecules*, 2006, **39**, 5330.

70. J. W. Chan, B. Yu, C. E. Hoyle and A. B. Lowe, *Chem. Commun.*, 2008, 4959.

71. R. Ranjan and W. J. Brittain, *Macromolecules*, 2007, **40**, 6217.

72. M. Kar, P. S. Vijayakumar, B. L. V. Prasad and S. S. Gupta, *Langmuir*, 2010, **26**, 5772.

73. P. Paoprasert, J. W. Spalenka, D. L. Peterson, R. E. Ruther, R. J. Hamers, P. G. Evans and P. Gopalan, *J. Mater. Chem.*, 2010, **20**, 2651.
74. T. Zhang, Y. Wu, X. Pan, Z. Zheng, X. Ding and Y. Peng, *Eur. Polym. J.*, 2009, **45**, 1625.
75. C. Li, J. Hu, J. Yin and S. Liu, *Macromolecules*, 2009, **42**, 5007.
76. N. Li and W. H. Binder, *J. Mater. Chem.*, 2011, **21**, 16717.
77. W. H. Binder, D. Gloger, H. Weinstabl, G. Allmaier and E. Pittenauer, *Macromolecules*, 2007, **40**, 3097.
78. E. Drockenmuller, I. Colinet, D. Damiron, F. Gal, H. Perez and G. Carrot, *Macromolecules*, 2010, **43**, 937.
79. T. Zhang, Z. Zheng, X. Ding and Y. Peng, *Macromol. Rapid Commun.*, 2008, **29**(21), 1716–1720.
80. R. Ranjan and W. J. Brittain, *Macromol. Rapid Commun.*, 2008, **29**, 1104.
81. R. Ranjan and W. J. Brittain, *Macromol. Rapid Commun.*, 2007, **28**, 2084.
82. Y. Tran and P. Auroy, *J. Am. Chem. Soc.*, 2001, **123**, 3644.
83. D. Hritcu, W. Müller and D. E. Brooks, *Macromolecules*, 1999, **32**, 565.
84. O. Prucker and J. Rühe, *Macromolecules*, 1998, **31**, 592.
85. T. von Werne and T. E. Patten, *J. Am. Chem. Soc.*, 2001, **123**, 7497.
86. M. Husseman, E. E. Malmström, M. McNamara, M. Mate, D. Mecerreyes, D. G. Benoit, J. L. Hedrick, P. Mansky, E. Huang, T. P. Russell and C. J. Hawker, *Macromolecules*, 1999, **32**, 1424.
87. R. Jordan, A. Ulman, J. F. Kang, M. H. Rafailovich and J. Sokolov, *J. Am. Chem. Soc.*, 1999, **121**, 1016.
88. M. D. K. Ingall, C. H. Honeyman, J. V. Mercure, P. A. Bianconi and R. R. Kunz, *J. Am. Chem. Soc.*, 1999, **121**, 3607.
89. R. Jordan and A. Ulman, *J. Am. Chem. Soc.*, 1998, **120**, 243.
90. B. Zhao and W. J. Brittain, *Macromolecules*, 2000, **33**, 8813.
91. M. Ejaz, S. Yamamoto, K. Ohno, Y. Tsujii and T. Fukuda, *Macromolecules*, 1998, **31**, 5934.
92. J.-B. Kim, M. L. Bruening and G. L. Baker, *J. Am. Chem. Soc.*, 2000, **122**, 7616.
93. M. Manuszak Guerrini, B. Charleux and J.-P. Vairon, *Macromol. Rapid Commun.*, 2000, **21**, 669.
94. C. Perruchot, M. A. Khan, A. Kamitsi, S. P. Armes, T. von Werne and T. E. Patten, *Langmuir*, 2001, **17**, 4479.
95. K. N. Jayachandran, A. Takacs-Cox and D. E. Brooks, *Macromolecules*, 2002, **35**, 6070.
96. M. Matsuo, Y. Sugiura, T. Kimura and T. Ogita, *Macromolecules*, 2002, **35**, 6070.
97. W. Huang, J.-B. Kim, M. L. Bruening and G. L. Baker, *Macromolecules*, 2002, **35**, 1175.
98. D. M. Jones and W. T. S. Huck, *Adv. Mater.*, 2001, **13**, 1256.
99. D. Bontempo, N. Tirelli, G. Masci, V. Crescenzi and J. A. Hubbell, *Macromol. Rapid Commun.*, 2002, **23**, 417.
100. S. Edmondson, V. L. Osborne and W. T. S. Huck, *Chem. Soc. Rev.*, 2004, **33**, 14.

101. J. N. Kizhakkedathu and D. E. Brooks, *Macromolecules*, 2003, **36**, 591.
102. J. N. Kizhakkedathu, R. Norris-Jones and D. E. Brooks, *Macromolecules*, 2004, **37**, 734.
103. J. N. Kizhakkedathu, K. R. Kumar, D. Goodman and D. E. Brooks, *Polymer*, 2004, **45**, 7471.
104. M. Zhang, L. Liu, H. Zhao, Y. Yang, G. Fu and B. He, *J. Colloid Interface Sci.*, 2006, **301**, 85.
105. D. A. Shipp, *J. Macromol. Sci. Part C: Polym. Rev.*, 2005, **45**, 171.
106. J. Chiefari, Y. K. Chong, F. Ercole, J. Krstina, J. Jeffery, T. P. T. Le, R. T. A. Mayadunne, G. F. Meijs, C. L. Moad, G. Moad, E. Rizzardo and S. H. Thang, *Macromolecules*, 1998, **31**, 5559.
107. C. Schilli, M. G. Lanzendörfer and A. H. E. Müller, *Macromolecules*, 2002, **35**, 6819.
108. G. Moad, E. Rizzardo and S. H. Thang, *Aust. J. Chem.*, 2005, **58**, 379.
109. G. Moad, Y. K. Chong, A. Postma, E. Rizzardo and S. H. Thang, *Polymer*, 2005, **46**, 8458.
110. J. Chiefari, R. T. A. Mayadunne, C. L. Moad, G. Moad, E. Rizzardo, A. Postma and S. H. Thang, *Macromolecules*, 2003, **36**, 2273.
111. Y. K. Chong, T. P. T. Le, G. Moad, E. Rizzardo and S. H. Thang, *Macromolecules*, 1999, **32**, 2071.
112. S. Perrier and P. Takolpuckdee, *J. Polym. Sci.: Polym. Chem.*, 2005, **43**, 5347.
113. A. Favier and M.-T. Charreyre, *Macromol. Rapid Commun.*, 2006, **27**, 653.
114. K. Bridger, D. Fairhurst and B. Vincent, *J. Colloid Interface Sci.*, 1979, **68**, 190.
115. B. Vincent, *Chem. Eng. Sci.*, 1993, **48**, 429.
116. C. Li and B. C. Benicewicz, *Macromolecules*, 2005, **38**, 5929.
117. C. Li, J. Han, C. Y. Ryu and B. C. Benicewicz, *Macromolecules*, 2006, **39**, 3175.
118. P. S. Chinthamanipeta, S. Kobukata, H. Nakata and D. A. Shipp, *Polymer*, 2008, **49**, 5636.
119. P. Espiard and A. Guyot, *Polymer*, 1995, **36**, 4391.
120. H. B. Sunkara, J. M. Jethmalani and W. T. Ford, *Chem. Mater.*, 1994, **6**, 362.
121. S. A. Asher, J. Holtz, L. Liu and Z. Wu, *J. Am. Chem. Soc.*, 1994, **116**, 4997.
122. E. Bourgeat-Lami and J. Lang, *J. Colloid Interface Sci.*, 1998, **197**, 293.
123. C. Barthet, A. J. Hickey, D. B. Cairns and S. P. Armes, *Adv. Mater.*, 1999, **11**, 408.
124. Y. Zhao and S. Perrier, *Macromolecules*, 2006, **39**, 8603.
125. J. Nicolas, Y. Guillaneuf, C. Lefay, D. Bertin, D. Gigmes and B. Charleux, *Prog. Polym. Sci.*, 2013, **38**, 63.
126. V. Ladmiral, T. Morinaga, K. Ohno, T. Fukuda and Y. Tsujii, *Eur. Polym. J.*, 2009, **45**, 2788.
127. Z. Xu, C. Wang, W. Yang, Y. Deng and S. Fu, *J. Magn. Magn. Mater.*, 2004, **277**, 136.

128. Y. Deng, L. Wang, W. Yang, S. Fu and A. Elaissari, *J. Magn. Magn. Mater.*, 2003, **257**, 69.
129. H. Macková, D. Králová and D. Horák, *J. Polym. Sci. Polym. Chem.*, 2007, **45**, 5884.
130. G. Pibre, L. Hakenholz, S. Braconnot, H. Mouaziz and A. Elaissari, *e-Polymers*, 2009, No. 139.
131. A. Elaissari, M. Rodrigue, F. Meunier and C. Herve, *J. Magn. Magn. Mater.*, 2001, **225**, 127.
132. O. Yamada, T. Matsumoto, M. Nakashima, S. Hagari, T. Kamahora, H. Ueyama, Y. Kishi, H. Uemura and T. Kurimura, *J. Virol. Methods*, 1990, **27**, 203.
133. F. Mallet, G. Oriol, C. Mary, C. Verrier and B. Mandrand, *Biotechniques*, 1995, **18**, 678.
134. A. Elaissari, *e-Polymers*, 2005, No. 028.
135. F. Mallet, C. Hebrard, J. M. Livrozet, O. Lees, F. Tron, J. L. Touraine and B. Mandrand, *J. Clin. Microbiol.*, 1995, **33**, 3201.
136. F. Mallet, C. Hebrard, D. Brand, E. Chapuis, P. Cros, P. Allibert, J. M. Besnier, F. Barin and B. Mandrand, *J. Clin. Microbiol.*, 1993, **31**, 1444.
137. M. H. Charles, M. T. Charreyre, T. Delair, A. Elaissari and C. Pichot, *STP Pharma Sci.*, 2001, **11**, 251.
138. A. Elaissari, *Macromol. Symp.*, 2005, **229**, 47.
139. J.-C. Olivier, M. Taverna, C. Vauthier, P. Couvreur and D. Baylocq-Ferrier, *Electrophoresis*, 1994, **15**, 234.
140. P. Van Dulm and W. Norde, *J. Colloid Interface Sci.*, 1983, **91**, 248.
141. H. Kawaguchi, K. Fujimoto and Y. Mizuhara, *Colloid Polym. Sci.*, 1992, **270**, 53.
142. A. Elaissari, L. Holt, F. Meunier, C. Voisset, C. Pichot, B. Mandrand and C. Mabilat, *J. Biomater. Sci. Polym. Ed.*, 1999, **10**, 403.
143. A. Elaissari and V. Bourrel, *J. Magn. Magn. Mater.*, 2001, **225**, 151.
144. K. Satoh, A. Iwata, M. Murata, M. Hikata, T. Hayakawa and T. Yamaguchi, *J. Virol. Methods*, 2003, **114**, 11.
145. R. Veyret, A. Elaissari, P. Marianneau, A. A. Sall and T. Delair, *Anal. Biochem.*, 2005, **346**, 59.
146. A. Arkhis, A. Elaissari, T. Delair, B. Verrier and B. Mandrand, *J. Biomed. Nanotechnol.*, 2010, **6**, 28.

CHAPTER 10

Computer Simulations of Soft Nanoparticles and Their Interactions with DNA-Like Polyelectrolytes

SERGE STOLL

Institute Forel, University of Geneva, Earth and Environmental Science Section, 10 route de Suisse, CP 416, 1290 Versoix, Switzerland
Email: serge.stoll@unige.ch

10.1   Introduction

The design of functional molecular architectures and the formation of electrostatic complexes at the nanometre scale have emerged as a major and novel research area for biological and biomedical applications.[1–8] In particular, the controlled association between polyelectrolytes (*i.e.* charged polymers) and solid or soft nanoparticles or macroions such as micelles,[9] globular proteins,[10] dendrimers[11] and membranes,[12] using electrostatic driving binding mechanisms, has emerged as a promising means to stabilize, destabilize, assemble and control the chemical and biological reactivity of suspensions containing nanoparticles.[13] In particular, nanoparticles, especially those having dimensions below 100 nm, have been proved to be very promising owing to their unique properties with regard to their size (diffusive properties) and surface properties (chemical reactivities). In addition, many biopolymers, such as DNA, are also polyelectrolytes and the formation

RSC Nanoscience & Nanotechnology No. 34
Soft Nanoparticles for Biomedical Applications
Edited by José Callejas-Fernández, Joan Estelrich, Manuel Quesada-Pérez and Jacqueline Forcada

of assemblies with nanoparticles,[14] proteins,[15] membranes[16] and vesicles,[17] for example, are expected to play critical roles in biological regulation processes with important potential applications in therapeutic delivery systems.[18]

However, in most situations and since nanoparticle and colloids suspensions are thermodynamically unstable, they have to be made stable or at least metastable for long periods.[19] Usually, an energy barrier is created by the formation of charges at the surface of the nanoparticles of interest. Flexible soft polymer chains can also be used to create a protective steric layer around the nanoparticles.[20] As a result, electric and/or steric stabilization is ensured to avoid irreversible aggregation during the different steps of handling (synthesis, functionalization, storage) and in application areas where they are expected to be used. This is particularly important with respect to stability problems in biological fluids that exhibit significant ionic strength and consist of mixtures containing other nanoparticles, colloids, macroions, multivalent ions, *etc.* For biomedical applications, magnetic nanoparticles have, for example, to be stabilized both in suspension and under physiological conditions, by the adsorption of suitable polyelectrolytes, which can provide both electrostatic and steric stabilization, on the nanoparticle surfaces.[21]

In addition, for given applications, the nanoparticle surface properties must be controlled. Different criteria then can be considered, such as particle size, particle dissolution, size distribution, surface polarity, presence of surface reactive groups which are pH dependent, hydrophilic *versus* hydrophobic balance of the surface and properties of the deposited polyelectrolyte corona at the nanoparticle surfaces.[22–24]

It should be noted that the interaction between polyelectrolytes and the nanoparticle surfaces is still today a complex topic. The long-range attractive and/or repulsive character of electrostatic interactions between polyelectrolytes and nanoparticles, solution chemistry, chemical composition of the different compounds, geometry and concentration of both polyelectrolytes and nanoparticles, and also competitive adsorption and aggregation processes, *etc.*, give these solutions very specific and labile properties which are only partially understood and hence difficult to control.[25,26] Hence so far, little is known at present about the use of polyelectrolytes with nanoparticles regarding all the parameters and variables that need to be considered. Therefore, there is an urgent need for the understanding of the dynamics and structure of such complexes at the molecular level. Future advances in the design of such functional molecular nanostructures composed both of nanoparticles and polyelectrolytes must be based on a better and detailed understanding of nanoparticle–polyelectrolyte interactions at the molecular level, the resulting structures and long-term structural stability.[27–32] The coupling of the experimental techniques used in soft condensed matter (static and dynamic light scattering techniques, nanoparticle tracking analysis, electrophoretic measurements, thermodynamic calorimetric analysis, *etc.*) with detailed computer simulations is expected to improve considerably the

understanding and design of functional nanoparticles and the final structural behaviour (*e.g.* stability) in solution.

Computer simulations provide a very valuable approach to gain an insight
into the understanding of the structures of nanoparticle complexes and their
interactions with their surroundings. Even simple questions such as the
effects of size variation on the properties of the nanoparticle complexes,
surface charge changes and corona structure can be difficult to answer
experimentally unless precise control over the building process is
achieved.[13,33]

Computer modelling is able to test independently the influence of various
parameters such as pH, temperature, nanoparticle charge, building block
properties and specific environments such as extreme temperatures and
physiological conditions. It can also provide at the atomistic level more
detailed information about the structures of these complex than is possible
with experimental measurements. In this chapter, we attempt to give the
reader an overview of problems that can be addressed and show how
simulations can therefore act as an efficient tool to explore the large parameter space for complex nanostructures where experiments cannot give
definitive answers.

We start by briefly discussing some of the computational methods that are
used and have been developed to model the structure of individual nanoparticles, systems containing nanoparticles and nanoparticle interactions
with their surroundings by describing both their theoretical basis and their
advantages and disadvantages. This brief methodological overview is followed by the presentation of one computer simulation technique; coarse-
grained Monte Carlo simulation. As a prototypical system, we consider the
problem of the interaction of polyelectrolyte chains with oppositely charged
nanoparticles, discuss different situations and try to isolate the potential
influence of experimental parameters.

10.2 Computer Simulation Techniques

Investigations dealing with nanoparticle systems cover a wide range of time
and length scales,[34–38] from femtosecond dynamics including all atomic
details to real-time macroscopic phenomena (Figure 10.1). The wide range of
timescales is introduced by the short-time Brownian motion and the long-
time hydrodynamic behaviour of the solvent whereas the range of length
scales is related to the size separation between nanoparticles and the assemblies that they can form, large polymer chains, gels and solvent molecules. Since nanoparticles and the structures they form cannot be easily
treated with quantum mechanical methods, it is more appropriate to use
'classical' molecular approaches. Molecular approaches are commonly used
in studies relating to the formation and evolution of individual nanoparticles. Classical computer simulation methods include the Monte Carlo
(MC), molecular dynamics (MD) and Brownian dynamics (BD) methods.[39] In
general, the MC method is the easiest to implement but it should be keep in

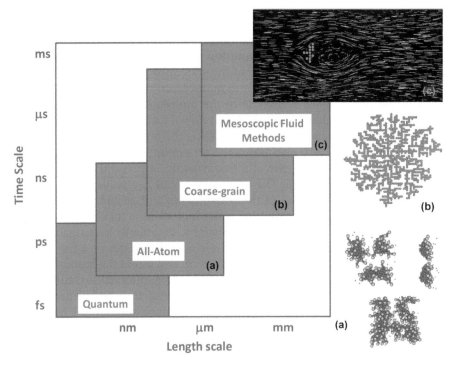

Figure 10.1 Illustration of different simulation techniques and corresponding temporal and spatial scales. (a) Surfactant micelle formation process considering an all-atom description;[52] (b) fractal structure resulting from the aggregation of individual nanoparticles;[46] (c) hydrodynamic behaviour of a fractal structure in a fluid.[46]

mind that it only provides information on systems in their equilibrium state. The MD method, on the other hand, is more complex but capable of simulating both non-equilibrium and equilibrium systems and gives kinetic or time-related information. In computer simulations, the instantaneous, systematic forces acting on each atom or particle, in the case of coarse-grain methods, are usually given in the form of interaction potentials (described in a force field) and where the energy is calculated as the sum over individual particles, pairs, triplets, *etc.*[40] Force fields are typically parameterized from a combination of experimental results and quantum calculations. Potential energy functions are additive and usually contain on the one hand bonding terms describing covalent interactions between atoms and on the other hand non-covalent terms to describe long-range interactions. Bonding terms represent a molecular deformation from an arbitrary reference geometry. Non-bonding terms describe van der Waals interactions, hydrogen bonding, electrostatic attractions and/or repulsions. A common description of these interactions includes a combination of Lennard–Jones potentials and Coulomb's law.

10.2.1 Molecular Dynamics

Based on Newton's equations of motion, this well-established method[41,42] allows the determination of the evolution of simple molecules and systems and the time-dependent behaviour of the system. During MD simulations, a system undergoes conformational and momentum changes so that different parts of the phase space accessible to the molecules can be explored. The conformational search capability of MD is one of its most important uses. By providing several mechanisms for controlling the temperature and pressure of simulated systems, MD also allows the generation of statistical ensembles from which various energetic, thermodynamic, structural and dynamic properties can be calculated. The advantage of MD is a realistic description of the microscopic dynamics of the constituents and permits, for example, the investigation of the transport of small molecules across bilayers, membrane channels, *etc.* Current MD simulations are able to investigate membrane patches of a few tens of nanometres over timescales of a few tens of nanoseconds. The largest scale simulations carried out at the atomistic level are also able to study lipid–protein or lipid–DNA interactions and to investigate channels through a bilayer lipid membrane. It should be noted that the bilayer structures in almost all of these atomistic simulations has to be preassembled because the timescale of self assembly from a dispersed mixture would greatly exceed the simulation timescale. The most time-consuming part of MD is the calculation of the force acting on each atom. The main difficulty with such an approach is that it cannot be used to model the behaviour of realistic processes containing nanoparticles (such as nanoparticle aggregation). A considerable simulation time would be required to follow the motion of the solvent. As a result, the application of MD to soft-matter systems is nowadays relatively limited.

10.2.2 Brownian Dynamics

In solutions containing different species, such as dispersions of nanoparticles, polymer chains, polyelectrolytes and individual molecules such as salt particles, the timescale characterizing the motion of each species can differ by several orders of magnitude. To simulate the dynamics of such systems when the fast motion of the small species is not relevant and where MD methods would require too demanding computer resources, approaches such as the Brownian dynamics (BD) method can be used.[43,44] In BD, the fast motion of the smaller particles is not considered and their impact on the large objects is represented by a combination of random and frictional forces. Here the Newtonian equations of motion are replaced by Langevin-type equations of motion. As a result, BD can be considered as a coarse-grain method including the viscous aspects of the surrounding solvent, thermal fluctuations and the dynamic correlation between the particles. As they do not generally account for the electrostatic and van der Waals effects, they are not always appropriate when strong electrostatic effects play a role, which is often the case. However, the computational cost of BD is so important that one cannot treat more than a few hundred Brownian particles. It should be noted that dissipative particle dynamics (DPD) is also an alternative[45,46] to classical MD and includes hydrodynamic and Brownian fluctuations. Here fluid molecules are not explicitly represented but groups of molecules called dissipative particles interact with each other, exchange momentum and move randomly. However, the method is computationally expensive, since the number of particles of interest is usually much smaller than the number of dissipative particles.

10.2.3 Monte Carlo Simulations

Monte Carlo simulations are used in many areas of mathematics, physics and chemistry.[47-49] MC simulations contain one random variable and use a sort of random path *via* a statistical sampling of multiple replicates. Here the particle momenta and velocities are not involved. A reliable pseudo-random number generator is also critical for the success of MC simulations and they are used to sample the conformational space and based on a Markov chain sequence of trials for which the result of successive trials depends on the immediate predecessor. This approach is successfully used to describe building processes between polymers, micelles formation, gel behaviour and the complexation process between nanoparticles and polyelectrolyte chains. MC simulation scheme can be used both at the atomistic level and at the coarse-grain level.[50-54]

10.2.4 Coarse-Grain Models

Although MD simulations have become more applicable to experimental systems with the development of increasingly accurate, specific force fields, they remain limited to nanometre length scales owing to the computational demand of calculating the interactions amongst all the atoms and in particular the van der Waals and electrostatic energy functions. For nanoparticles that are immersed in an aqueous solvent, collective phenomena reaching beyond tens of nanometres in size and a few microseconds, a different strategy must be adopted. Such restrictions have therefore resulted in the development of coarse-grain (CG) models, which provide the ability to compute and investigate the properties of materials such as nanoparticles, proteins and membranes, lipid interactions, self-assembly and phase transitions between self-assembled morphologies at the mesoscale level.[55-59] This was achieved by reducing the level of detail in the system representation. For example, each monomer constituting a polymer chain is represented by a single spherical coarse-grain unit. Hydrophobic and hydrophilic components are just linear monomer units or beads, making the parameterization easier. On the scale of the beads, the solvent can be viewed as a hydrodynamic continuum, characterized by its dielectric constant, viscosity and temperature. Since nanoparticles and polyelectrolyte systems can exhibit dimensions up to a few hundred nano-metres, coarse-grain models are ideal candidates in modelling soft matter-related systems. The challenge here consists in including just enough details so as to capture the physical chemistry that is relevant to investigate the problem of interest. This is usually done by (i) using multiscale modelling in which the various levels of treatment are coupled and transferred into one another and/or (ii) starting with a simple model and incrementally increasing its complexity by including additional interactions until the phenomenon of interest is captured.

As mentioned in the Introduction, this chapter offers an illustrative ex-ample of the application of computer simulation within coarse-grain models to soft matter issues: the complexation between polyelectrolyte chains and oppositely charged nanoparticles; the interested reader can find in the sci-entific literature other examples of biotechnological relevance,[60,61] many of them straightforwardly related to some of the soft nanoparticles analysed in this book.

This is particularly true for the swelling of micro- and nanogels. The usual model of a microgel is developed from an idealized representation of a polymer network commonly used in simulations of gels, in which monomer units and cross-linkers are modelled as spheres.[61,63] The network is also characterized by the numbers of nodes (cross-linkers), the number of chains connected to the nodes (functionality) and the number of neutral and charged monomer units per chain. Molecules such as N,N'-methylenebis-acrylamide can be considered as tetrafunctional cross-linkers. For that rea-son, a diamond-like topology (usual in simulations of gels) is also assumed. As commented on before, simulating the solvent explicitly is prohibitively time consuming. Nevertheless, an effective solvent-mediated hydrophobic attractive interaction between non-polar monomer beads is required in aqueous solutions to capture the thermo-shrinking behaviour of some temperature-sensitive nanoparticles, such as poly(N-isopropylacrylamide)-based micro- and nanogels.[62]

Even within this level of coarse graining (and without an explicit solvent), it should be pointed out that the number of particles required for the simulation of a whole microgel could be extremely high. In fact, simulating microgel particles with diameters greater than a few tens of nanometres can become an unassailable task. For some purposes, however, one could then simulate a small and inner piece of the microgel assuming that the rest of the gel behaves in the same manner and surface effects are negligible. In other words, we would simulate an infinite network, replicating a small piece of it (the simulation cell). This would permit us to examine computing-intensive properties, such the volume fraction or the osmotic pressure. Also. some of these properties could be related to extensive quantities (*e.g.* the diameter of the nanoparticle) if a reference state is perfectly known.[62] If the network is small enough, the nanogel can be explicitly simulated including the ionic atmosphere around it.[64]

10.2.5 Coarse-Grain Monte Carlo Model of Polyelectrolyte and Nanoparticle Complex Formation

In this section, a method for modelling the complexation process between a strong polyelectrolyte chain and one nanoparticle is discussed. The coarse-grain approach presented here has been demonstrated to be effective and reliable in studying specific phenomena such as the polyelectrolyte corona at the nanoparticle surface, nanoparticle surface charge modifications and calculation of adsorption/desorption limits. Quantitative comparisons with experiments were also made possible by providing microscopic insight into polyelectrolyte–nanoparticle system properties such as the evolution of the nanoparticle surface charge.

10.2.5.1 The Model

A pearl necklace coarse-grain model is used to generate off-lattice three-dimensional polyelectrolyte chains. They are represented as a succession of N freely jointed hard spheres and each sphere is considered to be a physical monomer of radius $\sigma_m = 3.57$ Å. The fraction of ionized monomers α is then adjustable by placing negative charges (equal to -1) at the centre of the monomers. The bond length is constant and equal to the Bjerrum length $l_B = 7.14$ Å. The nanoparticle is represented as a uniformly charged sphere with a variable radius σ_p so as to obtain full insight into nanoparticle curvature effects. The solvent is treated as a dielectric medium with a relative dielectric permittivity constant ε_r taken as that of water at 298 K, *i.e.* 78.5. The total energy E_{tot} ($k_B T$ units) for a given conformation is the sum of repulsive electrostatic interactions between the polyelectrolyte monomers and attractive electrostatic interactions between the polyelectrolyte and the nanoparticle E_{el}. Excluded volume interactions E_{ev} are also considered to include both monomer and particle excluded volumes. Van der Waals interactions can also be added into the model so as to take into account polyelectrolyte or nanoparticle hydrophobicity and non-electrostatic forces. To consider implicitly the solvent ionic strength and potential-related changes and reduce the simulation time, a Debye–Hückel approach is used to describe the electrostatic interactions. In the Debye–Hückel model, all pairs of charged monomers ij within the polyelectrolyte interact with each other *via* a screened Debye–Hückel long-range potential:

$$u_{el}(r_{ij}) = \frac{z_i z_j e^2}{4\pi \varepsilon_r \varepsilon_0 r_{ij}} \exp(-\kappa r_{ij}) \qquad (10.1)$$

where z_i is the amount of charge on unit i, e the elementary charge, ε_0 the dielectric permittivity of the vacuum, ε_r the dielectric permittivity of the solution, κ the inverse Debye screening length and r_{ij} the distance between the charged monomers.

Monomers interact with the particle according to a Verwey–Overbeek potential:

$$u'_{\text{el}}(r_{ij}) = \frac{z_i z_j e^2}{4\pi\varepsilon_r\varepsilon_0 r_{ij}} \frac{\exp\left[-\kappa(r_{ij} - \sigma_{\text{p}})\right]}{1 + \kappa\sigma_{\text{p}}} \tag{10.2}$$

where σ_{p} represents the nanoparticle radius.

Since free ions are not included explicitly in the simulations, their overall effects on monomer–monomer and monomer–particle interactions are described *via* the dependence of the inverse Debye screening length κ^2 on the electrolyte concentration according to

$$\kappa^2 = 1000 e^2 N_{\text{A}} \sum_i \frac{z_i^2 C_i}{\varepsilon_0 \varepsilon_r k_{\text{B}} T} \tag{10.3}$$

where N_{A} is Avogadro's number, C_i the ionic concentration, k_{B} the Boltzmann constant and T the absolute temperature.

The polyelectrolyte intrinsic chain stiffness can also be considered in the model and adjusted by a square-potential with variable amplitude to vary its strength. This gives the bending energy:

$$E_{\text{tor}} = \sum_{i=2}^{N} k_{\text{ang}}(\alpha_i - \alpha_0)^2 \tag{10.4}$$

where $\alpha_0 = 180°$, α_i is the angle achieved by three consecutive monomers $i - 1$, i and $i + 1$ and k_{ang} defines the strength of the angular potential or chain stiffness.

10.2.5.2 The Monte Carlo Procedure

Monte Carlo simulations are performed according to the Metropolis algorithm in the canonical ensemble (Figure 10.2). Successive 'trial' chain configurations are generated to obtain a reasonable sampling of low-energy conformations. After applying elementary movements which are randomly selected, the Metropolis selection criterion is employed either to select or to reject the move. If the change in total energy ΔE resulting from the move is negative, then the move is selected. If ΔE is positive a Boltzmann factor p:

$$p = \exp\left(\frac{-\Delta E}{k_{\text{B}} T}\right) \tag{10.5}$$

is computed and a random number z with $0 \leq z \leq 1$ is generated. If $z \leq p$, the move is selected. When $z > p$, the trial configuration is rejected and the previous configuration is retained and considered as a 'new' state in calculating ensemble averages. This conformation is the one that is perturbed in the next step. The perturbation process is continued a specified number of times (a typical run requires several million perturbations) until the conformation is energy minimized and equilibrated. To generate new conformations, the

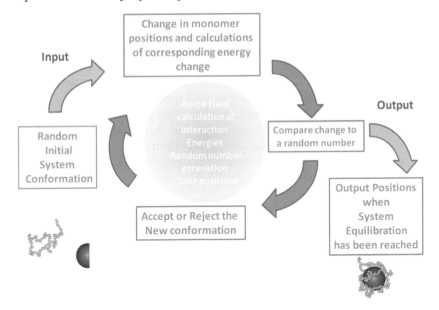

Figure 10.2 Schematic Monte Carlo simulation algorithm.

monomer positions are randomly modified by specific movements. These movements include three 'internal' or elementary movements (end–bond, kink–jump and crankshaft), the pivot and the reptation.[64] The use of all these movements is very important to ensure the ergodicity of the system and also the convergence toward minimized conformations.

One important challenge and problem is to allow the energy of the complex structure to be minimized gradually without trapping the structure in a local energy minimum. This problem is of particular importance when compact conformations have to be achieved or when large polyelectrolyte chains are considered because a few monomer–monomer contacts can lead to the formation of 'irreversible' bonds that freeze the complex structure. Some MC refinements are therefore necessary to overcome the formation of structures in local minima and increase the chances of success when sampling new conformations. Anneal Monte Carlo can be used to minimize the complex structures gradually by altering the temperature from an initial value to a final value and *vice versa*, to increase the chances of success when sampling conformations. However, as this method requires a large amount of CPU time, it should be used for the formation of dense structures only.

To investigate the formation of polyelectrolyte–nanoparticle complexes, the central monomer of the chain is initially placed at the centre of a large three-dimensional spherical box and the particle is randomly placed in the cell. The polyelectrolyte and the oppositely charged particle are then allowed to move (a random motion is used to move the particle). After each calculation step, the coordinates of both the particle and monomers are translated in order to replace the central monomer of the polyelectrolyte in the middle of the box. It should be noted that the chain has the possibility to diffuse further away and leave the particle surface during a simulation run (so the polyelectrolyte desorption process can be investigated). After relaxing the initial conformation through 10^6 cycles (equilibration period), chain properties are then calculated and recorded every 1000 cycles. Owing to the large number of possible situations to be investigated with regard to the changes in polyelectrolyte intrinsic rigidity, size and charge of the nano-particle and ionic strength, the application of this model with respect to the actual processor speeds has currently limited the chain length to 100 monomer units (200 for the isolated polyelectrolytes).

10.3 Monte Carlo Simulations of Complex Formation Between Nanoparticles and Polyelectrolyte Chains

10.3.1 Complexation Between a Strong Negatively Charged Polyelectrolyte Chain and an Oppositely Charged Nanoparticle

Using the previously described coarse-grain MC approach, we focus first on the influence of the polyelectrolyte length N. As the ionic strength C_i is also expected, *via* screening effects, to play a key role in controlling the electro-statics, including chain conformation *and* polyelectrolyte–nanoparticle interaction energies, we also focus on its influence.

10.3.1.1 Influence of Ionic Strength and Polyelectrolyte Chain Length on Complex Structures

It can be clearly seen from Table 10.1 that no polyelectrolyte adsorption is observed when $C_i \geq 1$ M, *i.e.* at high ionic strength. Attractive surface–polymer interactions in high ionic strength conditions are not strong enough to overcome the entropy loss of the polymer due to its confinement near the particle. This is an important point to consider for the potential application of such nanoparticle complexes in a physiological medium because of the high ionic strength of the latter. It is also important to observe that long chains are found preferentially adsorbed on the nanoparticle surface because less en-tropy is lost compared with the short chains, while they gain approximately the same (total) adsorption energy. When the polyelectrolyte is adsorbed on the nanoparticle, its conformation or corona structure is expected to be dif-ferent from that in solution. When $N = 25$, the adsorption approaches that on a nearly planar surface. In this case, the chain can fully spread on the surface with dimensions close to its dimensions in a free solution. When chain length is increased to $N = 100$, the polyelectrolyte continues to wrap around the nanoparticle to optimize the number of contacts. The conformation of the polyelectrolyte is then dictated by the particle size and is subject to the highest level of deformation (the maximum is observed here when $N = 140$).

Table 10.1 Equilibrated conformations between fully charged polyelectrolyte chains and oppositely charged nanoparticles of identical sizes. Polyelectrolyte length adjustment indicates that large polyelectrolytes are partially adsorbed at the nanoparticle surface and form an extended tail in solution. Also at high ionic strength the polyelectrolyte is not adsorbed at the nanoparticle surface owing to the strong reduction of the electrostatic interactions between the polyelectrolyte and the nanoparticle.

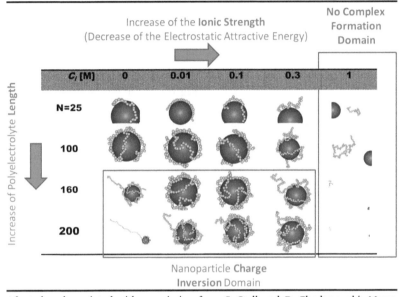

Adapted and reprinted with permission from S. Stoll and P. Chodanowski, *Macromolecules*, 2001, **34**, 2320. Copyright 2001 American Chemical Society.

On increasing chain length further, both the excluded volume and electrostatic repulsions between the monomers prevent any additional adsorption on the surface *via* the formation of an extended tail in solution. On the other hand, monomer adsorption is again promoted by increasing the ionic strength, which reduces the electrostatic repulsions between the monomers at the nanoparticle surface. The results presented here clearly demonstrate that the polyelectrolyte corona and amount of adsorbed polyelectrolyte are not simple monotonic functions of N and C_i. in particular when the polyelectrolyte is large enough to form a tail in solution.

10.3.1.2 *Impact of Polyelectrolyte Intrinsic Rigidity*

The behaviour of polyelectrolyte–nanoparticle complexes is usually described by considering the ionic concentration of the solution, polyelectrolyte length and charge linear density and nanoparticle charge and diameter. Nonetheless, an essential parameter is the polyelectrolyte chain flexibility, which includes both chain stiffening due to electrostatic monomer–monomer repulsions and stiffness of the underlying chain backbone, which is also called intrinsic flexibility.[65] Both effects control the polyelectrolyte persistence length, which is equal to 50 nm in the case of DNA. To achieve a comprehensive description of DNA-like–nanoparticle or polysaccharide-nanoparticle assemblies, equilibrated conformations of complex formation between a polyelectrolyte with variable flexibility and nanoparticles as a function of C_i and k_{ang} are presented in Table 10.2. No adsorption is observed when $C_i \geq 1$ M whereas adsorption is always observed when $C_i \leq 0.1$ M. When $C_i = 0.3$ M, Table 10.2 clearly demonstrates that adsorption is strongly controlled by the value of the polyelectrolyte persistence length. As shown here, adsorption is promoted by (i) decreasing the chain stiffness (and subsequently the energy required to confine the semi-flexible polyelectrolyte at the particle surface) and (ii) decreasing the ionic concentration (thus increasing the electrostatic attractive interactions between the monomer and the particle surface that we consider as the driving force for the adsorption). Because of charge screening in the high-salt regime, monomer–nanoparticle interactions are not large enough to overcome the polyelectrolyte

Table 10.2   Equilibrated conformation of complexes formed between nanoparticles (of identical sizes) and polyelectrolytes having different flexibilities (but with identical contour length). Typical structures are obtained such as tennis ball and solenoid conformations. The increase in the ionic strength and chain rigidity are expected to promote polyelectrolyte desorption. Solenoid conformations are only achieved for a rigid polyelectrolyte and when electrostatic coupling between the polyelectrolyte and the nanoparticle is strong.

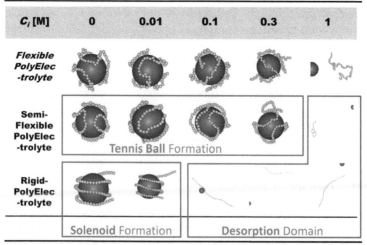

Adapted and reprinted with permission from S. Stoll and P. Chodanowski, *Macromolecules*, 2002, **35**, 9556. Copyright 2002 American Chemical Society.

confinement near the nanoparticle and therefore no complex formation is observed. When the chain is semi-flexible, 'tennis ball' conformations are achieved, whereas when rigid chains are considered, the intrinsic flexibility forces the polyelectrolyte to adopt solenoid conformations similarly to the packaging of DNA in chromatin, a fundamental element of biology in which a DNA of approximately 150 base pairs is tightly wrapped around a small protein complex to form the nucleosome.

10.3.1.3 Nanoparticle Size and Surface Charge Effects

We focus now on the influence of curvature effects related to the nanoparticle size variation. Indeed, depending on the relative size of the nanoparticle compared with the polyelectrolyte contour length, important changes in the polyelectrolyte corona structure are expected. Examples of equilibrated conformations of polyelectrolyte–nanoparticle complexes as a function of the ionic concentration C_i and nanoparticle/monomer size ratio σ_p/σ_m are presented in Table 10.3. The polymer length is constant and equal to $N = 100$ monomer units. The central point charge Q of the nanoparticle is adjusted so as to keep a constant surface charge density equal to $+100$ mC m^{-2}. As shown in Table 10.3, three distinct domains may be defined. Small nanoparticles do not permit the adsorption of all the polyelectrolyte monomers owing to the confinement energy of the chain. As a result, extended tails are formed in solution and the corona in very extended. When the nanoparticle radius is large enough, the polyelectrolyte collapses on it to form classical 'tennis ball'-like conformations. The corona is formed of trains and loops. On further increasing nanoparticle size, polyelectrolytes can spread to the same extent as on a flat surface with a corona essentially composed of trains. Finally, on increasing the ionic concentration, the charged polyelectrolyte starts to desorb. When screening is important, no adsorption is observed.

From a more quantitative point of view, when the polyelectrolyte is adsorbed at the particle surface, its conformation is expected to be different from that in solution without the influence of the nanoparticles. To gain an insight into the extent of change, the mean square radius of gyration $\langle R_g^2 \rangle$ of the polyelectrolyte can be calculated from the simulation data and represented as a function of C_i for different σ_p/σ_m values (Figure 10.3). It is clearly demonstrated here that maximum polyelectrolyte deformation is achieved when $C_i = 0.1$ M. Below this concentration, the electrostatic repulsion between the monomers is strong enough to limit the adsorption process through the formation of extended tails in solution. When $\sigma_p/\sigma_m = 8$, maximum deformation is achieved whatever the ionic concentration. The nanoparticle surface area is large enough that the polyelectrolyte is able to wrap around the sphere to optimize the number of contacts. When $\sigma_p/\sigma_m > 8$, the conformation and size of the polyelectrolyte are now dictated by the size of the particle. On increasing the particle size further, the nearly planar surface limit is now reached. In this case, the chain can spread fully on the surface with dimensions close to those in solution.

Table 10.3 MC equilibrated conformations of polyelectrolyte–particle complexes as a function of the ionic strength C_i and nanoparticle size ratio. The polyelectrolyte contour length is constant $(N=100)$, as is the nanoparticle surface charge density. Depending on the size of the particle, three regions may be defined. Small particles do not permit the adsorption of full polyelectrolyte owing to the importance of the confinement energy of the chain. As a result, extended tails are formed in solution. When the particle radius is large enough, the polyelectrolyte collapses on it to form 'tennis ball'-like conformation. Then, with further increase in particle size, polyelectrolytes can fully spread on the flat nanoparticle surface.

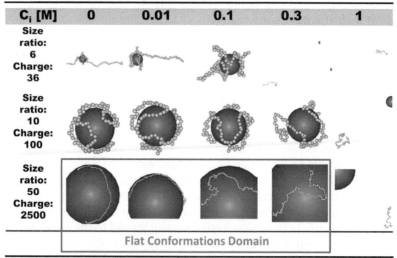

Adapted and reprinted with permission from P. Chodanowski and S. Stoll, *Journal of Chemical Physics*, 2001, **115**, 4951. Copyright 2001 AIP Publishing LLC.

10.3.1.4 Nanoparticle Charge Neutralization, Charge Inversion and Overcharging

An important issue when dealing with nanoparticle assemblies and applications is related to the final charge of the complex, which is expected to control the stability of the complex in solution. Moreover, the surface charge, which can be determined experimentally using electrophoretic measurements, constitutes one of the few experimental parameters allowing direct comparison between simulations and experiments. In Figure 10.4, we calculated as a function of the polyelectrolyte length N the variation of the number of adsorbed monomers N^{ads} in the salt-free case. With increasing size of the polyelectrolyte, several key results are found: (i) the chain is fully collapsed on the charged nanoparticle provided that the complex is undercharged; (ii) when $N>Q$ (*i.e.* the number of charged monomers is greater than the nanoparticle charge), more monomers adsorb on the particle surface than is necessary to

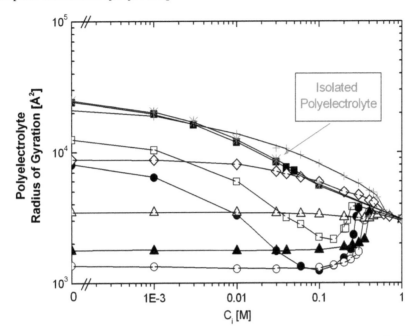

Figure 10.3 Mean square radius of gyration of the polyelectrolyte chain $\langle R_g^2 \rangle$ ($\mathring{A}^2$) as a function of ionic strength and polyelectrolyte nanoparticle relative size ratio σ_p/σ_m ($N=100$). $\sigma_p/\sigma_m = 2$ ($\blacksquare$), 5 ($\square$), 6 ($\bullet$), 8 ($\bigcirc$), 10 ($\blacktriangle$), 15 ($\triangle$), 25 ($\diamondsuit$), 50 (+), isolated ($\times$). The size of the nanoparticle and strength of the electrostatic interactions result in a complex and non-monotonic behaviour of the polyelectrolyte dimensions and resulting corona. Maximum polyelectrolyte compaction is achieved when $\sigma_p/\sigma_m = 8$.
Adapted and reprinted with permission from P. Chodanowski and S. Stoll, *Journal of Chemical Physics*, 2001, **115**, 4951. Copyright 2001 AIP Publishing LLC.

neutralize it and, as a result, the complex is overcharged. Moreover, accumulation of monomers close to the surface continues up to a critical value (here $N=151$); (iii) beyond that critical number of monomers, a protruding tail in solution appears, in good agreement with theoretical models predicting a first-order transition at this point followed by a small decrease in N^{ads} with N to reach rapidly a plateau value. It should be noted that tail formation is also observed in cells when considering DNA–histone complexes owing to the excess charge remaining from DNA chains.

We now consider the overcharging issue in the presence of added salt. All calculations were performed with $C_i \leq 0.03$ M, *i.e.* before the desorption process takes place. The number of collapsed monomers as a function of N and C_i is reported in Figure 10.5. By increasing the screening of the electrostatic interactions, it is clearly demonstrated that the polyelectrolyte monomer electrostatic confinement effect at the nanoparticle surface is reduced, thus allowing the adsorption of a greater number of monomers on

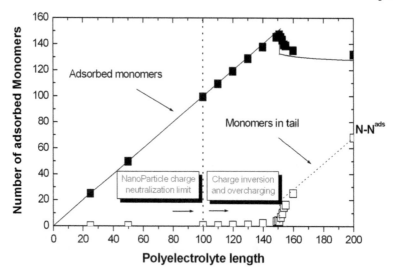

Figure 10.4 Number of adsorbed monomers N^{ads} and monomers in tails $(N - N^{ads})$ as a function of the polyelectrolyte contour length N. The number of adsorbed monomers is found to increase linearly and then stabilize after nanoparticle charge inversion. The additional monomers form an extended tail in solution to expel the excess charge.

Adapted and reprinted with permission from S. Stoll and P. Chodanowski, *Macromolecules*, 2001, **34**, 2320. Copyright 2001 American Chemical Society.

the nanoparticle surface. In addition to the increase in charge inversion with increase in C_i, Figure 10.5 also demonstrates that the position of the first-order transition is increased with increase in ionic strength.

10.3.1.5 Complexation Process Between One Polyelectrolyte and Several Nanoparticles

The main purpose of this section is to extend the model to systems involving several nanoparticles.[66] Two parameters are adjusted here, the surface charge of the nanoparticles and the ionic strength.

As shown in Table 10.4, dense aggregates are achieved with such flexible and strong polyelectrolytes. Here the polyelectrolyte chains act as a polymeric glue for the nanoparticles. It is found that, at low nanoparticle surface charge density, by decreasing the ionic strength complex formation is promoted, with regard to the number of nanoparticles attached to the polyelectrolyte chains, because of the increase in electrostatic attractions between the charged nanoparticles and the polyelectrolyte chain. On the other hand, when highly charged nanoparticles are considered, the best conditions for the formation of aggregates are obtained at high ionic strength so as to limit the effect of the electrostatic repulsions between the nanoparticles within the aggregate. The results presented here again

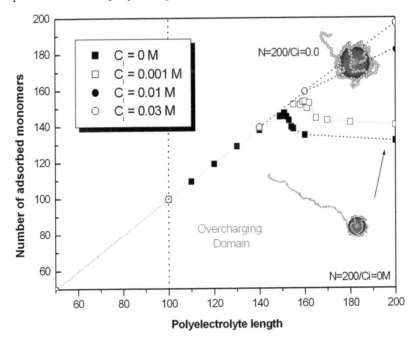

Figure 10.5 Number of collapsed monomers N^{ads} as a function of the total number of monomers in the chain N for different values of the ionic strength. It is found that by increasing the ionic strength, more monomers can be adsorbed at the nanoparticle surface.
Adapted and reprinted with permission from S. Stoll and P. Chodanowski, *Macromolecules*, 2001, **34**, 2320. Copyright 2001 American Chemical Society.

demonstrate the non-linear behaviour of such complexes with regard to the ionic strength variation of the solution and indicate that the best conditions for complex formation are obtained with intermediate ionic strength values.

10.3.2 Complex Formation Between Nanoparticles and Weak Polyelectrolyte Chains

Owing to the importance of solution pH for the acid–base properties of polyelectrolytes and nanoparticles (and effects on polyelectrolyte degree of ionization and nanoparticle surface charge), it is critically important to understand the effects of pH on polyelectrolyte conformation and complex formation.[67] It is important to note that, because of their connectivity along the polyelectrolyte chain, charged monomers interact strongly between themselves and their acid–base properties are different from those of ideal systems. In the case of a polyacid, the total amount of charge increases with increase in the pH of the solution. However, the electrostatic interactions oppose the deprotonation. Hence one has to consider in the models an

Table 10.4 Equilibrated conformations of fully charged flexible polyelectrolyte in the presence of several nanoparticles *versus* ionic concentration C_i and nanoparticle surface charge. Four surface charge densities, $\sigma = +10$, $+25$, $+50$ and $+100$ mC m^{-2}, are considered. Non-adsorbed nanoparticles are not presented. Dense aggregates are mainly obtained and the number of nanoparticles is limited both by the nanoparticle surface charge (controlling the electrostatic repulsive interactions between the nanoparticle) and ionic strength, which plays a non-linear role depending on the nanoparticle surface charge.

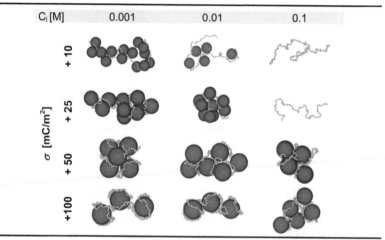

Reprinted with permission from S. Ulrich, M. Seijo, A. Laguecir and S. Stoll, *J. Phys. Chem. B*, 2006, **110**, 20954. Copyright 2006 American Chemical Society.

apparent dissociation constant K that is related to the solution pH and the polyelectrolyte degree of ionization α *via* the Henderson–Hasselbach equation:

$$pK = pH - \log\left(\frac{\alpha}{1-\alpha}\right) \qquad (10.6)$$

The differences in the acid–base properties of connected monomers in chains and isolated monomers are then given by

$$\Delta pK = pK - pK_0 = pH - pK_0 - \log\left(\frac{\alpha}{1-\alpha}\right) \qquad (10.7)$$

where pK_0 is the negative logarithm of the dissociation constant of a monomer in the absence of electrostatic interactions. ΔpK will largely depend on polyelectrolyte length, complex formation and the presence of the nanoparticles, and also on the ionic concentration of the solution because of the influence of screening effects between charges. Again, ionic strength is expected to play a central role here. To understand such effects better, the MC coarse-grain model has to be modified so as to include pH variations. After a given number of Monte Carlo steps to equilibrate the polyelectrolyte conformation, a monomer is chosen at random and, depending on the solution pH, its charge state is switched on or off. The energy change, ΔE, which determines the probability of accepting the new charge, is the sum of the change in electrostatic interaction ΔE_c and a term that corresponds to the change in free energy of the intrinsic association reaction of a monomer:

$$\Delta E = \Delta E_c \pm k_B T(\text{pH} - \text{p}K_0) \times \ln 10 \qquad (10.8)$$

Regarding the $\pm$ symbol in eqn (10.8), when a monomer is deprotonated a minus sign is used and when the monomer is protonated a plus sign is required. the grand canonical simulations, the chemical potential is fixed, hence the difference $\text{pH} - \text{p}K_0$ is an input parameter and, after energy minimization, the degree of ionization α is measured.

At this point, titration curves, *i.e.* the degree of dissociation of the polyelectrolyte as a function of pH, can give useful information on the complex formation between one nanoparticle and a weak anionic polyelectrolyte and can point out the importance of several competing effects.

As shown in the titration curve in Figure 10.6, (i) attractive interactions between the charged polyelectrolyte and the nanoparticle change the acid–base properties of the polyelectrolyte by promoting chain ionization and (ii) the ionic concentration decreases the attractive interaction between the polyelectrolyte and the nanoparticle but promotes polyelectrolyte ionization. The equilibrated structures in Table 10.5 also indicate that owing to charge mobility, polyelectrolyte charges accumulate at the nanoparticle surface, suggesting that annealed polyelectrolyte (polyelectrolyte chains having mobile charges) are expected to bind more strongly than quenched (fixed) polyelectrolytes with equivalent charge density, in good agreement with experimental findings.

10.3.3 Effect of Solution Chemistry (Presence of Di- and Trivalent Ions) on Complex Formation

Multivalent ions, which are naturally occurring in biological or environmental systems, are expected to induce important conformational changes to polyelectrolytes, dendrimers, gels, membranes, *etc.*, *via* local and very specific interactions.[68,69] Such specific effects cannot be explained by an implicit ion representation and consequently the model has to be refined. If free ions must be explicitly present then salt particles (both anions and cations) should be described by impenetrable hard spheres to take into account the ion excluded volume effect and ion mobility. For consistency, it is also important to consider the counterions of both the nanoparticle and polyelectrolyte chain. In our extended coarse-grain model, both the nanoparticle and polyelectrolyte counterions and also salt anions and salt cations have fixed radii of 2 Å whereas di- and trivalent salt cations have radii of 2.5 Å. Counterions and salt particles have a permanent charge on their centre of $+1$ for polyelectrolyte counterions, -1 for nanoparticle counterions, $+1$, $+2$ or $+3$ for salt cations and -1 for salt anions. Since the polyelectrolyte

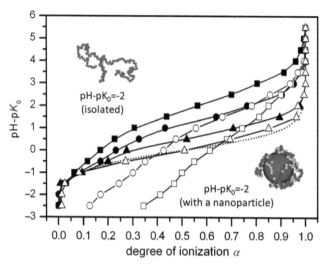

Figure 10.6   Titration curves of a weak flexible isolated polymeric acid ($N=160$) at various ionic strengths. In the absence of nanoparticles and at different salt concentrations, charge screening is found to control the polyelectrolyte degree of ionization ($C_i = \blacksquare$, 0.001 M; $\bullet$, 0.01 M; $\blacktriangle$, 0.1 M). In the presence of one nanoparticle (surface charge density 100 mC m^{-2}, the acid–base properties of the polyelectrolyte are profoundly modified ($C_i = \square$, 0.001 M; $\bigcirc$, 0.01 M; $\triangle$, 0.1 M). By adding an oppositely charged nanoparticle, the resulting complex formation changes the acid–base properties of the polyelectrolyte chain significantly by promoting monomer deprotonation. Blue beads represent uncharged monomers and yellow beads charged monomers.
Adapted and reprinted with permission from S. Ulrich, M. Seijo and S. Stoll, *Curr. Opin. Colloid Interface Sci.*, 2006, **11**, 268. Copyright 2006 Elsevier.

linear charge density is pH dependent, each monomer is a titrating site which can be neutral or negatively charged (-1). In the simulation box, all pairs of charged objects interact with each other *via* a full Coulombic electrostatic potential, which is positive or negative when repulsive or attractive interactions are considered and is defined as

$$u_{ij}\left(r_{ij}\right) = \begin{cases} \infty, & r_{ij} < R_i + R_j \\[2mm] \dfrac{z_i z_j e^2}{4\pi\varepsilon_0\varepsilon_r r_{ij}}, & r_{ij} \geq R_i + R_j \end{cases} \tag{10.9}$$

where $z_{i,j}$ represent the charges carried by monomers, the nanoparticle, counterions or salt particles, r_{ij} is the distance between them (centre to centre) and $R_{i,j}$ are their radii. Thus the total energy E_{tot} for a given conformation, which also includes hard core interactions, is given by the sum of the whole pairwise potentials u_{ij}. In the following examples, during the MC simulations, only the polyelectrolyte chain charges will vary with pH. Hence

Table 10.5 Weak polyelectrolyte–nanoparticle complex conformations considering a weak anionic polyelectrolyte (with $N = 160$) and an oppositely charged nanoparticle (radius $= 35.7$ Å). All pairs of charged monomers and nanoparticles interact with each other *via* a screened Debye–Hückel potential. The adsorption/desorption limit, conformation and number of monomers in the train depend on ionic concentration, nanoparticle surface charge density and pH of the solution. Blue beads represent uncharged monomers and yellow beads charged monomers.

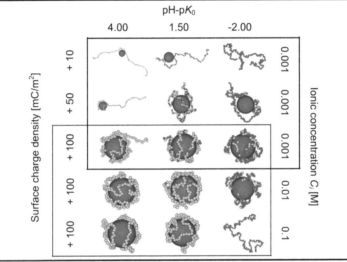

Adapted and reprinted with permission from S. Ulrich, M. Seijo and S. Stoll, *Curr. Opin. Colloid Interface Sci.*, 2006, **11**, 268. Copyright 2006 Elsevier.

the acceptance of each protonation/deprotonation step of the polyelectrolyte monomers is related to the MC Metropolis selection criterion already discussed. Each monomer charge is switched on or off depending on whether the monomer is neutral or charged, and oppositely charged counterions are randomly inserted or removed when charges appear on or disappear from the chain, respectively, to keep the system electrostatically neutral. The addition of explicit counterions results from the presence of an alkali such as NaOH. As a result, during the titration processes, the simulation box is coupled to a proton bath to establish a constant pH. For a given $pH - pK_0$ value, an energy stabilization equilibration period of 2.5×10^5 MC steps is achieved, followed by a production period of 7.5×10^5 steps where observables such as the chain degree of ionization α, polyelectrolyte radius of gyration, number of monomers in trains, tails and loops or the monomer radial distribution function around the nanoparticle, in addition to salt particle and counterion positions, are recorded to calculate ensemble averages.

10.3.3.1   Complex Formation in the Presence of a Monovalent Salt

The case of an isolated weak polyelectrolyte chain and one nanoparticle surrounded by explicit monovalent counterions and monovalent salt particles at $C_i = 1 \times 10^{-3}$ M is discussed first. Both pH and nanoparticle surface charge densities ranging from 25 to 100 mC m^{-2} have been adjusted. As shown in Table 10.6, polyelectrolyte deprotonation is induced by the increase of pH $-$ pK_0 values, which concomitantly promotes the adsorption of the polyelectrolyte on the nanoparticles. The nanoparticle effective charge is also controlled by the presence of its counterions, salt anions and polyelectrolyte monomers, whereas attractive interactions between monomers with counterions, salt cations and nanoparticles affect the polyelectrolyte linear charge density. It is found that on increasing pH $-$ pK_0, the polyelectrolyte chain moves closer to the nanoparticle surface, therefore releasing the nanoparticle counterions in the bulk. This indicates that entropy is expected to play a role here. When the polyelectrolyte linear charge density becomes higher than the nanoparticle bare charge (overcharging situation), polyelectrolyte counterions and salt cations are attracted again around the complex so as to reduce the electrostatic energy of the system (Table 10.6, *e.g.* when $\alpha = 1$ and $\sigma = 25$, 50 mC m^{-2}). At a given pH $-$ pK_0 value, the increase in the nanoparticle surface charge promotes polyelectrolyte deprotonation due to an increase in the attractive interactions between monomers and the nanoparticles. The results presented here clearly indicate that in some circumstances the co- and counterions also play key and subtle roles in the complex formation.

10.3.3.2   Effect of Salt Valency

The case of an isolated weak flexible polyelectrolyte $(N = 100)$ and a nanoparticle at $\sigma = 50$ mC m^{-2} surrounded by explicit monovalent counterions and also mono-, di- or trivalent salt particles $(C_i = 1 \times 10^{-3}$ M$)$ is now discussed. Table 10.7 presents the equilibrated conformations of one polyelectrolyte and one nanoparticle at various pH $-$ pK_0 values and salt valencies. It is clearly shown that increasing the salt valency results in higher polyelectrolyte degree of ionization α for a given pH $-$ pK_0 value. This is due to stronger specific monomer–salt cation associations and screening efficiency. Therefore, polyelectrolyte and nanoparticle complexation is in direct competition with the association of salt cations around polyelectrolyte chains. With increase in pH $-$ pK_0, the associated nanoparticle counterions and salt anions are progressively released into the bulk and replaced by the charged monomers of the polyelectrolyte. When $\alpha > 0.5$ and with a monovalent salt, polyelectrolyte counterions and salt cations are not distinguishable and the association is then unspecific. The case with divalent salt results in a specific association of the salt cations with the polyelectrolyte due to stronger monomer–salt attractive interactions. Indeed, by

Table 10.6 Weak flexible polyelectrolyte in the presence of one nanoparticle surrounded by explicit monovalent counterions and monovalent salt particles. Neutral and negatively charged monomers are represented by grey and yellow spheres, polyelectrolyte and nanoparticle counterions by purple and blue beads and salt particles by cyan spheres. Nanoparticle surface charge densities σ range from 25 to 100 mC m^{-2}. The degree of ionization α is also presented in the inset.

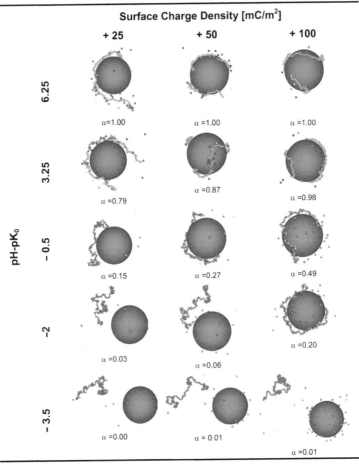

considering a trivalent salt, the adsorption of the trivalent cations around the polyelectrolyte chains when $\alpha > 0.5$ strongly influences the final conformations of the nanoparticle–polyelectrolyte complex. As a result, a subtle competition between salt cation interactions leads to the formation of locally

Table 10.7 Equilibrated conformations of a weak, flexible, negatively charged polyelectrolyte chain and one positively charged nanoparticle ($\sigma = +50$ mC m^{-2}). Each monomer can be neutral (grey sphere) or carry one negative charge (yellow sphere). Chain and nanoparticle monovalent counterions are purple and blue. Cases with a mono-, di- and trivalent salt (cyan spheres) are considered. The polyelectrolyte adsorption is promoted by an increase in the chain linear charge density. When $\alpha > 0.5$, the association between monomers and salt cations occurs more strongly with higher salt valencies, hence promoting the chain deprotonation and the competition with the polyelectrolyte–nanoparticle complexation.

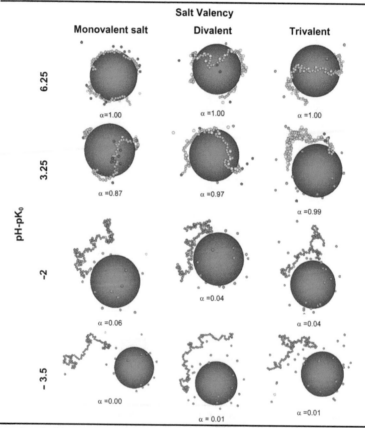

Reprinted with permission from F. Carnal and S. Stoll, *J. Phys. Chem. B*, 2011, **115**, 12007. Copyright 2011 American Chemical Society.

collapsed polyelectrolyte segments, which are desorbed from the nanoparticle surface (for instance, $\mathrm{pH} - \mathrm{p}K_0 = 3.25$ and $z_i = 3$). Therefore, the overall complexation process is found to be less efficient with a trivalent salt than with a mono- or divalent salt.

10.4   Conclusions and Outlook

The effects of ionic strength, nanoparticle size and surface charge, poly-electrolyte length, persistence length, pH and the presence of multivalent ions on the complex formation between a polyelectrolyte and oppositely charged nanoparticles have been investigated using a coarse-grain MC approach. It is clearly shown that MC simulations constitute a rewarding and invaluable approach successively to increase the model refinement and to isolate the relevant factors that control the polyelectrolyte conformation in solution and at the surface of the nanoparticles. From this model, adsorption limits, which are important for potential applications, distribution of ions, final complex charge, influence of pH and complex stability with regard to the solution ionic strength can thus be investigated at least in a qualitative way to address the formulation of polyelectrolyte–nanoparticle mixtures and guide new experiments.

From the simulation data presented here, important results on the complex formation between nanoparticles and polyelectrolytes can be derived in particular with regard to the influence of ionic strength, which is a key parameter in biological applications. Our simulations point out the importance of two competing effects when the ionic concentration increases; on the one hand, by decreasing the electrostatic repulsions between the charged polyelectrolyte substantial amounts of monomers can be attached to the nanoparticle surface; however, on the other hand, the electrostatic attraction between the nanoparticle and the monomer becomes less important, giving the monomers and the polyelectrolyte the opportunity to leave the particle surface. Therefore, the best conditions for complex formation will be obtained at intermediate ionic strengths.

Adsorption of charged polymers on nanoparticles is controlled not only by ionic concentration but also by particle diameter. Surface curvature effects clearly limit the amount of adsorbed monomers; large particles allow the polyelectrolyte to spread on the surface whereas small particles limit the number of adsorbed monomers that may be attached to it. When small particles are considered, the low-salt regime is dominated by polyelectrolyte monomer–monomer repulsions, forcing the polyelectrolyte to form extended tails in opposite directions. When the particle size is of the order of or larger than the radius of gyration of the charged polymers, polyelectrolytes can wrap fully around the nanoparticle.

MC results also demonstrate that the complexation between a polyelectrolyte and a charged sphere can lead to overcharging when the polyelectrolyte size is large enough. In the case of added salt, our simulations indicate the importance of charge inversion with increase in ionic concentration.

Chain stiffness influences the amount of adsorbed monomers, the monomer distribution at the particle surface and adsorption/desorption limits. The amount of adsorbed monomers has a maximum value for the semi-flexible chains in the low salt concentration regime. In such conditions, the polyelectrolyte is strongly adsorbed at the particle surface as a solenoid and the confinement energy does not contribute to the formation of tails in solution. When the intrinsic stiffness of the chain is small, tennis ball-like conformations are achieved. In contrast, when rigid chains are considered, the polyelectrolyte becomes tangential to the particle.

Polyelectrolyte adsorption is promoted by increasing its degree of ionization and hence $pH - pK_0$ values when weak polyelectolytes are considered, and/or decreasing the ionic concentration to promote electrostatic attractive effects.

This synthesis of simulation data reported here is a preliminary step towards an even more precise modelling of the problem to gain an insight into the behaviour of more concentrated polymer and nanoparticle solutions (systems with several chains). A simple model involving one chain interacting with one particle has been described, but it can be extended to more concentrated systems involving several chains (and/or nanoparticles). Also, the model and related concepts could be extended to different systems, including full soft systems such as gels, star polymers and dendrimers. We hope that the observations presented in this chapter will be particularly useful for the rational design of systems containing polyelectrolytes, ions and nanoparticles.

Acknowledgements

Many of the ideas presented here are based on the results of graduate and postgraduate students over the years, in particular P. Chodanowski, S. Ulrich, M. Seijo and F. Carnal. I am also grateful to Manuel Quesada Pérez for his critical reading of, comments on and input to this chapter. This research was supported over the years by the Swiss National Foundation (SNF), Swiss Commission pour la Technologie et l'Innovation (CTI) and the University of Geneva.

References

1. J. M. Lehn, *Supramolecular Chemistry: Concepts and Perspectives*, Wiley-VCH, Weinheim, 1997.
2. G. Decher, *Science*, 1997, **277**, 1232.
3. X. Zhang, H. Chen and H. Zhang, *Chem. Commun*, 2007, 1395.
4. R. Haag, *Angew. Chem. Int. Ed.*, 2004, **43**, 278.
5. Q. A. Pankhurst, J. Connolly, S. K. Jones and J. Dobson, *J. Phys. D: Appl. Phys.*, 2003, **36**, R167.
6. S. Wang, N. Mamedova, N. A. Kotov, W. Chen and J. Studer, *Nano Lett.*, 2002, **2**, 817.
7. J. M. Nam, C. C. Thaxton and C. A. Mirkin, *Science*, 2003, **301**, 1884.
8. Y. Zhang, N. Kohler and M. Zhang, *Biomaterials*, 2002, **23**, 1553.

9. D. W. McQuigg, J. I. Kaplan and P. L. Dubin, *J. Phys. Chem.*, 1992, **96**, 1973.
10. C. L. Cooper, P. L. Dubin, A. B. Kayitmazer and S. Turksen, *Curr. Opin. Colloid Interface Sci.*, 2005, **10**, 52.
11. C. L. Cooper, A. Goulding, A. B. Kayitmazer, S. Stoll, S. Turksen, S. Ulrich, S.-I. Yusa, A. Kumar and P. L. Dubin, *Biomacromolecules*, 2006, 7, 1025.
12. L. Ouyang, R. Malaisamy and M. L. Bruening, *J. Membr. Sci.*, 2008, **310**, 76.
13. S. Ulrich, M. Seijo and S. Stoll, *Curr. Opin. Colloid Interface Sci.*, 2006, **11**, 268.
14. H. J. Chung, C. M. Castro, H. Im, H. Lee and R. Weissleder, *Nat. Nanotechnol.*, 2013, **8**, 369.
15. M. L. Bulyk, E. Gentalen, D. J. Lockhart and G. M. Church, *Nat. Biotechnol.*, 1999, **17**, 573.
16. J. J. Cockburn, N. G. A. Abrescia, J. M. Grimes, G. C. Sutton, J. M. Diprose, J. M. Benevides, G. J. Thomas, J. K. H. Bamford, D. H. Bamford and D. I. Stuart, *Nature*, 2004, **432**, 122.
17. R. S. Dias, B. Lindman and M. G. Miguel, *J. Phys. Chem. B*, 2002, **106**, 12600.
18. S. Parveen, R. Misra and S. K. Sahoo, *Nanomed.: Nanotechnol. Biol. Med.*, 2012, **8**, 147.
19. R. J. Hunter, *Introduction to Modern Colloid Science*, Oxford University Press, Oxford, 1993.
20. A. Elaissari (ed.), *Colloidal Nanoparticles in Biotechnology*, Wiley, Hoboken, NJ, 2008.
21. A. Elaissari (ed.), *Colloidal Biomolecules, Biomaterials and Biomedical Applications*, Marcel Dekker, New York, 2004.
22. M. Lundqvist, J. Stigler, G. Elia, I. Lynch, T. Cedervall and K. A. Dawson, *Proc. Natl. Acad. Sci. U. S. A.*, 2008, **105**, 14265.
23. C. Ge, J. Du, L. Zhao, L. Wang, Y. Liu, D. Li, Y. Yang, R. Zhou, Y. Zhao, Z. Chai and C. Chen, *Proc. Natl. Acad. Sci. U. S. A.*, 2011, **108**, 16968.
24. P. L. Rodriguez, *Science*, 2013, **339**, 971.
25. R. A. Sperling and W. J. Parak, *Philos. Trans. R. Soc. London, Ser. A*, 2010, **368**, 1333.
26. C. Fang, N. Bhattarai, C. Sun and M. Zhang, *Small*, 2009, **5**, 1637.
27. M. Pavlin and V. B. Bregar, *Dig. J. Nanomater. Biostruct.*, 2012, 7, 1389.
28. A. Hartland, J. R. Lead, V. I. Slaveykova, D. O'Carroll and E. Valsami-Jones, *Nat. Educ. Knowledge*, 2013, **4**, 7.
29. J. Buffle, K. J. Wilkinson, S. Stoll, M. Filella and J. Zhang, *Environ. Sci. Technol.*, 1998, **32**, 2887.
30. D. Palomino, C. Yamunake, P. Le Coustumer and S. Stoll, *J. Colloid Sci. Biotechnol.*, 2013, **2**, 2.
31. D. Palomino and S. Stoll, *J. Nanopart. Res.*, 2013, **15**, 1428.
32. F. Loosli, P. Le Coustumer and S. Stoll, *Water Res.*, 2013, **47**, 6052.
33. J.-P. Chapel and J.-F. Berret, *Curr. Opin. Colloid Interface Sci.*, 2012, **17**, 268.

34. S. Wang and E. E. Dormidontova, *Soft Matter*, 2011, 7, 4435.

35. Q. H. Zeng, A. B. Yu and G. Q. Lu, *Prog. Polym. Sci.*, 2008, **33**, 191.

36. G. Allegra, G. Raos and M. Vacatello, *Prog. Polym. Sci.*, 2008, **33**, 683.

37. D. Fritz, K. Koschke, V. A. Harmandaris, N. F. van der Vegt and K. Kremer, *Phys. Chem. Chem. Phys.*, 2011, **13**, 10412.

38. R. B. Ross and S. Mohanty (eds), *Multiscale Simulation Methods for Nanomaterials*, Wiley, Hoboken, NJ, 2008.

39. D. Frenkel and B. Smit, *Understanding Molecular Simulation: from Algorithms to Applications*, Academic Press, New York, 2001.

40. M. P. Allen and D. J. Tildesley, *Computer Simulation of Liquids*, Oxford University Press, Oxford, 1989.

41. D. C. Rapaport, *The Art of Molecular Dynamics Simulation*, Cambridge University Press, Cambridge, 2004.

42. M. Nedyalkova, S. Madurga, S. Pisov, I. Pastor, E. Vilaseca and F. Mas, *J. Chem. Phys.*, 2012, **137**, 174701.

43. A. Satoh, *Introduction to Practice of Molecular Simulation: Molecular Dynamics, Monte Carlo, Brownian Dynamics, Lattice Boltzmann and Dissipative Particle Dynamics*, Elsevier, Amsterdam, 2011.

44. M. Seijo, S. Ulrich, M. Filella, J. Buffle and S. Stoll, *Environ. Sci. Technol.*, 2009, **43**, 7265.

45. R. D. Grootand and P. B. Warren, *J. Chem. Phys.*, 1997, **107**, 4423.

46. B. Chopard, H. Nguyen and S. Stoll, *Math. Comput. Simul.*, 2006, 72, 103.

47. D. P. Landau and K. Binder, *A Guide To Monte Carlo Simulations in Statistical Physics*, Cambridge University Press, Cambridge, 2005.

48. R. Y. Rubinstein and D. P. Kroese, *Simulation and the Monte Carlo Method*, 2nd edn, Wiley, Hoboken, NJ, 2008.

49. S. Ulrich, M. Seijo, F. Carnal and S. Stoll, *Macromolecules*, 2011, **44**, 1661.

50. F. A. Escobedo and J. J. de Pablo, *J. Chem. Phys.*, 1996, **104**, 4788.

51. M. Quesada-Pérez, J. G. Ibarra-Armenta and A. Martín-Molina, *J. Chem. Phys.*, 2011, **135**, 094109.

52. C. Arnold, S. Ulrich, S. Stoll, P. Marie and Y. Holl, *J. Colloid Interface Sci.*, 2011, **353**, 188.

53. F. Carnal and S. Stoll, *J. Phys. Chem. A*, 2012, **116**, 6600.

54. S. Ulrich, A. Laguecir and S. Stoll, *J. Nanopart. Res.*, 2004, **6**, 595.

55. K. Kremer, *Soft Matter*, 2009, **5**, 4357.

56. S. Diez Orrite, S. Stoll and P. Schurtenberger, *Soft Matter*, 2005, **1**, 364.

57. S. O. Nielsen, C. F. Lopez, G. Srinivas and M. L. Klein, *J. Phys.: Condens. Matter*, 2004, **16**, R481.

58. M. Müller, K. Katsov and M. Schick, *Phys. Rep.*, 2006, **434**, 113.

59. J. Shillcock and R. Lipowsky, *Biophys. Rev. Lett.*, 2007, **18**, 33.

60. A. V. Dobrynin, *Curr. Opin Colloid Interface Sci.*, 2008, **13**, 376.

61. J. Ramos, A. Imaz, J. Callejas-Fernandez, L. Barbosa-Barros, J. Estelrich, M. Quesada-Perez and J. Forcada, *Soft Matter*, 2011, 7, 5067.

62. M. Quesada-Perez, J. Ramos, J. Forcada and A. Martin-Molina, *J. Chem. Phys.*, 2012, **136**, 244903.

63. M. Quesada-Perez and A. Martin-Molina, *Soft Matter*, 2013, **9**, 7086.

64. S. Ulrich, A. Laguecir and S. Stoll, *J. Chem. Phys.*, 2005, **122**, 094911.
65. S. Ulrich, A. Laguecir and S. Stoll, *Macromolecules*, 2005, **38**, 8939.
66. S. Ulrich, M. Seijo, A. Laguecir and S. Stoll, *J. Phys. Chem. B*, 2006, **110**, 20954.
67. S. Ulrich, M. Seijo and S. Stoll, *J. Phys. Chem. B*, 2007, **111**, 8459.
68. F. Carnal and S. Stoll, *J. Phys. Chem. B*, 2011, **115**, 12007.
69. F. Carnal, S. Ulrich and S. Stoll, *Macromolecules*, 2010, **43**, 2544–2553.

Subject Index